Sixth Edition

PROBLEM SOLVING AND PROGRAMMING CONCEPTS

Maureen Sprankle

Professor Emeritus, College of the Redwoods

Prentice
Hall

Upper Saddle River, New Jersey
Columbus, Ohio

Library of Congress Cataloging-in-Publication Data

Sprankle, Maureen.
 Problem solving and programming concepts / Maureen Sprankle.--6th ed.
 p. cm.
 ISBN 0-13-048268-4
 1. Computer programming. 2. Problem solving--Data processing. I. Title.

 QA76.6 .S684 2003
 005.1--dc21

2002032659

Editor in Chief: Stephen Helba
Assistant Vice President and Publisher: Charles E. Stewart, Jr.
Assistant Editor: Mayda Bosco
Production Editor: Alexandrina Benedicto Wolf
Production Supervision: Custom Editorial Productions, Inc.
Design Coordinator: Diane Ernsberger
Cover Designer: Ali Mohrman
Cover Image: Jeremy Stanton
Marketing Manager: Ben Leonard
Production Manager: Matthew Ottenweller

This book was set in Times Roman by Custom Editorial Productions, Inc., and was printed and bound by the Banta Company. The cover was printed by Phoenix Color Corp.

Earlier edition ©1989 by Science Research Associates.

Pearson Education Ltd.
Pearson Education Australia Pty, Limited
Pearson Education Singapore, Pte. Ltd.
Pearson Education North Asia Ltd.
Pearson Education Canada, Ltd.
Pearson Educación de Mexico, S.A. de C.V.
Pearson Education—Japan
Pearson Education Malaysia, Pte. Ltd.
Pearson Education, *Upper Saddle River, New Jersey*

10 9 8 7 6 5 4 3 2 1
ISBN: 0-13-048268-4

Preface

Knowledge of problem solving and programming concepts is necessary for those who develop applications for users. Unfortunately, many students have greater difficulty with problem solving than they do with the syntax of computer languages. The art of programming is learning multiple techniques and applying those techniques to specific problems. When students learn basic programming and problem-solving techniques, they can then concentrate on the syntax when learning specific languages. These techniques may be presented in a separate class on problem solving or with a first language course that concentrates on problem solving. This approach tends to decrease students' frustration and improves their success rates.

This book is intended for a one-semester introductory course for programming majors. It can serve as a primary text or as a supplement. Although this book is written for students who have little or no computer experience, those who have studied a computer language can benefit from the generic presentation of the material.

The text provides a step-by-step progression of ideas with detailed explanation and many illustrations, from the basics of mathematical functions and operators to the design and use of techniques such as codes, arrays, pointers, other data structures, database concepts, and object-oriented programming concepts. The text uses problem-solving tools including problem analysis charts, interactivity charts, IPO charts, algorithms, and flowcharts to design a solution to a problem. The appendices present additional tools, including Universal Modeling Languages concepts, Nassi-Schneiderman charts, and Warnier-Orr diagrams. *Putting It All Together* sections illustrate a complete solution for a given problem, using the concepts previously presented. In some cases, an earlier solution is updated to incorporate more sophisticated techniques. Throughout the text, problems presented for solution are typical of the business world and provide excellent experience for students. These problems can then be presented in a language course so that students can finish the solution on the computer.

Organization of This Text

Unit One, Introduction to Problem Solving and Programming, presents basic concepts of problem solving, an introduction to how problems are solved on computers, mathematical concepts required for problem solving using a computer, and steps for analyzing

a problem and designing an appropriate solution. Unit Two, Logic Structures, presents basic concepts of programming, including local and global variables, parameters, and four basic logic structures. The four basic logic structures are sequential, decision, loop, and case logic structures. Unit Two presents programming techniques using these programming logic structures. Unit Three, Data Structures, presents the concepts of arrays, sorting techniques, search methods, stacks, linked lists, binary trees, and databases. Unit Four, Object-Oriented Programming, presents basic concepts in the design of a solution using object-oriented languages. Unit Five, File Processing is extremely important for a COBOL student. Database tools are replacing traditional techniques of working with large amounts of data. However, there are still many COBOL programs in government and industry that are being maintained. Unit Six, Problem Solving for Application Software, is presented for the programmer who will be using these applications to write reports and create budgets and who will need to combine applications to create a document or presentation of a problem. The unit presents concepts of word processing, design, spreadsheets, and information processing. Appendixes present formulas commonly used in business, sort and merge methods, ASCII and EBCDIC codes, blank forms, other problem-solving tools, and concepts of UML.

Acknowledgments

I am indebted to those who reviewed the manuscript and offered suggestions and constructive comments. In particular, I thank Philip Davis, DelMar College, Texas; Richard Windley, ECPI College of Technology, Virginia; and Costas Vassiliadis, Ohio University.

In addition I wish to thank Mayda Bosco and Alex Wolf at Prentice Hall and Megan Smith-Creed at Custom Editorial Productions, Inc., for their support and help throughout the production process.

I would also like to thank Jeremy Stanton, a student at College of the Redwoods, for the cover image.

Dedication

To my husband, Norm; my three daughters, Bonnie, Heidi, and Tamara; my two sons-in-law, Gary and Todd; and my good friends Ann and Craig—for their continuing love, support, and understanding.

Contents

UNIT ONE

INTRODUCTION TO PROBLEM SOLVING AND PROGRAMMING

Chapter 1

General Problem-Solving Concepts

Overview

Problem Solving in Everyday Life
Types of Problems
Problem Solving with Computers
Difficulties with Problem Solving

Objectives

When you have finished this chapter, you should be able to:

1. Describe the difference between heuristic and algorithmic solutions to problems.
2. List and describe the six problem-solving steps to solve a problem that has an algorithmic solution.
3. Use the six problem-solving steps to solve a problem.

Problem Solving in Everyday Life

People make decisions every day to solve problems that affect their lives. The problems may be as unimportant as what to watch on television or as important as choosing a new profession. If a bad decision is made, time and resources are wasted, so it's important that people know how to make decisions well. There are six steps to follow to assure the best decision. These steps, the six steps in problem solving, include the following:

1. *Identify the problem.* The first step toward solving a problem is to identify the problem. In a classroom situation, most problems have been identified for you and given to you in the form of written assignments or problems out of a book. However, when you are doing problem solving outside of the classroom, you need to make sure you identify the problem before you start solving it. If you don't know what the problem is, you cannot solve it.

2. *Understand the problem.* You must understand what is involved in the problem before you can continue toward the solution. This includes understanding the knowledge base of the person or machine for whom you are solving the problem. If you are setting up a solution for a person, then you must know what that person knows. A different set of instructions might have to be used depending on this knowledge base. For example, you would use a more detailed set of instructions to tell someone how to find a restaurant in your city if he or she has a limited knowledge of the city than if he or she knows it well. When you are working with a computer, its knowledge base is the limited instructions the computer can understand in the particular language or application you are using to solve the problem. Knowing the knowledge base is very important since you cannot use any instructions outside of this base. You also must know your own knowledge base. You cannot solve a problem if you do not know the subject. For example, to solve a problem involving calculus, you must know calculus; to solve a problem involving accounting, you must know accounting. You must be able to communicate with your client and be able to understand what is involved in solving the problem.

3. *Identify alternative ways to solve the problem.* This list should be as complete as possible. You might want to talk to other people to find other solutions than those you have identified. Alternative solutions must be acceptable ones. You could go from Denver to Los Angeles by way of New York, but this would probably not be an acceptable solution to your travel needs.

4. *Select the best way to solve the problem from the list of alternative solutions.* In this step, you need to identify and evaluate the pros and cons of each possible solution before selecting the best one. In order to do this, you need to select criteria for the evaluation. These criteria will serve as the guidelines for evaluating each solution.

5. *List instructions that enable you to solve the problem using the selected solution.* These numbered, step-by-step instructions must fall within the knowledge base set up in step 2. No instruction can be used unless the individual or the machine can understand it. This can be very limiting, especially when working with computers.

6. *Evaluate the solution.* To evaluate or test a solution means to check its result to see if it is correct, and to see if it satisfies the needs of the person(s) with the problem. (When a person needs a piece of furniture to sleep on, buying him a cot may be a *correct* solution, but it may not be very satisfactory.) If the result is either incorrect or unsatisfactory, then the problem solver must start the process again.

If any of these six steps are not completed well, the results may be less than desired.

People solve problems daily at home, or work, or wherever they go. Problems at home include such things as what to cook for dinner, which movie to see this evening, which car to buy, or how to sell the house. At work, the problems might involve dealing with fellow employees, work policies, management, or customers. The better the decisions an employee can make, the more valuable that person will be to the company. In each case, the six steps in problem solving can be followed. Most people use them without even knowing it.

Take the problem of what to do this evening.

1. Identify the problem. How do the individuals wish to spend the evening?

2. Understand the problem. With this simple problem, also, the knowledge base of the participants must be considered. The only solutions that should be selected are ones that everyone involved would know how to do. You probably would not select as a possible solution playing a game of chess if the participants did not know how to play.
3. Identify alternatives.
 a. Watch television.
 b. Invite friends over.
 c. Play games.
 d. Go to the movies.
 e. Play miniature golf.
 f. Go to the amusement park.
 g. Go to a friend's party.
 The list is complete only when you can think of no more alternatives.
4. Select the best way to solve the problem.
 a. Weed out alternatives that are not acceptable, such as those that cost too much money or do not interest one of the individuals involved.
 b. Specify the pros and cons of each remaining alternative.
 c. Weigh the pros and cons to make the final decision. This solution will be the best alternative if all the other steps were completed well.
5. Prepare a list of steps (instructions) that will result in a fun evening.
6. Evaluate the solution. Are we having fun yet?

By going through these steps, the problem solver can be assured that he or she has arrived at the best possible solution and will achieve the desired results.

Types of Problems

Problems do not always have straightforward solutions. Some problems, such as balancing a checkbook or baking a cake, can be solved with a series of actions. These solutions are called **algorithmic solutions.** Once the alternatives have been eliminated, for example, and once one has chosen the best among several methods of balancing the checkbook, the solution can be reached by completing the actions in steps. These steps are called the **algorithm.** The solutions of other problems, such as how to buy the best stock or whether to expand the company, are not so straightforward. These solutions require reasoning built on knowledge and experience, and a process of trial and error. Solutions that cannot be reached through a direct set of steps are called **heuristic solutions.**

The problem solver can use the six steps for both algorithmic and heuristic solutions. However, in step 6, evaluating the solution, the correctness and appropriateness of heuristic solutions are far less certain. It's easy to tell if your completed checkbook balance is correct and satisfactory, but it's hard to tell if you've bought the best stock. With heuristic solutions, the problem solver will often need to follow the six steps more than once, carefully evaluating each possible solution before deciding which is best. Furthermore, this same solution may not be correct and satisfactory at another time, so the problem solver may have to reevaluate and resolve the same problem later. The stock that did well in January may do poorly in June.

Most problems require a combination of the two kinds of solutions.

Problem Solving with Computers

In this book, the term **solution** means the instructions listed during step 5 of problem solving—the instructions that must be followed to produce the best results. **Results** means the outcome or the completed computer-assisted answer. **Program** means the set of instructions that make up the solution after they have been coded into a particular computer language.

Computers are built to deal with algorithmic solutions, which are often difficult or very time-consuming for humans. People are better than computers at developing heuristic solutions. Solving a complicated calculus problem or alphabetizing 10,000 names is an easy task for the computer, but the problem of how to throw a ball or how to speak English is not. The difficulty lies in the programming. How can problems such as how to throw a ball or speak English be solved in a set of steps that the computer can understand?

The field of computers that deals with heuristic types of problems is called artificial intelligence. Artificial intelligence enables a computer to do things like build its own knowledge bank and speak in a human language. As a result, the computer's problem-solving abilities are similar to those of a human being. Artificial intelligence is an expanding computer field.

Until computers can be built to think like humans, people will process most heuristic solutions and computers will process many algorithmic solutions. Therefore, this book will deal only with algorithmic solutions. Heuristic problem solving can help determine alternative solutions. However, for computer use, they must be transformed into an algorithmic format.

Difficulties with Problem Solving

People have many problems with problem solving. Some have not been taught how to solve problems. Others are afraid to make a decision for fear it will be the wrong one. Often, when people go through the problem-solving process, they complete one or more of the steps inadequately. They may not define the problem correctly or may not generate a sufficient list of alternatives. When choosing the best alternative, they may eliminate good alternatives or list the pros and cons too hastily. They may not use a logical sequence of steps in their solution, or they may focus on details before the framework for the solution is in place. Finally, they may incorrectly or haphazardly evaluate the solution.

The problem-solving process is not easy. It takes practice and time to perfect, but in the long run the process proves to be of great benefit.

When solving problems on the computer, one of the most difficult tasks for the problem solver is writing the instructions. Take the task of deciding which number is the largest from a group of three numbers. Almost anyone can immediately tell which is the largest, but many cannot explain the steps they followed to arrive at it. Most people will say, "I can't explain how I know, I just know it!" This explanation is not good enough for the computer. The computer is a tool that will perform only tasks that the user can explain.

The computer has a specific system of communication that programmers and users must learn. This system demands that no step in the solution to a problem be left unstated and that all steps be in the proper order. You must assume the computer knows nothing except what you tell it, and think of it as an ignorant but efficient aid to problem solving.

Summary

The six steps in problem solving lead to the best possible solution to a problem. They include 1) identifying the problem, 2) understanding the problem, 3) identifying alternative solutions, 4) selecting the best solution, 5) preparing a list of instructions to solve the problem by the chosen solution, and 6) evaluating the solution. Solutions to problems are classified as algorithmic or heuristic. Algorithmic solutions are reached in a series of steps. Heuristic solutions are attained through trial and error. Algorithmic solutions are easier to define for computer use than heuristic ones.

Several things can go wrong in the problem-solving process. Sometimes the problem solver does not define the problem correctly or does not complete one or more of the steps. At other times, the list of alternatives is not sufficient or the list of instructions to the computer is incorrect. Sometimes one gets lost in details and overlooks the general framework, or does not test the solution. If the problem-solving process is incomplete, the solution will not produce the desired results.

The steps in problem solving can be applied to problems in daily life as well as to problems put on computer. Good problem-solving techniques enable you to look at a problem logically and unemotionally, saving time and other resources.

New Terms

algorithm	program
algorithmic solution	results
heuristic solution	solution

Questions

1. What are the six steps of problem solving?
2. What is an algorithmic solution to a problem?
3. Name three current problems in your life that could be solved through an algorithmic process. Explain why each of these problems is algorithmic in nature.
4. What is a heuristic solution to a problem?
5. Name three current problems in your life that might be solved through a heuristic approach. Explain why each of these problems is heuristic in nature.
6. Name three problems that might arise at home, at school, or in a business that could be solved more efficiently with computer assistance. Do these problems require an algorithmic or heuristic solution? Why?

Problems

1. Complete the six problem-solving steps to solve one of the problems you listed in Question 3. Follow the form outlined as follows:
 Step 1: Identify the problem.

(Problem 1 continued on next page)

Step 2: Understand the problem.
 a. Comments about the problem to aid in understanding it.
 b. Description of the knowledge base (this list would include what you would be expected to know to follow the solution).

Step 3: Identify alternative solutions.

Solution	Pros	Cons
a.		
b.		
c.		
.		
.		
.		

Step 4: Select the best solution.
 Why did you select this solution?

Step 5: List a set of numbered step-by-step instructions to attain the solution.
 1.
 2.
 3.
 .
 .
 .

Step 6: Test the solution.
 Does this solution work?
 If not, how might you change the solution so it will work?

2. For each of the following tasks, write a set of numbered, step-by-step instructions (a solution) so complete that another person could perform the task without asking questions. Define the knowledge base of this person by listing what you expect the person to know in order to follow your directions. For example, for task a (below) make a cup of cocoa, the knowledge base might include such things as knowledge of milk or water, refrigerator, pan, spoon, cocoa, cup, range top or microwave, and so forth.
 a. Make a cup of cocoa.
 b. Sharpen a pencil.
 c. Walk from the classroom to the student lounge, your dorm, or the cafeteria.
 d. Start a car (include directions regarding what to do if the car doesn't start).
 e. Get a glass of water from your kitchen.

3. Develop the knowledge base for the following problems:
 a. balancing your checkbook
 b. driving a car
 c. repairing a car
 d. building a house
 e. calculating the cost of using your car for a month

4. Write a solution to the problem of finding the largest number out of three numbers. List the specific steps that would enable another person to find the largest among three numbers presented.

5. This problem is about Otto the robot. Otto is not very smart; in fact, his knowledge base includes only 15 instructions. However, he can follow a set of instructions and do what he is told, as long as they are part of the 15 instructions in his knowledge base. He executes instructions in sequential order; that is, when he is finished with

one instruction, he executes the next one in order. He does only what he is told to do. In the following sections you will find Otto's instruction set, assumptions, an example solution, and problems.

Otto's knowledge base includes the following instructions:

Body Instructions
1. Stand up.
2. Sit down.
3. Take a step (take one step forward—must be standing).
4. Turn (turn right 90 degrees—must be standing).
5. Raise arms (to a forward position perpendicular to body).
6. Lower arms (to a side position).

Arithmetic Instructions
Otto has one number in his memory that he can add one to or subtract one from. This number is zero when the execution of the instruction begins. This number is considered a variable because the value of the memory location may change, in contrast to a constant that does not have a memory location.

7. Add one (adds one to a number in his memory).
8. Subtract one (subtracts one from a number in his memory).
9. Goto *s* (jumps to instruction step *s*—*s* is a known number—and continues to execute instructions in sequential order starting at *s*. *s* may be an instruction number before or after the present instruction).

Example:
20. Goto 23
 This means go to instruction 23 and execute from there.

Conditional Instructions
Otto has the ability to make three different types of tests and follow one of two sets of one or more instructions according to the yes or no results for each test. In this way, Otto can make a decision as to the course of action he will take. A test would be set up in the following way (20 is the instruction number in the numbered step-by-step solution to the problem):

Example of a test instruction:
20. Test: (the test)
 yes: set of instructions when the result of the test is yes.
 no: set of instructions when the result of the test is no.

When there are many instructions in the conditional instructions set, it is better to use the Goto instruction to transfer to where the instructions for yes and no are. (See Example 2 of a test instruction.)

Example 2 of a test instruction:
20. Test: (the test)
 Yes: Goto 21
 No: Goto 24
21. yes instruction 1
22. yes instruction 2, etc.
23. go to 26
24. no instruction 1
25. no instruction 2, etc.
26. continuation of other instructions

(Problem 5 continued on next page)

10. Test: Are your hands touching anything?
 yes:
 no:
11. Test: Are your hands touching a door?
 yes:
 no:
12. Test: Is the number in your memory zero?
 yes:
 no:

Repeat Instruction

13. Repeat x times [set of instructions](repeats bracketed instructions x number of times—x is a known number, which is considered a constant. You may not use the number created in the memory location as this value).

 Example of a repeat instruction:
 18. Repeat five times [take step].
 This will instruct Otto to take five steps.

Other Instructions

14. Open door (door will close after Otto walks through it).
15. Stop (halts execution of instructions).

Assumptions:

 a. Otto will start in a seated position.
 b. Otto will end in a seated position in the same chair.

Example Problem Solution

Write a set of instructions to direct Otto to walk forward five steps and back five steps.

Solution:
1. Stand up.
2. Repeat five times [take step].
3. Repeat two times [turn].
4. Repeat five times [take step].
5. Repeat two times [turn].
6. Sit down.

Another solution (combining instructions 2–5 above into new instruction 2):
1. Stand up.
2. Repeat two times: [a. repeat five times [take step]] [b. repeat two times [turn]].
3. Sit down.

Problems (tasks):

Write a set of numbered, step-by-step instructions to direct Otto to perform the following tasks:

a. Direct Otto to walk in a square that is three steps on each side. All turns are to be right-hand turns. How many solutions can you determine?
b. Direct Otto to walk in a square that is three steps on each side. All turns are to be left-hand turns.
c. Direct Otto to walk to a wall that is any distance forward and then walk back again. Otto will be able to touch both the wall and the back of the chair when his hands are raised. The wall may be any number of whole steps in front of Otto. (*Hint:* This solution requires the use of one of the test instructions.)

d. Direct Otto to walk to a wall that is any distance forward and then walk back again. Otto will be able to touch the wall when his hands are raised; however, the chair height is lower than his hands. Otto will not be able to touch the back of the chair on the way back to the chair.

e. Write one set of instructions that will direct Otto through any maze with its entrance and exit on the same side. The maze has no dead ends and no intersections. From Otto's starting position, the entrance is two steps forward. When Otto is facing the entrance, the exit is three steps to his right. Figure 1.1 shows three example mazes for you to test your solution. Do not make separate solutions for each maze. Your solution should allow Otto to walk through any such maze as those shown in Figure 1.1 (*Hint:* Test for a wall and a door at every step.)

f. Modify the solution in task e (above) to accommodate a change in the place of the exit. Your solution should allow for the exit to be any place on the perimeter of the maze. Figure 1.2 shows three example mazes for you to test your solution. Again, do not write separate solutions for each maze. Your solution should allow Otto to walk back to his chair from any exit, such as those in Figure 1.2.

g. Can you modify the solution in task e to accommodate dead ends? If not, what instructions would you add to Otto's instruction set to accomplish this task?

h. Can you modify the solution in task e to accommodate intersections? If not, what instructions would you add to Otto's instruction set to accomplish this task?

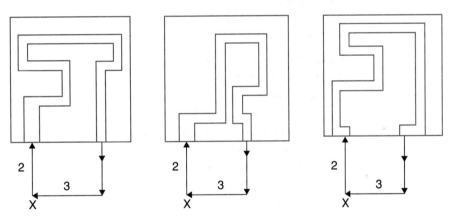

Figure 1.1 Example Mazes for Problem 5e

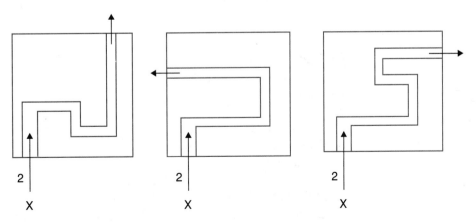

Figure 1.2 Example Mazes for Problem 5f

6. How would you modify Otto's instruction set (knowledge base) so as to be able to write a solution for the following tasks?
 a. Direct Otto to walk in a triangle three steps on a side.
 b. Direct Otto to walk to a table and pick up a book.
 c. Instruct Otto to detect an object in his path.

7. Write an instruction set, problems, and solutions for your own robot. Name your robot. Have another student in your class try to develop a solution for one of your problems using your instruction set.

WHAT'S WRONG WITH THIS?

What is wrong with the following sets of instructions for Otto? How would you correct them?

1. a. Stand up.
 b. Take step.
 c. Are your hands touching anything?

 Yes: Goto 2.
 No: Goto 4.
 d. Turn.
 e. Turn.
 f. Sit down.
 g. Stop.

2. a. Stand up.
 b. Repeat 3 times [Take step].
 c. Turn.
 d. Goto 2
 e. Sit down.
 f. Stop.

Chapter 2

Beginning Problem-Solving Concepts for the Computer

Overview

Constants and Variables
 Rules for Naming and Using Variables

Data Types
 Numeric Data
 Character Data—Alphanumeric Data
 Logical Data
 Other Data Types
 Rules for Data Types
 Examples of Data Types

Functions

Operators

Expressions and Equations

Objectives

When you have finished this chapter, you should be able to:

1. Differentiate between variables and constants.
2. Differentiate between character, numeric, and logical data types.
3. Identify operators, operands, and resultants.
4. Identify and use functions.
5. Identify and use operators according to their placement in the hierarchy chart.
6. Set up and evaluate expressions and equations using variables, constants, operators, and the hierarchy of operations.

Although problems that arise in daily life are of many types, problems that can be solved on computers generally consist of only three: 1) computational, problems involving some

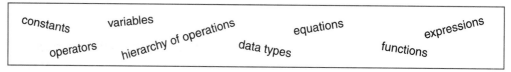

Figure 2.1 Important Concepts to Learn

kind of mathematical processing; 2) logical, problems involving relational or logical processing, the kinds of processing used in decision making on the computer; and 3) repetitive, problems involving repeating a set of mathematical and/or logical instructions. This chapter explains some computer fundamentals and demonstrates ways to set up expressions and equations to solve these types of problems on the computer. End users (users of prewritten software), as well as programmers, need to know these computer fundamentals. See Figure 2.1. In this chapter, both end users and programmers will be referred to as programmers.

Two of the most fundamental concepts that you will learn in this chapter are the constant and the variable. A programmer takes the data, the unorganized facts, and the information, the organized facts, relevant to a problem and defines them as constants or variables. They are the building blocks of the equations and expressions that ultimately make up solutions to computer problems. The programmer defines each constant and variable in a problem solution as a particular data type, such as numeric or character.

Other concepts that are essential to developing computer solutions to problems are operators and functions. Operators are the many signs and symbols that show relationships between the constant and variables in the expressions and equations that make up the solution. The programmer has to know all of the many operators and how to use them. The order in which operators are processed is determined by a hierarchy that programmers need to know as well.

Operators are combined with constants and variables to create expressions and equations. Expressions and equations are used in instructions that are the building blocks of the solution. Functions are sets of instructions that are so commonly used that they are built into a computer language or application, saving the programmer the trouble of writing them.

These are key concepts. Without an understanding of how the computer uses and defines data, without knowing what the operators are, and without knowing how to use these concepts to construct expressions and equations, a programmer or a user of applications software cannot effectively use the computer to solve problems.

When you study this chapter, it may help to keep these pointers in mind:

- Take each topic as it is presented, and learn the concepts pertaining to it.
- Understand the examples.
- Complete the questions and problems at the end of the chapter.
- Don't skip sections.
- Don't feel that something is too hard before you make an effort to understand it.
- Take each problem one step at a time.
- Don't skip steps.
- Don't assume anything.
- Don't be afraid to read a passage over again. Once may not be enough!

Constants and Variables

The computer uses constants and variables to solve problems. They are data used in processing. A **constant** is a value, that is, a specific alphabetical and/or numeric value, that never changes during the processing of all the instructions in a solution. Constants can be any type of data—numeric, alphabetical, or special symbols. (Data types are explained in more detail later.) Each constant is either part of the instructions in the solution or a piece of data (a datum) to be entered during processing. In some programming languages and applications, constants can be named. In this case, the constant is given a location in memory and a name. During the execution of the program, this constant is given a value and then is referred to by its name. Once the constant is given a value, it cannot be changed during the execution of the program. For example, because the value of PI does not change, it would be a constant and defined within the program. This constant may be given a name, but the only way to change the value of the constant is to change the program.

In contrast, the value of a **variable** may change during processing. A programmer must give a name to each variable value used in a solution. The programmer uses a variable name as a reference name for a specific value of the variable. In turn, the computer uses the name as a reference to help it find that value in its memory. The computer sets up a specific memory location to hold the value of each variable name found in a program. Variables can be any data type, just as constants can. For instance, consider the cost of a pair of shoes. This data item should be given a variable name because the cost of a pair of shoes may change during the processing of the program or during multiple executions of the program. The variable name should be consistent with what the value of the variable represents. In this case, the name of the variable is SHOE_COST because it represents the cost of the pair of shoes:

SHOE_COST

56.00

However, if the cost of the shoes changes during the next execution of the program, then the value of SHOE_COST will change but the variable name will not:

SHOE_COST

35.00

Notice in Table 2.1 that when the value of a constant or a variable contains alphanumeric data—that is, numbers and/or special symbols—it is surrounded by quotation marks. These marks indicate to the computer that the value is a datum, and not a variable name. The computer must have some way to distinguish between the two. Also notice that the variable names stand for specific data. The variable name CASH has a sum of money as its value. The variable name CITY has a city name as its value. Finally, notice that there are no blank spaces in variable names. A space indicates the end of a variable name for the computer. If you need a space in a variable name, use the underline character, for example LAST_NAME. You would not use a dash, as a dash is a subtraction symbol. The programmer decides on the variable name.

Table 2.1 Constants and Variables on the Computer

Constants	Variables
Rules: Constants cannot be changed. Examples: Value [25] Value [−1.5] Value ["ARCATA"] Value ["95521"] Named Constants Rules: A constant cannot be changed after it is initially given a value. Storage location given a name. Referred to by the given name. Example: PI [3.142857]	Rules: Storage locations are given names. Values of the contents for name variable locations can be changed. Referred to by variable name in the instructions. Examples: Variable Name—AGE Value [25] Variable Name—CASH Value [83.59] Variable Name—CITY Value ["EUREKA"] Variable Name—ZIPCODE Value ["95501"]

In a solution that calculates payroll for a company, the name of the company would be a constant since it does not change. The employee name, the hours, and the rate of pay would be variables because the values of these items change for each employee. If there were no such thing as a variable, the programmer would have to write a separate set of instructions for each employee. It is far more efficient to have one rather than a thousand programs to process payroll for 1,000 employees.

The rules for naming variables differ from language to language and application to application. However, the general rule is to name the variable as near to its meaning as possible: PAYRATE for the rate of pay, HOURS for the number of hours worked during the pay period, and the like. Some languages and applications have character-length limitations or other restrictions for names (called reserved words), so adjustments have to be made as necessary. In this book, there will be no character-length limitations for variable names.

It is important to understand the difference between the name of a variable and the value of a variable. The *name* is the label the computer uses to find the correct memory location; the *value* is the contents of the location. The computer uses the variable name to find the location; it uses the value found at the memory location to do the processing.

It is also important to be consistent in the use of variable names because the computer will go only to the location with the specified name, whether or not it is the one intended by the user. For example, if you use HOURS for hours worked, then you must use HOURS at all times in referring to hours worked not HRS or Hours. If the computer cannot find a memory location by the specified name, it will either name a new memory location and yield an incorrect result, or it will return an error message indicating there is no location as referenced and then cease execution.

Table 2.2 Incorrect Variable Names

Data Item	Incorrect variable names	Problem	Corrected variable name
Hours Worked	Hours Worked	Space between words	Hours_Worked
Name of client	CN	Does not define data item	Client_Name
Rate of Pay	Pay-Rate	Uses a mathematical operator	Pay_Rate
Quantity per customer	Quantity/customer	Uses a mathematical operator	Quantity_Per_Customer
6% sales tax	6%_sales_tax	Starts with a number	Six_percent_sales_tax or Sales_tax
Client address	Client_address_for_client_of_XYZ_corporation_in_California	Too long	Client_address
Variable name Introduced as Hours	Hrs	Inconsistent name	Hours
Variable name Introduced as Hours	Hours_worked	Inconsistent name	Hours

Rules for Naming and Using Variables, see Table 2.2

1. Name a variable according to what it represents, that is, HOURS for hours worked, PAYRATE for rate of pay, and so on. Create as short a name as possible but one that clearly represents the variable.
2. Do not use spaces in a variable name. If a space is needed, use the underline character.
3. Start a variable name with a letter.
4. Do not use a dash (or any other symbol that is used as a mathematical operator) in a variable name. The computer will recognize these symbols as mathematical operators, turn your variable into two or more variables, and treat your variable as a mathematical expression.
5. After you have introduced a variable name that represents a specific data item, this exact variable name must be used in all places where the data item is used. For example, if the data item *hours worked* has the variable name of HOURS, HOURS must be used consistently. You may not use HRS or HOURSWORKED to represent the same data item. If you do, the computer views these variables as new and different data items and will assign a new memory location to the new name.
6. Be consistent when using upper- and lowercase characters. In some languages HOURS is a different variable name than Hours.

Data Types

To process solutions, the computer must have data. **Data** are unorganized facts. They go into the computer as input and are processed by the program. What is returned to the user is output, or **information.** This information is printed in the form of reports. For example, when a computer calculates the balance of a checkbook, data are the checks, the deposits, and the bank charges. The information from the processing is shown on the balance sheet (see Figure 2.2).

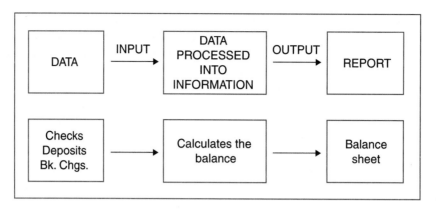

Figure 2.2 Procesing Data—How a Computer Balances a Checkbook

The data the computer uses are of many different types. Computers must be told the **data type** of each variable or constant. The most common data types are numeric, character, and logical. A few languages and applications also use the date as a data type. Other languages allow the programmer to define data types.

Numeric Data

Numeric data include all types of numbers (see Table 2.3). Numeric is the only data type that can be used in calculations. The subtypes of numeric data include *integers* and *real numbers*. Integers are whole numbers, such as 5,297 or −376. They can be positive or negative. Real numbers, or floating point numbers, are whole numbers plus decimal parts. In problem solving on the computer, programmers use integers when there is no reason for using partial numbers, as when they are designing counters, expressions used for counting things, such as inventory items or people. A real number can be expressed in scientific notation, such as 2.3E5 or 5.4E–3. In scientific notation, the E stands for *times 10 to the power of*. Therefore, 2.3E5 is the same number as 2.3×10^5, or

Table 2.3 Data Types and Their Data Sets

Data Type	Data Set	Examples
Numeric: INTEGER	All whole numbers	3580 −46
Numeric: REAL	All real numbers (whole + decimal)	3792.91 4739416.0 0.00246
CHARACTER (surrounded by quotation marks)	All letters, numbers, and special symbols	"A" "a" "M" "z" "k" "1" "5" "7" "8" "0" "+" "=" "(" "%" "$"
STRING (surrounded by quotation marks)	Combinations of more than one character	"Arcata" "95521" "707-444-5555"
LOGICAL	TRUE FALSE	TRUE FALSE

230,000.0, and 5.4E–3 is the same number as 5.4×10^{-3}, or 0.0054. A number expressed in scientific notation is always considered a real data type. The computer does not use commas in a number during the processing of a calculation, only when the number is formatted as output for the user.

Numeric data are used in business for values, such as rate of pay, salary, tax, or price, that have calculations performed on them. Numbers such as an account number or a zip code, which would not have calculations performed on them, would not be designated by the programmer as numeric data.

Each data type has a *data set*, the set of symbols necessary to specify a datum as a particular data type. The data set for the numeric data type includes all base 10 numbers, the plus sign ($+$), and the negative sign ($-$). The data set for the integers includes all whole numbers within the limitations of the computer or the language, and for the real numbers, all whole numbers with decimal parts within the limitations of the computer or the language, including zero as a whole number and zero as a decimal part.

Character Data—Alphanumeric Data

The **character data** set, sometimes called alphanumeric data set, consists of all single-digit numbers, letters, and special characters available to the computer—a, A, Z, 3, #, &, and so forth—placed within quotation marks. An uppercase letter is considered a different character from a lowercase letter. The ASCII (American Standard Code for Information Interchange) character set contains 256 characters. The first 128 comprise a standard set (see Appendix C for the ASCII and EBCDIC data codes), and the second 128 differ with each computer. Characters cannot be used for calculations even if they consist of only numbers. When more than one character is put together, the computer considers this item a *string*—derived from *a string of characters*. Some languages do not differentiate between characters and strings. All character data are considered string data.

Character and string data can be compared and arranged in alphabetical order in the following way. The computer gives each character a number. This numeric representation is necessary because the computer works only with numbers. The numbers are compared to see which is larger and are then arranged in ascending numeric order. Since B has a larger number representing it than A, B is placed after A. C would follow B by the same method, and so forth. String data can be tested in the same way to alphabetize names, cities, and the like. Banana is larger than Apple because B has a larger number representing it than A. Joan is larger than James since the letter o is attributed a larger number than a. Uppercase letters have smaller numeric representations than lowercase letters. It is important to bear this difference in mind when you are comparing letters to see if they are equal. Languages and applications that differentiate between lower case and upper case are called case sensitive. When they are not case sensitive, upper and lower case letters are treated the same. You can use functions before the comparison to insure that data are not case sensitive by making all letters either upper or lower case.

Character data or string data can be joined together with the $+$ operator in an operation called **concatenation.** When two pieces of character data are joined, the concatenation results in the second being placed at the end of the first, as in "4" + "4" = "44" (not "8"). Concatenation could be used to join a first name with a last name or to join pieces of data together to make a code for an inventory item. The concatenation operator varies with each language or application.

Most data in business—names, account numbers, addresses, cities, states, telephone numbers, zip codes—are string data. As a rule, items that would not have mathematical computations performed on them should be designated string data types. There

is another reason for designating a zip code as string data. If you are using numeric data, there is no way to hold onto leading zeros, that is, the zeros at the front of a number, such as 00987. As a result, a zip code on the East Coast could not be entered correctly as a numeric datum. The leading zeros could be preserved only with string data.

Logical Data

Logical data consist of two pieces of data in the data set—the words TRUE and FALSE. (Some languages accept *yes*, T, and Y for TRUE, and *no*, F, and N for FALSE as part of the data set.) These are used in making yes-or-no decisions. For example, logical data type might be used to check someone's credit record; TRUE would mean her credit is okay, and FALSE would mean it's not okay. A home accounting system could use logical data type for check returns; TRUE would mean the check has been returned, and FALSE would mean it has not. There are many such uses for logical data. They are discussed further in later chapters. Remember the logical data set True and False are *not* string data.

Other Data Types

There are other data types available to most programmers, such as the date data type and user-defined data types. The date data type is a number for the date that is the number of days from a certain date in the past such as the first day of the 20th century or January 1, 1940. The use of this data type for the date allows the user to subtract one date from another date to calculate the number of days between dates. This simplifies the calculation when the dates cross over years or months. The date is printed in the date format instead of a number. The date data type is a numeric data type as you can perform mathematical calculations on any date.

Programmers may define their own data types, such as data types that have as their data set soda pop brands, or types of cars, or computer components. The user must specify the items in the data set for each user-defined data type. If the data item is not contained within the data set, it is not part of the set. For example, if the programmer does not include HARD DISK as part of the data set for a user-defined data type of computer parts, then the computer does not recognize a HARD DISK as a computer part.

This book will use the date data type in a few places; however, user-defined data types will not be further discussed or used.

Rules for Data Types

1. The data that define the value of a variable or a constant will most commonly be one of three data types: numeric, character (including character string), or logical.
2. The programmer designates the data type during the programming process. The computer then associates the variable name with the designated data type.
3. Data types cannot be mixed. For example, string data cannot be placed in a variable memory location that has been designated as numeric, and vice versa. When the computer expects a certain data type, the user must use that type or the computer will return an error message.
4. Each of the data types uses what is called a *data set* (see Table 2.2). The numeric data uses the set of all base 10 numbers, the plus sign ($+$), and the

negative sign (−); the character type uses the set of all characters available to the computer; the logical data type uses the set of data consisting of the words TRUE and FALSE. The use of any data outside the data set results in an error.

5. Any numeric item that must be used in calculations resulting in a numeric result must be designated as numeric data type. All other numbers should be designated as character or character-string data types, even if data are all numbers, as in zip codes.

Examples of Data Types

Table 2.4 illustrates some common uses for numeric, character (including character string), and logical data types. Data are drawn from everyday life and business, and are

Table 2.4 Examples of Data Types

Data	Data Type	Explanation
The price of an item: 7.39, 12.98	Numeric: real	The price of an item would be used in calculations. The price is money and needs decimals.
An account number: "A2453," "2987"	Character string	An account number consists of alphanumeric or simply numeric data. It is not used for calculations.
A quantity: 12389	Numeric: integer	A quantity is used for calculations. It is an integer because it normally is a whole number.
The name of a company: "Smith Corp."	Character string	A name of something is alphabetical and therefore would be character string data type.
A credit check: TRUE, FALSE	Logical	The credit check of a customer would be a choice of two answers such as *yes,* it is ok, or *no,* it is not ok. Therefore, it should be logical data.
A zip code: "95521" "76548" "00538"	Character string	A zip code would be a character string because there are no calculations to be done on a zip code, and the leading zeros need to be retained.
A date: 01/23/87 or "03/14/87"	Date or character string	If the date data type is available then it is used; if not, then it has to be a string because it is alphanumeric.
A date: 187259	Numeric: integer	A calendar date is the number of days from a given date; such as, the first day of the 20th century. It can be used for calculations.
Social Security Number "333-33-333"	Character string	A SSN is string data because it is alphanumeric and is not used for calculations.

examples of the kinds of data that are commonly used in solving various types of problems on the computer. To the right of the data examples are data types and explanations of why each type is appropriate to the data.

Functions

Functions are small sets of instructions that perform specific tasks and return values. They are usually built into a computer language or application. Functions are used as parts of instructions in a solution. Because they are basic tasks that are used repeatedly in the problem-solving process, by using them a programmer or user can shorten the problem-solving time and improve the readability of the solution. Each language or application has a set of functions within it. This set varies with the language, the application, and the computer. Most languages allow programmers to write their own functions. Libraries of functions can be added to many languages.

The form of a function is the name of the function followed by an open parenthesis, followed by the data needed to perform the function and concluded by a closed parenthesis:

<p style="text-align:center">functionname(data)</p>

The value of the result of the function is returned in the name of the function.

Functions use data. These data are listed as part of the function and are called **parameters.** Functions normally do not alter the parameters. Take the square root function, SQRT(N). This function will calculate the square root of N. SQRT is the name of the function. N is the data needed to calculate the square root and, therefore, it is the parameter. Parameters are surrounded by parentheses. The maximum function, MAX(N1, N2, N3), will find which of three numbers is the largest. The name of the function is MAX. The parameters, data surrounded by parentheses and needed to do the calculation, are N1, N2, and N3. Not all functions need parameters. The function RANDOM generates or calculates a random number. No data are needed to do this calculation, so there are no parameters. A parameter can be a constant, a variable, or an expression. (An expression is a calculation, such as interest/100, which has not been given a permanent memory location in the computer.) The names of the functions may vary from language to language and from application to application. This book will give a generic name to each function and list the parameters in parentheses.

Table 2.5 lists and defines a few of the basic functions that are found in many languages. Functions are unique to each language. As a programmer you need to investigate the functions used in the language in which you are writing your solution. These functions have been divided into classes.

1. *Mathematical functions*. Often used in science and business, mathematical functions calculate such things as square root, absolute value, or a random number. Other mathematical functions used primarily for scientific purposes have not been included in the table.
2. *String functions*. These are used to manipulate string variables. For example, they copy part of the string into another variable, find the length or the number of characters in the string, and so forth.
3. *Conversion functions*. These functions are used to convert data from one data type to another. For example, since character strings cannot be used in calculations, one of these functions would convert a string value to a numeric value.

4. *Statistical functions.* These functions are used to calculate things such as maximum values, minimum values, and so forth.
5. *Utility functions.* This class is very important in business programming because most reports require some use of utility functions. They access information outside the program and the language in the computer system. Examples of these include date and time functions.

Table 2.5 Functions *(continued on page 24)*

Function*	Definition	Example	Result
Mathematical Functions Used in Business			
SQRT(N)	Returns the square root of N.	SQRT(4)	2
ABS(N)	Returns the absolute value of N.	ABS(−3)	3
ROUND(N, n1)	Returns the rounded value of N to the n1 place.	ROUND (3.7259,2)	3.73
INTEGER(N)	Returns the closest whole number less than or equal to N.	INTEGER (5.7269)	5
RANDOM	Returns a random number between 0 and 1, but not 1. This number is mathematically generated. To find a number N between N1 and N2, inclusive, use the following formula: N = INTEGER (RANDOM * (N2 − N1 + 1)) + N1, where N1 is the smallest number and N2 is the largest number.	RANDOM	0.239768
SIGN(N)	Returns the sign of N: 1 when N is positive, 0 when N is zero, −1 when N is negative.	SIGN(7.39)	1
String Functions			
MID(S, n1, n2)	Returns a set of n2 characters starting at n1 in the string S.	MID(S, 3, 2) where S = "THOMAS"	"OM"
LEFT(S, n)	Returns a set of n characters on the left side of the string S.	LEFT(S, 3) where S = "THOMAS"	"THO"
RIGHT(S, n)	Returns a set of n characters on the right side of the string S.	RIGHT (S, 3) where S = "THOMAS"	"MAS"
LENGTH(S)	Returns the number of characters in the string S.	LENGTH(S) where S = "THOMAS"	6

*Definitions of symbols:

N is a numeric value—a constant, a variable, or an expression

S is a string value—a constant, a variable, or an expression

n, n1, n2 are integer values—a constant, a variable, or an expression

Table 2.5 *(continued from page 23)*

Function	Definition	Example	Result
Conversion Functions (change data type)			
VALUE(S)	Changes a string value into a numeric value.	VALUE ("57.39")	+57.39
STRING(N)	Changes a numeric value into a string value.	STRING (+57.39)	"57.39"
Statistical Functions			
AVERAGE(list)	Returns the average of a list of numbers.	AVERAGE (5, 3, 8, 6)	5.5
MAX(list)	Returns the maximum value from a list of numbers.	MAX (5, 3, 8, 6)	8
MIN(list)	Returns the minimum value from a list of numbers.	MIN (5, 3, 8, 6)	3
SUM(list)	Returns the sum of a list of numbers.	SUM (5, 3, 8, 6)	22
Financial Functions are listed in Chapter 20.			
Utility Functions			
DATE	Returns the current date from the system. The date may be in various forms: mm/dd/yy, day only, month only, year only, or Julian calendar.	DATE	09/15/98
TIME	Returns the current time from the system, The time may be in various forms: hh:mm:ss, seconds from midnight, or minutes from midnight.	TIME	9:22:38
ERROR	Returns control to the program when a systems error occurs.		

Table 2.5 is only a partial list of functions. There are other functions that are not universal or not used very much. The programmer should check the particular language or software package to find any additional functions provided there.

Operators

The computer has to be told how to process data. This task is accomplished through the use of operators. **Operators** are the data connectors within expressions and equations. They tell the computer how to process the data. They also tell the computer what type of processing (mathematical, logical, or whatever) needs to be done. The types of operators used in calculations and problem solving include mathematical, relational, and logical operators. Without these operators very little processing can be done.

The *operand* and the *resultant* are two concepts related to the operator. Operands are the data that the operator connects and processes. The resultant is the answer that results when the operation is completed. For example, in the expression 5 + 7, the + is the operator, 5 and 7 are the operands, and 12 is the resultant. Operands can be constants or variables. The data type of the operands and the resultant depends on the operator.

Mathematical operators include addition, subtraction, multiplication, division, integer division, modulo division, powers, and functions. The computer has a symbol for each of them (see Table 2.6). You are probably familiar with addition, subtraction, multiplication, division, and powers. However, you may not be familiar with integer and modulo division.

These two operations are related. In integer division, the resultant is the whole number in the quotient. In modulo division, the resultant is the whole number remainder. These two types of division are used in business to find the hours and minutes someone has worked given the total number of minutes, or the days and weeks worked given the total number of days. For example, if Jane Smith has worked 19 days during the month, then, assuming a 5-day work week, she has worked $19\backslash5 = 3$ weeks and

Table 2.6 Operators and Their Computer Symbols

Operator	Computer Symbol	Example	
Mathematical		Operation	Resultant
Addition	+	3.0 + 5.2	8.2
Subtraction	−	7.5 − 4.0	3.5
Multiplication	*	8.0 * 5.0	40.0
Division	/	9.0/4.0	2.25
Integer division	\	9\4	2
Modulo division	MOD	9 MOD 4	1
Power	^	3 ^ 2	9
Relational			
Equal to	=	5 = 7	FALSE
Less than	<	5 < 7	TRUE
Greater than	>	5 > 7	FALSE
Less than or equal to	<= (two key strokes)	5 <= 7	TRUE
Greater than or equal to	>= (two key strokes)	5 >= 7	FALSE
Not equal to	<> (two key strokes)	5 <> 7	TRUE
Logical			
Not	NOT	NOT TRUE	FALSE
And	AND	TRUE AND TRUE	TRUE
Or	OR	TRUE OR FALSE	TRUE

19 MOD 5 = 4 days. Therefore, she has worked 3 weeks and 4 days. When dividing 19 by 5 the result is 3 with a remainder of 4. The 3 is the resultant of the integer division and the 4 is the resultant of the modulo division. If the dividend is less than the divisor, such as 28 MOD 379, then the resultant is always equal to the dividend, in this case 28. The resultant of 28 \ 379 is zero because the dividend is less than the divisor. When you divide 28 by 379 your resultant is zero with a remainder of 28.

Relational operators include the following: equal to, less than, greater than, less than or equal to, greater than or equal to, and not equal to. A programmer uses relational operators to program decisions. The operands of a relational operator can be either numberic or character (a string); however, both operands must be of the same data type. The resultant of a relational operation is logical data type TRUE or FALSE. The programmer designs one action or set of actions that will follow when a relational expression is TRUE, and another action or set of actions that will follow when the expression is FALSE. The use of relational operators is the only way for the computer to make decisions.

For example, when a credit card customer's balance is *less than $500* (TRUE), then the customer can charge another purchase. When the balance is *not less than $500* (FALSE), then he cannot charge another purchase. The expression would be set up as BALANCE < 500. The operands are BALANCE and 500; the operator is <; the resultant is either TRUE or FALSE depending on the value of the balance. The programmer would use this expression in what is called a *decision instruction*.

Relational operators are also used to control repetitive instructions called *loops*. A set of instructions to enter data for a client, which would be repeated until the data are entered for every client, is one example of a loop. When the computer processes this type of instruction, the loop repeats until the resultant changes from TRUE to FALSE, or vice versa. In the example, the resultant would change when the computer can find no more client data to enter.

Logical operators are the third type of operator (see Table 2.7). Logical operators are used to connect relational expressions (decision-making expressions) and to perform operations on logical data. For example, a store might require a driver's license or a check-cashing card on file for a customer to cash a check. When the customer has a driver's license, the check can be cashed. When he or she has a check-cashing card, the check can be cashed. The expression is written as LICENSE OR CARD. LICENSE and CARD are the operands. They are logical data; that is, the value of each is TRUE or FALSE. The operator is OR. The resultant is TRUE or FALSE depending on the values of the operands. When one or both of the operands is TRUE, then the resultant is TRUE. When both of the operands are FALSE, then the resultant is FALSE.

The OR operator is one of three logical operators. The others are AND and NOT. When the AND operator is used, the resultant is TRUE only when both of the operands are TRUE. When either or both of the operands are FALSE, the resultant is FALSE. A programmer uses AND when two requirements must be TRUE in order for an action or set of actions to take place, such as when a store requires a customer to have *both* a driver's license *and* a check-cashing card to cash a check. The NOT operator is the only logical operator that requires only one operand. The resultant of the NOT operator changes in an operand from TRUE to FALSE, or from FALSE to TRUE. A programmer uses the NOT operator to change an operand to the opposite value. This operation is sometimes called *reversing* the value of the operand.

These mathematical, relational, and logical operators have a **hierarchy,** or precedence, an order in which their operations take place (see Table 2.8). To reorder the

Table 2.7 Definitions of the Logical Operators

NOT

A	Not A	When A is		The Resultant is
T	F	NOT TRUE	IS	FALSE
F	T	NOT FALSE	IS	TRUE

AND

A	B	A AND B	When A is		When B is		The Resultant is
T	T	T	TRUE	AND	TRUE	IS	TRUE
T	F	F	TRUE	AND	FALSE	IS	FALSE
F	T	F	FALSE	AND	TRUE	IS	FALSE
F	F	F	FALSE	AND	FALSE	IS	FALSE

OR

A	B	A OR B	When A is		When B is		The Resultant is
T	T	T	TRUE	OR	TRUE	IS	TRUE
T	F	T	TRUE	OR	FALSE	IS	TRUE
F	T	T	FALSE	OR	TRUE	IS	TRUE
F	F	F	FALSE	OR	FALSE	IS	FALSE

normal processing sequence, the programmer uses parentheses. The processing of the operands (as directed by the operators) always starts with the innermost parentheses and works outward, and processes from left to right. Each level of operators within a set of parentheses requires that the computer make another pass through the parentheses, until all levels have been cleared. This hierarchy is important to the programmer because the order of the operations determines the result of the expression. If the operations are not completed in the correct order, the result of the expression may be incorrect. For example, assuming a 40-hour work week and overtime pay at 1.5 times regular pay, the expression to calculate overtime pay would subtract 40 from the hours worked and multiply the result by the regular wage times 1.5. The expression would be written as

$$(\text{HOURS} -40) * \text{WAGE} * 1.5$$

As you can see from the hierarchy chart, multiplication is processed before subtraction, so parentheses must be added to tell the computer to do the subtraction first, before the multiplication. If the parentheses are not added, the result will be incorrect.

In this hierarchy table the integer and modulo divisions are on a separate level from real number multiplication and division. However, many languages place these four operators on the same level. When this is true, it is important to place parenthesis around the integer or modulo operations so that the integer and modulo divisions are

Table 2.8 Hierarchy of Operations

Order of Operations	Operand Data Type	Resultant Data Type
() Reorders the hierarchy; all operations are completed within the parentheses using the same hierarchy.		
1.　　Functions		
Mathematical Operators		
2.　　Power	Numeric	Numeric
3.　　\, MOD	Numeric	Numeric
4.　　*, /	Numeric	Numeric
5.　　+, −	Numeric	Numeric
Relational Operators		
6. =, <, >, <=, >=, <>	Numeric or string or character	Logical
Logical Operators		
7.　　NOT	Logical	Logical
8.　　AND	Logical	Logical
9.　　OR	Logical	Logical

executed before the multiplication and division. In the following example you will get two different results depending upon whether you execute the integer and modulo division on the same level or separate them with parenthesis to force different levels. Look at the equation:

$$F = 6 * 2 \setminus (6 + 2)$$

If the multiplication and integer division is on the same level the answer will be 1. When they are on different levels, where the integer division will be executed first, the answer will be 0. Remember the execution of operators on the same level is completed from left to right. To force the integer division to be executed first, resulting in a zero resultant, the equation would have to be changed to:

$$F = 6 * (2 \setminus (6 + 2))$$

Check with the language you are working with to find out whether integer and modulo divisions are on the same or different levels. In this book we will assume they are on the different levels. However, please be aware of the difference when starting to develop equations and expressions for the language you are learning.

Each type of operator requires a certain data type for operands and determines the data type of the resultant. Mathematical operators require numeric data as the operands

and have numeric resultants. When any of the relational or logical operators are used (whether or not they are used in conjunction with mathematical operators), the resultant is logical data. Relational operators use numeric, character, or string data types as the operands; however, the data type of the operands of an operator must be the same. The logical operators require logical data for operands.

Expressions and Equations

A knowledge of constants and variables, of the three data types, and of operators, operands, and resultants is not very valuable until you can use these concepts to create expressions and equations. The problem you are trying to solve may be calculating pay, including tax, Social Security, and medical deductions. It may be an inventory problem with reorder quantities and cash values. It may be a problem of calculating interest or payments on a loan, or of putting a mailing list in order. These tasks all require the use of different operators, and would be written as expressions and equations in order for the computer to process the data and arrive at a result. There are very few computer problems that do not use expressions and equations in their solutions.

Expressions and equations make up part of the instructions in the solution to a computer problem. An **expression** processes data, the operands, through the use of operators. For example, to find the number of square feet in a room you would multiply the length of the room by the width in the expression

$$LENGTH * WIDTH$$

An **equation** stores the resultant of an expression in a memory location in the computer through the equal ($=$) sign. The expression above would be used as part of an instruction in the equation

$$AREA = LENGTH * WIDTH$$

The resultant of the expression LENGTH * WIDTH would then be stored in a memory location called AREA.

As you can see in Table 2.9, expressions use no equal sign (except as a relational operator) and are used as part of an instruction, such as part of an equation. The resultant of an expression is not stored and, therefore, is not available for use at another time. An expression can use numeric, string, character, or logical data types.

Equations are often called *assignment statements* because the variable on the left-hand side of the equal sign is assigned the value of the expression on the right-hand side. The equal sign does not mean *equals*; instead, it means *replaced by* or *is assigned the value of*. This distinction is important because it is possible to have equations on the computer that are not allowed in a mathematics class, such as $N = N + 1$ (one is added to the old value of N to get a new value of N). This equation is the way the computer counts. There is only one variable on the left-hand side of the equal sign. On the right-hand side of the equal sign is an expression. The right-hand side is processed before the assignment is made (see Table 2.9 and the examples that follow).

For the beginning programmer, a big part of learning to solve problems on the computer is learning how to write and evaluate expressions and equations. (Recall that to *evaluate* means to test for correctness using actual data.) The following examples illustrate how to use the concepts you've learned in this chapter to write and evaluate expressions and equations.

Table 2.9 Expressions and Equations

Expressions	Equations
A + A and B are numeric. The resultant is numeric and is not stored.	= A + B C, A, and B are numeric. The resultant is stored in C.
A < A and B are numeric, character, or string. The resultant is logical and is not stored.	= A < B A and B are numeric, character, or string. The resultant is stored in C; C is logical.
A OR B A and B are logical. The resultant is logical and is not stored.	C = A OR B C, A, and B are logical. The resultant is stored in C.

Example 1: Setting Up a Numeric Expression

A programmer will often be given an expression to use in solving a problem. Assume the programmer has to modify the following mathematical expression for computer use:

$$X(3Y + 4) - \frac{4Y}{X + 6}$$

The appropriate computer expression would be the following:

$$X * (3 * Y + 4) - 4 * Y / (X + 6)$$

All variables, constants, and operators have to be on the same line. There must be an operator between variables and/or constants. For the computer, there are no assumed multiplication signs. Parentheses are added when the hierarchy has to be reordered or when parentheses would improve readability.

Follow these steps to complete the expression:

1. Write all parentheses, operands, and operators on a single line with the dividend first, followed by the divisor:

$$X (3Y + 4) - 4Y / X + 6$$

2. Insert all implied operators. The computer does not use any implied operators as used in mathematics. These usually include multiplication and power signs.

$$X * (3 * Y + 4) - 4 * Y / X + 6$$

3. Insert all parentheses where the hierarchy needs to be reordered. $X + 6$ needs to be calculated before completing the division, therefore parentheses need to be inserted around the $X + 6$.

$$X * (3 * Y + 4) - 4 * Y / (X + 6)$$

Example 2: Setting Up a Mathematical Equation

A mathematical equation might be given to the programmer in the following form:

$$Y + 3 = X(Z + 5)$$

The programmer has to modify the equation so that it is in the form of an assignment instruction:

$$Y = X * (Z + 5) - 3$$

Notice that there is only one variable (and no calculations) on the left side and an expression on the right side of the equal sign.

Follow these steps to complete the equation:

1. Write all parentheses, operands, and operators on a single line:

$$Y + 3 = X (Z + 5)$$

2. Use mathematical rules to complete the equation so that there is one variable on the left side of the equal sign. In this case, subtract 3 from each side:

$$Y + 3 - 3 = X (Z + 5) - 3$$
$$Y = X (Z + 5) - 3$$

3. Follow the steps in Example 1 to complete the equation.

$$Y = X * (Z + 5) - 3$$

Example 3: Setting Up a Relational Expression

A relational expression is used to make decisions. Given the expression

$$X \text{ is less than } Y + 5$$

the programmer would change its form to the following:

$$X < Y + 5$$

Because this is a relational expression, either a variable or a mathematical expression is on either side of the operator. X and Y would be the same data type, in this case, a numeric data type. The resultant would be a logical data type—TRUE or FALSE, depending on the values of X and Y.

Example 4: Setting Up a Logical Expression

Often a computer decision is based on more than one TRUE or FALSE operand. For example, in order to cash a check a customer must have a driver's license (A) or a check-cashing card (B) on file at the store. As is the case with all logical expressions, this case would require a logical operator, A OR B. The operands, A and B, are logical data type; it is either TRUE or FALSE that the customer has a driver's license, and TRUE or FALSE that she has a check-cashing card. The resultant is also a logical type: The customer can cash a check (TRUE or FALSE).

Example 5: Evaluating a Mathematical Expression

To find out if proposed solutions are correct, it is important for the programmer to evaluate, or test, all expressions and equations. Assume the programmer has written the-expression

$$5 * (X + Y) - 4 * Y/(Z + 6)$$

The programmer uses the following values to evaluate the expression:

$$X = 2 \qquad Y = 3 \qquad Z = 6$$

Notice the structure of the evaluation that follows. A dashed line leading to a circle indicates the operator. The solid lines indicate the operands. Following the illustration of the equation is a list of the operations and the resultant of each operation. The number of circles should equal the number of operators. Be sure the circles using other circles are below the circles they need to use, that is, 3 is lower than 1, 6 is lower than 3 and 5.

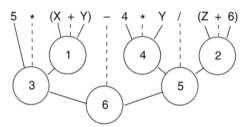

Operation	Resultant
1. X + Y	5
2. Z + 6	12
3. 5 * resultant of 1	25
4. 4 * Y	12
5. Resultant of 4 / resultant of 2	1
6. Resultant of 3 − resultant of 5	24

Illustrating the structure of the expression and listing the operations and results in this way simplifies the evaluation process for the programmer. The evaluation must be completed according to the hierarchy of operations. The operations within the parentheses would be completed first, moving from left to right—so X + Y would be completed first, and Z + 6 second. According to the hierarchy, the programmer would then complete the multiplications and the divisions from left to right. The division would be completed after the multiplications in this expression only because it is to the right. The final operation to be completed would be the subtraction.

This evaluation shows how to test a computer expression to see if it matches an equivalent mathematical expression. It gives the programmer a correct result that can be compared to the results manually calculated by the client, or by whoever requested the computer solution to the problem, so one can be sure that the solution is correct.

Example 6: Evaluating a Relational Expression

Assume the programmer has written the expression

$$A - 2 > B$$

The programmer uses the following values to evaluate the expression:

$$A = 6 \qquad B = 8$$

The following illustration shows the structure of the evaluation:

	Operation	Resultant
1.	A − 2	4
2.	Resultant of 1 > B	FALSE

According to the hierarchy of operations, the mathematical operation is completed first.

Example 7: Evaluating a Logical Expression

Assume the programmer has written the expression

$$A \text{ AND } B \text{ OR } C \text{ AND } A$$

The programmer uses the following values to evaluate the expression:

$$A = \text{TRUE} \qquad B = \text{FALSE} \qquad C = \text{TRUE}$$

	Operation	Resultant
1.	A AND B	FALSE
2.	C AND A	TRUE
3.	Resultant of 1 OR resultant of 2	TRUE

Example 8: Evaluating an Equation That Uses Both Relational and Logical Operators

Assume the programmer has written the following equation:

$$F = \text{NOT } (A < B) \text{ AND } (C \text{ OR } D)$$

The programmer uses the following values to evaluate the equation:

$$A = 4 \qquad B = 2 \qquad C = \text{TRUE} \qquad D = \text{FALSE}$$

The operations are completed in hierarchical order from left to right, as illustrated in the figure.

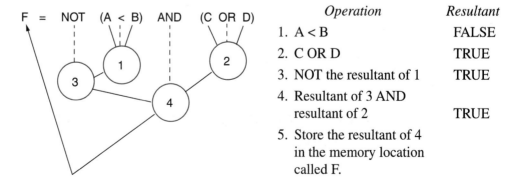

	Operation	Resultant
1.	A < B	FALSE
2.	C OR D	TRUE
3.	NOT the resultant of 1	TRUE
4.	Resultant of 3 AND resultant of 2	TRUE
5.	Store the resultant of 4 in the memory location called F.	

Example 9: Developing a Table of All Possible Resultants of a Logical Expression

It is often valuable for a programmer to figure out the resultants for all of the possible combinations of values of the variables in an expression. To create a table of all possible combinations, double the number of combinations from a table with one less unknown. For example:

■ One unknown—A. Two combinations: A can be either T or F.

■ Two unknowns—A and B. Four combinations: B can be either T or F for each value of A.

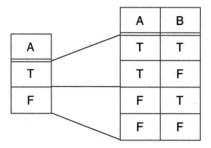

■ Three unknowns—A, B, and C. Eight combinations.

Given the expression

<div align="center">NOT A OR B</div>

the following table gives the resultants for all combinations of the values of A and B. The equation to find the number of combinations is 2^n, where n is the number of variables. For example, with two variables, A and B, there are four combinations; and with three variables, A, B, and C, there are eight combinations.

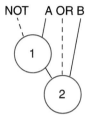

A	B	NOT A OR B
T	T	F
T	F	F
F	T	F
F	F	T

Example 10: Developing a Logical Expression from a Given Problem

Many times a programmer will be given the policies of a company and then be required to set up a logical expression from those policies. To do this, the programmer first needs to list the items on which the condition is dependent. Then the logical expression can be developed from this list.

> Problem: A large department store has its own charge card. The policy for a customer to charge an item is that the customer must have a valid charge card and either a balance of less than $500 or a charge of less than $50.

1. The first step is to list the items the charge is dependent on and their data types:
 a. the charge card (logical data type)
 b. the balance (numeric data type)
 c. the charge amount (numeric data type)
2. The second step is to write down these items as variables along with the conditions on each. Put them on one line, leaving space between, so that the logical operator can be added next. The resultant or the data type of each expression must be logical. Remember that the operators must match the data types of the operands (see Table 2.7). Do not put the logical operators in as of yet.

<p style="text-align:center">CHARGE_CARD BALANCE < 500 AMOUNT < 50</p>

3. The third step is to put in the logical operators and parentheses:

<p style="text-align:center">CHARGE_CARD AND (BALANCE < 500 or AMOUNT <50)</p>

Because the condition between the balance and the amount is one or the other, the OR operator is needed. The condition between the charge card and the other two items is both; this calls for the AND operator. Since the AND has precedence over the OR, the parentheses are added around the OR operation.

4. You have now created the logical expression. This expression can be used in various ways:

 a. In an assignment statement:

$$\text{OKTOCHG} = \text{CHARGE_CARD AND (BALANCE} < 500 \text{ OR}$$
$$\text{AMOUNT} < 50)$$

 b. In a decision statement (this will be explained in Chapter 6, Problem Solving with Decisions):

$$\text{IF CHARGE_CARD AND (BALANCE} < 500 \text{ OR AMOUNT} < 50)$$
$$\text{THEN PRINT "OKAY TO CHARGE"}$$
$$\text{ELSE PRINT "NOT OKAY TO CHARGE"}$$

Summary

This chapter explains fundamental concepts for solving problems on the computer, including constants, variables, data types, common functions, operators, expressions, and equations. It also shows the beginning programmer how to set up computer expressions and equations.

Constants are values that never change during processing, even if they are named constants; variables are values that do change. Constants and variables are assigned data types and are used in expressions and equations. The most common data types include numeric, character, and logical.

Expressions and equations use functions to process standard tasks. Most functions are part of the computer language. Expressions and equations use operators to process the data. These operators are of three types: mathematical, relational, and logical. By using operators and the hierarchy of operations, a programmer can design calculations and decisions, which are the key to solving problems on the computer.

New Terms

character data	information
concatenation	logical data
constant	logical operator
data	mathematical operator
data type	numeric data
equation	operator
expression	parameter
function	relational operator
hierarchy	variable

Questions

1. What is a constant? What is a variable?
2. Fill in the following table with the variable name and data type needed to solve an inventory problem for White Auto Supplies.

Data Item	Variable Name	Data Type
a. Name of vendor company		
b. Inventory item name		
c. Inventory number		
d. Quantity		
e. Price		
f. Address of company		
g. Date last ordered		
h. Reorder quantity		
i. Obsolete item (yes/no)		

3. Name the data type for each of the following constants. Explain your answer.
 a. 5.38
 b. "87654" *character string*
 c. TRUE *logic*
 d. "A"
 e. "707–434–5555"
 f. "New York"
 g. −389
 h. 2.45E6
 i. 48976.0
 j. FALSE

4. What is a function?

5. Why are functions used?

6. What is the difference between the operators /, \, and MOD? Give an example of each.

7. Find the result of the following operations:
 a. 5 + 4
 b. 10/2
 c. TRUE OR FALSE *true*
 d. 20 MOD 3 *2*
 e. 5 < 8
 f. 25 MOD 70 *25*
 g. "A" > "H" *False*
 h. NOT TRUE
 i. 25\70
 j. FALSE AND TRUE
 k. 20 * .5
 l. 35 < = 35
 m. 35/7
 n. FALSE OR FALSE
 o. TRUE AND TRUE
 p. 50 MOD 5

(Question 7 continued on next page)

q. $-35 < 67$
r. $4.0 \wedge 3$
s. $60 \backslash 9$
t. $35 < 35$
u. TRUE AND FALSE

8. Using the hierarchy chart (Table 2.8), list the order in which the following operations would be processed. (Remember: operations are processed left to right within a level in the hierarchy table.)
 a. $+, -, *$
 b. $/, \backslash, =$
 c. OR, $*, <$
 d. NOT, AND, $*$
 e. NOT, $>, +$
 f. AND, OR, NOT
 g. $<$, AND, $>, +$
 h. $*, \wedge, +$
 i. NOT, $+, \backslash$
 j. MOD, $\backslash, <$

9. Name the data type of the operands and the resultant of each of the following expressions or equations.

	Data Types of Operands	Data Types of Resultant
a. A * B		
b. D > R		
c. NOT C		
d. B AND F		
e. G = B		

10. What is the difference between an expression and an equation?

11. What does the equal sign mean in an equation? Is the meaning of the equal sign as a relational operator different from that in an assignment statement?

Problems

1. Evaluate the following equations, given the values A = 12, B = 3, C = 6, D = 2:
 a. $F = A + B/C - D \wedge 2$
 b. $F = (A + B)/C - D \wedge 2$
 c. $F = A + B/(C - D \wedge 2)$
 d. $F = (A + B) \text{MOD } C$
 e. $F = (A + B) \backslash D \wedge 2$

2. Write the following equations in computer form:

 a. $X = Y + 3Z - \dfrac{Z + Y}{Z - 3}$

b. $X = 5Y + \dfrac{3Z - 1}{4(3Z + 1) - Y}$

$X = (X - Y)^2$

c. $X = (X - Y)^2$

3. Is the = sign an assignment instruction or a relational operator in the following equations? Justify your answer.
 a. $A = B + 2 + C$
 b. $A - B = 2 + C$

4. Set up an equation to calculate the following (create your own variable names):
 a. The area of a room.
 b. The wall area of a room including windows and doors.
 c. The wall area of a room not including 2 windows and a door.
 d. The number of miles given a number of feet. (Use 5280 feet per mile.)
 e. The percent increase (or decrease) of a value given the beginning number and the ending number. How would the result differ between increase and decrease?
 f. The average of 5 numbers.
 g. Sale price of an item given an original price and a percentage discount.

5. Evaluate the following equations given $A = 5, B = 4, C = 3, D = 12$
 a. $E = A * B + D / C$
 b. $E = D \text{ MOD } A * B$
 c. $E = 5 * A \setminus D * (B + 1)$
 d. $E = D / B * ((A + 4) \setminus (C + 1))$

6. An employee came in to work and clocked in at MORNING_IN, clocked out at NOON_OUT for lunch, clocked back in at NOON_IN, and clocked out to go home at NIGHT_OUT (all in minutes since midnight). Set up equations to calculate the number of hours and the number of minutes the employee worked for the day. (You will develop two equations.)

7. Steve Thompson bought and charged an electric saw on DATE1 and made the first payment on DATE2. Write an equation to calculate how many days (DAYS) elapsed between the two dates.

$$DAYS =$$

8. Eureka Lumber Company gives a 2% discount if the remaining balance is paid within 10 days of purchase. Write a logical expression to verify whether or not a customer qualifies for this discount. Use the following variable names:

 BUYING_DATE: date the customer purchased the merchandise

 PAID_DATE: date the customer made the final payment

9. A part-time employee worked 20 hours in the first week and 15 hours in the second week of a two-week pay period. He is paid a weekly salary based on a 40-hour week. What is his full-time equivalent for the two weeks based on a 40-hour week (i.e., what percentage of full-time did he work)? Write a general equation that could be used to express and store the full-time equivalent of any hours worked per week.

$$FULL_TIME =$$

10. Evaluate the following equations, given $A = $ FALSE, $B = $ TRUE, $C = $ FALSE, $D = $ TRUE. (Include the structure of the order of processing—see page 34 for example.)
 a. $R = A \text{ AND } B \text{ OR } C \text{ AND } D$
 b. $R = \text{NOT } (A \text{ AND } B) \text{ OR NOT } (D \text{ AND } C)$

(Problem 10 continued on next page)

c. R = (A OR B) AND (D OR C)

d. R = NOT (A AND B OR C) AND (A OR B AND D)

e. R = C OR NOT (A AND D) AND (A OR B) OR NOT (A OR C)

11. Create a table that gives all possible answers for the following logical equations. (Include the structure of the order of processing—see page 34 for example.) Make clear how you set up the table.

a. R = A OR B

b. R = NOT A OR B

c. R = A AND B AND (B OR C)

d. R = NOT (A OR B) AND NOT (B OR C)

e. R = B AND NOT (A OR C) OR NOT (B AND C)

12. Evaluate the following for the values A = 5, B = 2, C = TRUE, D = FALSE. (Include the structure of the order of processing.)

a. R = A + 3 > B − 1 AND C OR D

b. R = NOT C OR D OR A − 3 < = B

13. Set up a logical expression for the following conditions. A company gives a bonus at the end of each fiscal year. For an employee to get a bonus, the following must be true:

a. The employee has been working at the company for more than six months with no negative reports.

b. The employee has earned more than $5,000 during the fiscal year.

14. Set up a logical expression for the following conditions. A retail store has this check-cashing policy:

a. The customer must have a driver's license.

b. When the check is for more than $50, the customer must have a check-cashing card on file.

Chapter 3

Programming Concepts

Overview

How the Computer Stores Data

Communicating with the Computer

Organizing the Problem

 Analyzing the Problem

 Developing the Interactivity (Structure) Chart

 Developing the IPO Chart

 Writing the Algorithms

 Drawing the Flowcharts

 Internal and External Documentation

Using the Tools

Testing the Solution

Coding the Program

Objectives

When you have finished this chapter, you should be able to:

1. Differentiate between fields, records, and files.
2. List and describe the six basic tools used in this book to aid in the development of a solution to a problem.
3. Use a Problem Analysis Chart to consolidate data for a problem.
4. Use an Interactivity Chart to designate the modules to be used in the solution of a problem.
5. Use an IPO chart to designate the input, processing, module number, and output for a solution of a problem.
6. Use algorithms and flowcharts to develop the instructions for each module in the solution of a problem.
7. Describe the importance of internal and external documentation.

Computers are only as good as their hardware, their software, and the people using them. Assuming the hardware has been engineered to accomplish a given job, the efficiency of the computer relies on the effectiveness of the programmer.

The computer must be told what to do. It needs a set of instructions, the solution, in order to process data in the correct sequence and reach the desired results. For greatest efficiency, the programmer or the user has to be able to easily read and interpret the solution. The solution will be most effective if the programmer follows certain steps and rules that have been developed over the years. The solution becomes a program when it is coded into a computer language during the final stages of programming.

Programmers have three resources that they need to optimize for cost effectiveness: computer memory, computer time, and programmer time. Early computers were much less useful and efficient than they are today. Programmers worked alone, and only when they left a company did their programs go to others. Since computers were slow and had small memories, conservation of computer time and memory was of utmost importance. For these reasons and the fact that the profession was new, programs were often hard to read and maintain. Over the years, both computers and programming methods have changed. Because programs are larger and more complex, and need to be completed in less time than ever before, today many programmers work together on the same program. As a result, readability is now what's most important. Since computers are much faster with much more memory, programmers may sacrifice these resources somewhat in favor of improved readability. Cost effectiveness, decrease in programmer frustration, and increase in programmer satisfaction are the rewards.

The type of problem the computer can solve best is one that can be solved algorithmically—that is, the problem can be solved with a single sequence of instructions. These instructions must be written in a way the computer can understand, and the computer executes them in the order presented in the program. If the instructions are written according to the appropriate rules and they correctly follow the syntax of the computer language, then the computer will complete the task. While this scrupulous adherence to the instructions is usually what the programmer wants, it is also, believe it or not, the most frustrating thing about the computer. If the instructions make sense to the computer, it will complete them. If the programmer has made an error in the instructions, the results are not likely to be what was intended. The computer does not reason, nor does it read minds; it only follows instructions.

How the Computer Stores Data

The computer stores data internally in memory locations. These data are found by the variable names used by a program. Each variable name is given a memory location, and each memory location can hold one and only one value at a time. When a user enters a new value into the variable location, the previous value is destroyed. These memory locations are temporary, as the internal memory is a volatile memory. When a program completes its instructions, and/or when the computer is turned off, the values stored in the internal memory are destroyed.

Data and instructions are temporarily stored in the computer's internal memory during the processing. When data, information, or programs have to be kept for future use, they are stored externally on an auxiliary storage medium such as a hard disk or

a floppy diskette in storage areas called **files.** There are basically two types of files: program files and data files. Program files contain the instructions to tell the computer what to do. This book helps the programming student create programs that will be saved on the auxiliary storage device as program files. Data files contain the data required to execute the program files. A data file consists of all the data pertaining to a subject, such as a payroll or a mailing list. Within each file are **records,** which contain all the data pertaining to one entity of the file, such as one employee's payroll data or one vendor's data. Within each record are **fields,** which contain one piece of data, such as a name, an address, or a pay rate.

A good analogy for the way a computer stores data externally is a filing cabinet. One drawer holding all the data on a company's employees is like a file on the computer. Each employee's folder in the file is a record. A field is one piece of data about the employee. For example, assume that the sales file for a company contains all the data needed to process orders for the week. Each record contains all the data from one sales receipt including quantity, item, price, name, and so forth. Each field contains one piece of the data from the sales receipt (Figure 3.1.)

Before a file can be used, the operating system, which is the manager of the computer, must open the files, preparing the computer to receive data from and send data to the auxiliary storage device. One of the tasks the operating system completes at this time is to set up a small temporary memory called a **buffer.** Data are transferred between the internal memory in the computer and the external memory on disk through the buffer. Normally one record or a group of records, called a *block* of records, is copied from memory location (internal or external) into the buffer and then copied to the other memory location. The use of buffers makes both external memory and input/output devices more efficient. At the conclusion of the program, the operating system also executes certain closing instructions when the user tells it to CLOSE the file. The user must OPEN and CLOSE every file or an error may occur. More about file processing is presented in Chapters 14, 15, 16, and 17.

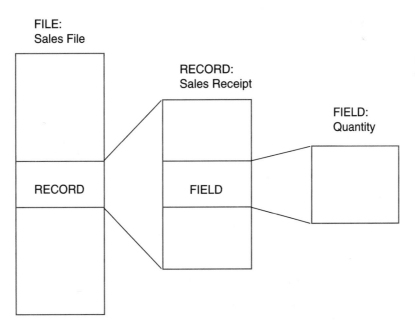

Figure 3.1 Files, Records, and Fields

Communicating with the Computer

Since the computer does not speak English as you do, you have to learn its system of communication. For the computer to "understand" a message, its system of communication requires instructions governed by specific rules. If the instructions you write are incorrect, the computer will give an error message, the wrong answer, or no answer at all. If the instructions are not properly sequenced, the computer will, nevertheless, execute them in the order given, and the result will be wrong.

The meaning of an instruction is essentially the same in any computer language or application. The differences between instructions from one language or application and another are in how they are set up. **Syntax** refers to the rules governing the computer operating system, the language, and the application. An error is called a **bug.** A bug must be found and corrected, a process called **debugging.** Many bugs are a result of syntax errors, but some are logic errors. You can find and correct most logic errors during the problem-solving process. You will find and correct syntax errors when you enter your program into the computer. All syntax errors must be corrected before you execute and test your program.

Although a set of instructions must be in a correct order to lead to the correct result, there may be several "correct" orders, just as there are several routes leading from New York to San Francisco. Whether or not the routes are equally efficient doesn't affect the result. When efficiency is important, then and only then should each route be examined and kept or discarded. People employ different logic and different ways of thinking. Two programmers may develop equally good solutions to a problem, but the solutions may look entirely different. There is nothing wrong with this. Computers are exact machines, but the people working with them are not.

Organizing the Problem

Certain organizational tools will help you learn to solve problems on the computer. The tools used in this book and illustrated in this chapter include the **problem analysis chart,** which shows a beginning analysis of the problem; the **structure chart** or **interactivity chart,** which shows the overall layout or structure of the solution; the **IPO chart,** which shows the input, the processing, and the output; the **algorithms,** which show the sequence of instructions comprising the solution; and the **flowcharts,** which are graphic representations of the algorithms. To analyze a problem and set up the most efficient solution, a programmer organizes the solution by using all or some of these tools. When the programmer does not use these tools during the problem-solving process, the solution takes longer to program, and the final program is less efficient, lacks readability, and increases programmer frustration.

Analyzing the Problem

To organize a solution, the programmer first has to understand and analyze the requirements of the problem. A good way to analyze a problem is to separate it into four parts, shown in the problem analysis chart (PAC) in Figure 3.2:

1. The given data
2. The required results
3. The processing that is required in the problem
4. A list of solution alternatives

Given Data	Required Results
Section 1: Data given in the problem or provided by the user. These can be known values or general names for data, such as price, quantity, and so forth.	Section 2: Requirements for the output reports. This includes the information needed and the format required.
Processing Required	Solution Alternatives
Section 3: List of processing required. This includes equations or other types of processing, such as sorting, searching, and so forth.	Section 4: List of ideas for the solution of the problem.

Figure 3.2 Problem Analysis Chart

(A blank problem analysis chart for student use appears in Appendix D, Figure D.1). As a problem unfolds (in a textbook or in the real world), the programmer can use this form to sort it out. The PAC allows the problem solver to eliminate the words and glean only the facts from the problem. Data, constants and variables, would be entered in the Given Data section. Variable data are the input values. Requirements for the output reports would be entered under Required Results. Any equations or other processing requirements would be listed in the Processing Required section. Finally, the programmer would write any other ideas that spring to mind concerning the solution in the Solution Alternatives section.

The following problem illustrates the use of the problem-solving tools. Calculate the gross pay of an employee. The formula to be used is

$$GROSSPAY = HOURS * PAYRATE$$

Figure 3.3 shows how the problem solver would fill in the PAC for this problem. The hours worked and the pay rate are the data given and are put into the Given Data box. The gross pay is what needs to be calculated and given to the user; therefore it is put into the Required Results box. The formula used is

$$GROSSPAY = HOURS * PAYRATE$$

It is put into the Processing Required box. The Solution Alternatives are as follows:

1. Define the hours worked and the pay rate as constants.
2. Define the hours worked and the pay rate as input values.

Given Data	Required Results
Hours Pay Rate	Gross Pay
Processing Required	Solution Alternatives
GROSSPAY = HOURS * PAYRATE	1. Define the hours worked and pay rate as constants. *2. Define the hours worked and pay rate as input values.

Figure 3.3 Problem Analysis Chart for the Payroll Problem

The second alternative will be used, as signified by the asterisk in the figure, since the program will not need to be changed in order to calculate the gross pay for another employee.

The problem analysis chart is an aid to clear thinking because it helps the problem solver to identify the essential data and information in a problem and to disregard the nonessentials. Most problems programmers work with are word problems. This is true whether they are working in a classroom situation or out on the job. The PAC helps to pare down the words.

Developing the Interactivity (Structure) Chart

The next step in organizing your solution is to divide the processing into subtasks called **modules** and then to connect these modules together to show the interaction of processing between the modules. Each module should contain the tasks to accomplish one function, such as entering data, printing results, or calculating results. There will be one module that controls the flow to most of the other modules. This will be called the Control or Main Module. Each module should contain no more than 20 instructions. This breakdown enables you to view a complex problem in simpler parts and to program smaller, simpler parts of a program rather than one large, complex program. This chart is called the interactivity chart or structure chart. In the rest of the text, we will refer to this chart as the interactivity chart since it really shows how the modules interact with each other.

Today's programmer will use two different ways of looking at a solution to a problem. Before the advent of the graphic user interface, all solutions were procedural in nature. Solutions were developed using a top-down method to develop a procedural program that was executed from the first lines of code to the end. The user had very little to do with the order of execution of the modules. The program was in complete control. However, with the advent of the graphic user interface, which allows the user to choose what to do next, programming changed from procedural to object-oriented programming. In object-oriented programming, execution is completed through an interactivity between the computer and the user. The user is in control, not the program. Modules are still used, and are basically the same. Within each module the execution is procedural. The difference lies in when each module will be executed. In procedural programming, there is a direct order of processing. In object-oriented programming, there is no preconceived order. The user directs the order of execution of the modules.

In developing the procedural interactivity chart, use the top-down method. The top-down method simply means that as you divide the problem into subtasks, you illustrate them in the order in which they will be processed from the top to the bottom of the chart. When using the top-down method, the module that encompasses the complete solution becomes what is called the *Control Module* because it controls the processing of all the data. The subtasks of this module are then located below it in the structure chart. The interactivity chart shows what will happen, not how it will happen.

As you can see in Figure 3.4, the interactivity chart shows the relationships of the parts of a problem to each other and to the whole. Notice that the programmer numbers the rectangles in a logical order from top to bottom and from left to right. This arrangement illustrates the order of processing. The programmer labels the top rectangle on the chart (the Control Module) with the lowest number (0000). The next level of rectangles starts with the number 1000 and increases from left to right by increments of 1000. The other sublevels increase by numbers that are one magnitude less than the numbers above them. That is, if the higher level of rectangles increases by increments of 1000, then the

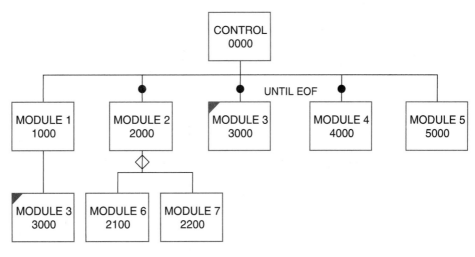

Figure 3.4 The Interactivity Chart

programmer would label the next level by increasing increments of 100, as in 2100, 2200, 2300, in the third row in Figure 3.4. The levels below that, if there are any, would increase by 10s, and so forth. The programmer marks any duplicate modules by darkening the upper left-hand corner and writes the name of the module inside the rectangle, giving the same name to identical tasks.

Notice in Figure 3.4 that Module 3 is found in two places, indicating that it is processed at two different points in the solution. Since it is the same module in both places, it carries the same name and number. The number is that of its top-most position in the interactivity chart. If both Module 3s are on the same level, the number should be that of its left-most position.

A module processes only those tasks below it and directly connected to it by a network of horizontal and vertical lines. In Figure 3.4, the Control Module processes Module 1, Module 2, Module 3, Module 4, and Module 5. The darkened circles indicate that the module is part of a set of modules that are processed many times—those in a loop. At the level below the Control Module, Module 1 processes Module 3, and Module 2 processes Module 6 and Module 7. However, only one of these modules will be processed since they are part of a decision. That is, the instructions in Module 2 and the appropriate data would lead the computer to choose one of the two modules. The diamond on the vertical line above the box indicates which modules are involved in a decision. Annotation may be placed beside the circle or diamond to indicate the condition. (Appendix D, Figure D.2 is an interactivity chart for student use.)

Figure 3.5 shows the interactivity chart for the payroll problem. The top rectangle is the Control Module, which processes the READ, the CALC, and the PRINT Modules.

The interactivity chart for the same solution completed as an object-oriented programming solution would show each of the subtask modules surrounding the control module (see Figures 3.6 and 3.7). Each module is given a number in a similar fashion as in the procedural solution. The connecting lines have arrows showing the direction of flow from one module to another. Connecting lines may connect with other lines when they go to the same place. Figure 3.6 shows the interactivity chart shown in Figure 3.4 reconstructed as an object oriented interactivity chart. Figure 3.7 shows the interactivity chart shown in Figure 3.5 reconstructed as an object-oriented interactivity chart. Object-oriented programming will be discussed further in Chapters 12 and 13.

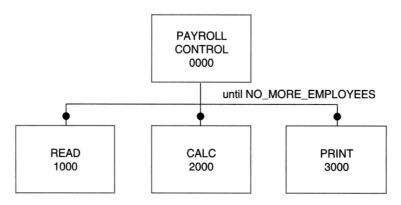

Figure 3.5 The Interactivity Chart for the Payroll Problem

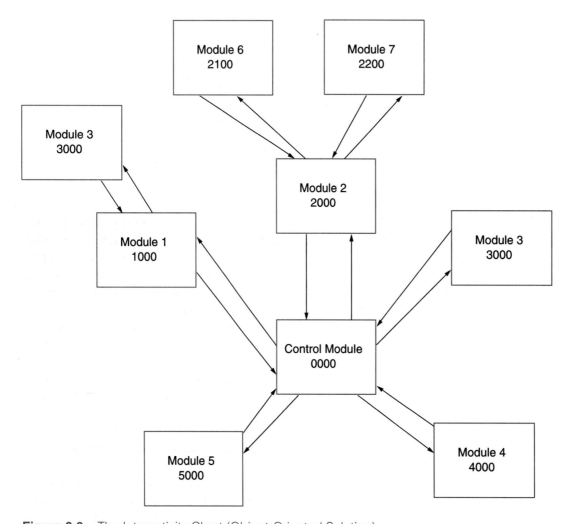

Figure 3.6 The Interactivity Chart (Object-Oriented Solution)

Developing the IPO Chart

The IPO (Input-Processing-Output) chart extends and organizes the information in the problem analysis chart. It shows in more detail what data items are input, what processing takes place on that data, and what information will be the end result, the output

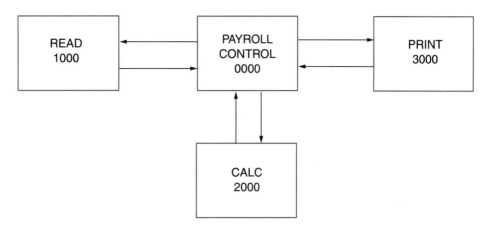

Figure 3.7 The Interactivity Chart for the Payroll Problem (Object-Oriented Solution)

Input	Processing	Module Reference Number	Output
All input data (from Section 1 of problem analysis chart)	All processing in steps (from Sections 3 and 4 of the problem analysis chart)	Module reference number from the interactivity chart	All output requirements (from Sections 1 and 2 of the problem analysis chart)

Figure 3.8 The IPO Chart

(see Figure 3.8). The IPO chart also shows where in the solution the processing takes place. The output is the first to be completed in an IPO chart. The next is the input, and the last is the processing. This may sound backwards, but it is really very logical. An analogy to completing an IPO chart would be planning a trip. The first thing you need to know is the destination—San Francisco, Paris, New York. This information is the output. Then you can plan what to pack, your means of transportation, and all the other essentials for starting out—the input. Finally, you plan the route—the processing. How can you know what to pack or what route to follow if you don't know where you're going?

The IPO chart has four sections: the Input, the Processing, the Module Reference Number, and the Output. The Input section contains all input data from the problem analysis chart. The input includes all given data from Section 1 on the problem analysis chart. The Processing section contains all processing, evident and implied, from Sections 3 and 4 of the problem analysis chart. The Module Reference Number section contains the number from the interactivity chart of the module in which each step in the processing is to be completed. The Output section includes all required output as designated by the problem and/or the user; it comes from Section 2, along with the needed input items from Section 1, of the problem analysis chart. (An IPO chart for student use appears in Appendix D, Figure D.3.)

To illustrate how to use an IPO chart, Figure 3.9 shows one for the payroll problem. The input, the processing, and the output are taken from Figure 3.3, the problem analysis chart for the payroll problem. The module reference numbers are taken from Figure 3.5, the interactivity chart for this problem. They show which module will perform each step in the processing. Notice also that these steps are numbered in the order to be processed.

Input	Processing	Module Reference Number	Output
Hours Worked Pay Rate	1. Enter Hours Worked 2. Enter Pay Rate 3. Calculate Pay 4. Print pay 5. End	1000 1000 2000 3000 0000	Gross pay

Figure 3.9 The IPO Chart for the Payroll Problem

Writing the Algorithms

After using the structure chart and the IPO chart, the next step in organizing a solution is for the programmer to develop sets of instructions for the computer, called algorithms (see Figure 3.10). **Pseudocode** may replace the algorithm in the development of a program when working in industry. Pseudocode is close to the actual language the programmer will be using. The algorithms we develop in this text are generic, and therefore we will not be able to use pseudocode. By using pseudocode, the programmer can start to write the actual code while finishing the final steps in the development of the program. Algorithms and pseudocode serve the same purpose.

To complete all of the algorithms needed to solve a problem, the programmer writes a separate set of instructions for each module in the structure chart. To be understood by the computer, the instructions have to be written according to certain rules that will be explained in future chapters throughout this book. Setting up the algorithms is probably the hardest part of problem solving on the computer. *The instructions cannot assume anything, cannot skip steps, must be executable one step at a time, and must be complete.* The modules are taken from the interactivity chart, and the processing is taken from the IPO chart. The number of instructions in an algorithm is determined by the way the programmer chooses to solve the problem. The algorithms pull the interactivity chart and the IPO chart together to give a logical step-by-step solution. Note in Figure 3.10, the control module uses an END since this is the end of the processing. The other modules use EXIT since the processing continues.

There are four algorithms and flowcharts for the payroll problem because there are four different modules in the structure chart. Figure 3.11 shows how the flowcharts and algorithms correspond for the payroll problem. Figure 3.12 shows the order of processing from module to module.

```
Control Module            Name of Module (list of parameters)

1. INSTRUCTION            1. INSTRUCTION

2. INSTRUCTION            2. INSTRUCTION

3. ..                     3. ..

4. ..                     4. ..

   ..                        ..

—. end                    —, exit
```

Figure 3.10 The Form of an Algorithm

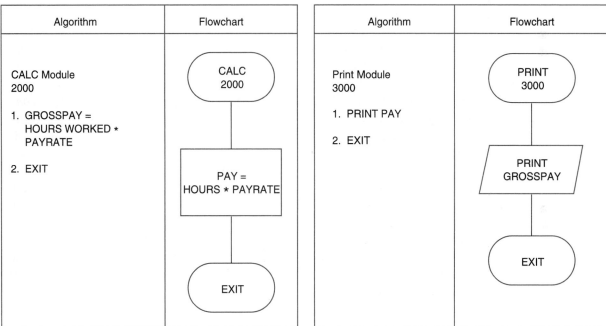

Figure 3.11 The Algorithms and Flowcharts for the Payroll Problem

Drawing the Flowcharts

From the algorithms the programmer develops the flowcharts, graphic representations of the algorithms. The algorithms and the flowcharts are the final steps in organizing a solution. Using them, the programmer can test the solution for bugs and go on to code the problem into a computer language for entry into the computer. A flowchart will show errors in logic not readily visible in the other charts. Also, a set of data can be tested easily using a flowchart.

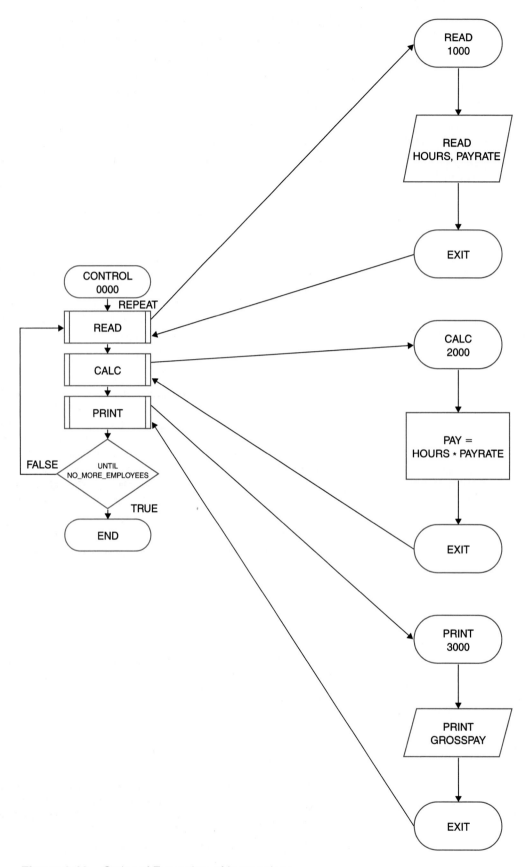

Figure 3.12 Order of Execution of Instructions

There are flowchart symbols for use with various types of processing. Figure 3.13 shows and explains some general flowchart symbols. There are many specific symbols, such as printer output, monitor output, and so forth, which are used in systems flowcharts. These are discussed later in this book in Chapters 14, 15, 16, and 17.

A flowchart shows the flow of the processing from the beginning to the end of a solution. Each block in a flowchart represents one instruction from an algorithm. *Flowlines* indicate the direction of the data flow. Most blocks have one or more *entrances*, flowlines directing the flow of the data into the block. Most blocks have only one *exit*, flowlines directing the data out of the block, since, in most cases, data can flow to only one other block. The exception to this rule is a block representing a *decision*

Flowchart Symbol	Explanation
Flowlines	Flowlines are indicated by straight lines with optional arrows to show the direction of data flow. The arrowhead is necessary when the flow direction might be in doubt. Flowlines are used to connect blocks by exiting from one and entering another.
Start End/Stop/Exit	Flattened ellipses indicate the start and the end of a module. An ellipse uses the name of the module at the start. The end is indicated by the word *end* or *stop* for the top or Control Module and the word *exit* for all other modules. A start has no flowlines entering it and only one exiting it; an end or exit has one flowline entering it but none exiting it.
Processing	The rectangle indicates a processing block, for such things as calculations, opening and closing files, and so forth. A processing block has one entrance and one exit.
I/O	The parallelogram indicates input to and output from the computer memory. An I/O block has one entrance and only one exit.
Decision	The diamond indicates a decision. It has one entrance and two and only two exits from the block. One exit is the action when the resultant is TRUE and the other exit is the action when the resultant is FALSE.

Figure 3.13 Flowchart Symbols *(continued on page 56)*

Flowchart Symbol	Explanation
Process Module	Rectangles with lines down each side indicate the process of modules. They have one entrance and only one exit.
COUNTER A B S Automatic Counter Loop	The polygon indicates a loop with a counter. The counter starts with A (the beginning value)and is incremented by S (the incrementor value) until the counter is greater than B (the ending value). COUNTER is a variable. A, B, and S may be constants, variables, or expressions.
On-Page Connectors* Off-Page Connectors*	Flowchart sections can be connected with two different symbols. The circle connects sections on the same page, and the home base plate connects flowcharts from page to page. Inside these two symbols the programmer writes letters or numbers. The on-page connector uses letters inside the circle to indicate where the adjoining connector is located. An A connects to an A, a B to a B, etc. The off-page connectors use the page number where the next part or the previous part of the flowchart is located. This allows the reader to easily follow the flowchart. On- and off-page connectors will have either an entrance or an exit.

*These connectors should be used as little as possible. They should be used to enhance readability. Overuse decreases readability and produces a cluttered effect.

Figure 3.13 *(continued from page 55)*

instruction, an instruction that enables the computer to make one of two choices. A decision block has two exits, one for each choice.

Besides the decision instructions, Figure 3.13 refers to another type of instruction that you will learn to use later, but that deserves a brief explanation here. A *loop* enables the computer to perform a task repeatedly during the processing of a solution. An *automatic counter loop* is a type of loop that enables the computer to count in increments of a given value; the increment can be 1, 2, 5, or whatever. That given value is the *incrementor* referred to in Figure 3.13.

Figure 3.14 shows a flowchart illustrating Otto the Robot walking to a wall and back again. Notice the use of the different flowchart symbols.

The following are some of the rules for drawing flowcharts:

1. You should write the instructions inside the blocks.
2. If there is something you need to remember, you can write a note beside a block. Test values of variables also can be placed beside flowchart blocks. This makes the flowchart an **annotated flowchart.**

Algorithm:

Start
1. Stand up.
2. Raise hands.
3. Are hands touching anything?
 Yes: GOTO 7.
 No: GOTO 4.
4. Add.
5. Take step.
6. GOTO 3.
7. Turn.
8. Turn.
9. Lower hands.
10. Is number in memory zero?
 Yes: GOTO 14.
 No: GOTO 11.
11. Subtract.
12. Take step.
13. GOTO 10.
14. Turn.
15. Turn.
16. Sit down.
17. Stop.

Flowchart:

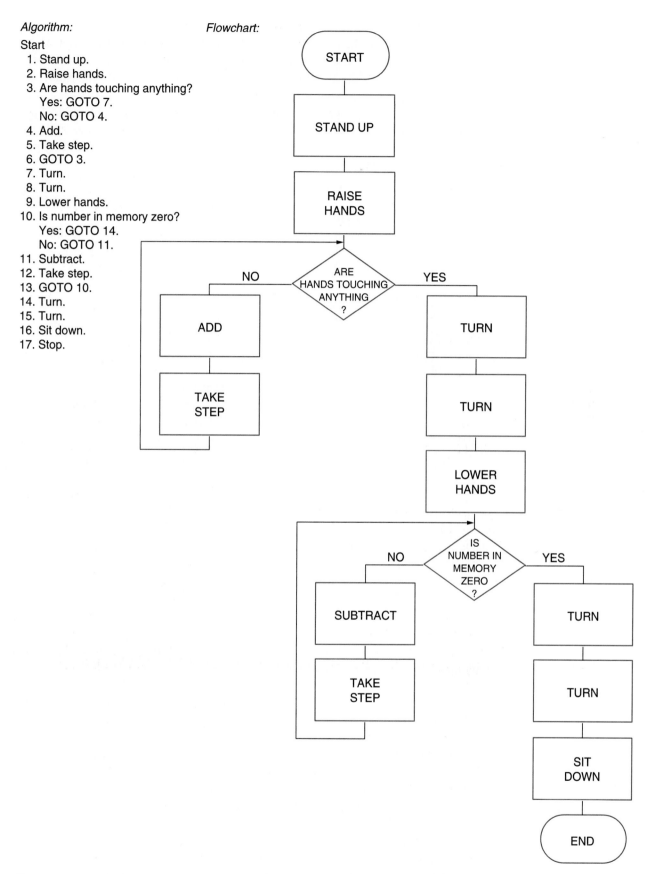

Figure 3.14 An Algorithim Flowchart for Otto the Robot (Problem 5d in Chapter 1)

3. A flowchart always starts at the top of the page and flows to the bottom. If you need more than one page for a flowchart, start another column on the same page or go on to another page. On- and off-page connectors are used to connect parts of the flowchart. A flowchart should not flow up, or sideways, or all over the page (see Figure 3.14).

4. Use a flowchart template to draw the symbols and a straight edge to draw the flowlines. When you use a template, the symbols are the same size and your flowchart will be neater and easier to read. You may use a flowcharting program to complete your flowcharts.

5. Make the blocks big enough to write instructions so they can be easily read.

6. Put the module number and name from the interactivity chart in the upper right-hand corner of the page for quick reference to the correct module.

7. Have plenty of paper on hand since the final copy of the flowchart normally will not be the first draft.

8. Use a pencil with a large eraser.

It is helpful to keep the algorithm and the flowchart of a module on the same page. Figure 3.15 shows a form (also in Appendix D, Figure D.4) you can use to develop your algorithms and flowcharts. Notice the last four sections of the form. The first is the Annotation section. Here you put notes about the algorithms and the flowcharts including test data, information about variables, things to remember, and so forth. The Test section is used to test the solution with sample data. The other two sections, Internal Documentation and External Documentation, are for notes to refer to later when you are completing the internal and external documentation. The last four sections of Figure 3.15 are optional but helpful in many cases.

Internal and External Documentation

Internal documentation consists of remarks written with the instructions to explain what is being done in the program. **External documentation** is made up of the manuals or help menus written about the solution. Documentation of your program is very important.

Good programming techniques require a program to be easily read by another programmer. The majority of the time, a program is passed from one programmer to another programmer. Proper internal documentation ensures that the new programmer will be able to learn about the program in the least amount of time. Internal documentation should include such things as people who have worked on the program, an abstract of the program, variable usage, notes about the development of the program, and anything else that will enable the new programmer to understand this program in the least amount of time. Internal documentation is written as the program is coded. Be careful not to overdo the internal documentation. Write enough information to help readability, but not so much that the instructions are lost in the remarks.

External documentation is for the user of the program. The end user will use the program to help solve problems. Proper external documentation ensures that the end user will have good instructions in order to learn to use this program in the least amount of time. External documentation should also allow the end user to find answers to questions arising while using the program. Good external documentation increases productivity for the end user. External documentation should include such things as tutorials, input definitions, usage instructions, installation instructions, command explanations, and anything else that will allow the end user to learn to use this program in the least amount of time. External documentation may be printed or electronic, such as help menus.

Algorithm	Flowchart	Annotation	Test	Internal Documentation	External Documentation

Figure 3.15 Algorithm and Flowchart Form

Using the Tools

The problem-solving tools used in this book are the problem analysis chart, the structure chart or interactivity chart, the IPO chart, the algorithm, and the flowchart. There is much duplication in these tools. Because of this duplication, you will find that after the first few weeks you will drop the use of some of these tools. However, for the first few problems, you should use all of the tools. In this way you will know which ones help you the most. You should end up using at least four of the six tools:

1. The PAC or the IPO chart
2. The structure chart or interactivity chart
3. The coupling diagram
4. The algorithm or the flowchart

Other tools are presented in Appendix E.

Testing the Solution

When a solution is complete, it is important to test it to make sure it meets the requirements of the user, and to check for errors in logic or in the setup of the expressions and equations. To test a solution, the programmer selects test data, a set of values for the input data, and works them through every step in the solution. It is important to select test data carefully, so the correctness of the results can be checked with as much accuracy as possible. The values of the input data should be capable of checking logic and calculations thoroughly. For example, to test the expression X/Y, the test data should be numbers other than 1. Since 1/1 is 1, and 1 * 1 is also 1, this set of data would not accurately test the expression.

A solution should never be assumed correct. If a bug is detected, the solution has to be modified to correct it, and then the testing process started again at the beginning. Logic errors, such as incorrect loop conditions or wrong conditions in a decision statement, are the hardest to find and correct. It takes only a few logic errors to produce a program that is unreadable and a programmer who is frustrated. It is far easier to correct the solution in the problem-solving stages of program development than after the program is put into the computer.

Coding the Program

Once you have completed the steps described so far, you have finished analyzing the problem and organizing the solution. This book stops with testing the solution. The next step is writing the solution of the problem into a computer language—that is, *coding* the solution. You will learn to code solutions when you take language and applications courses.

If the programmer follows the proper steps in developing the solution, there should be few logic errors in the program, and the testing and coding should go quickly. Most bugs will be syntax errors or will stem from a misunderstanding of the original problem. Often the process will appear shorter and easier if you skip these problem-solving steps and code directly into the computer. The experienced programmer knows the fallacy of this shortsighted reasoning; this approach is almost always more time consuming, and the resulting program is less readable and less efficient.

Summary

The efficiency of the computer relies on the effectiveness of the programmer or the user. The programmer must write solutions that are clear, organized, and correct. A solution is a sequence of instructions for solving a problem. A program is a solution after it has been coded into a computer language. Data are stored internally in memory locations according to variable names and externally on a magnetic disk. External data are stored in sections called files, which are divided into records, which are further divided into fields. The file must be opened before accessing the data and closed when the user has finished with it.

The steps in organizing the solution to a problem start with problem analysis, a process whereby you separate a problem into four parts, including given data, required results, processing needs of the problem, and solution alternatives. The problem analysis chart is an aid to completing this first step. Second, the interactivity chart is developed, showing the breakdown of the whole problem into subtasks called modules. The IPO chart is the third step; it shows the input, the processing, and the output. The algorithms are developed fourth, showing in detail the sequence of instructions in the solution. Finally, from the algorithms the programmer draws the flowcharts, showing graphic representations of the algorithms. After completing these steps, the programmer can test, code, and implement the solution.

New Terms

algorithm	interactivity chart
annotated flowchart	internal documentation
buffer	IPO chart
bug	module
debugging	problem analysis chart
external documentation	pseudocode
field	record
file	structure chart
flowchart	syntax

Questions

1. What is a program?
2. Why is problem analysis important?
3. Define a database, a file, a record, and a field. Give an example of each.
4. What happens when a programmer opens or closes a file?
5. Would each of the following be classified as a file, a record, or a field? Explain your answer.
 a. a pay rate
 b. all student data at a university
 c. one salesperson's data

(Question 5 continued on next page)

d. a telephone number

e. a quantity

f. an employee's personnel data

g. an inventory item number

h. a price

i. information on one class

j. a sales receipt

6. What are the tools of problem solving described in this chapter?

7. What are the relationships of the problem-solving tools to each other?

8. How would each of the flowchart symbols be used? Include an example in your explanation.

Problems

1. Develop a PAC for a solution to balance your checkbook. Include in the PAC various solutions for solving this problem.

2. List the various modules that you might include in the solution in Problem 1. Put these modules together in an interactivity chart.

3. Draw the flowchart for your solution to the problem in Chapter 1 problem 5e.

4. Draw a flowchart for the algorithm you developed in Chapter 1 problem 5f.

5. After completing problem 3 or 4, list 3 advantages to using an algorithm and 3 advantages to using a flowchart. Justify your answer. You may want to discuss the answer to this problem with other classmates.

Unit One

Supplementary Exercises

1. Write a set of numbered step-by-step instructions to plan a trip to San Francisco. Be sure to take into account the knowledge base of those who are going.
2. Write a set of numbered step-by-step instructions to plan a summer vacation to a place of your choice.
3. Write a set of instructions for Otto to walk in a square figure 8.
4. Define the data types of the following data items. Justify your answers.
 a. sum of money
 b. telephone number
 c. zip code
 d. quanitity of shoes
 e. check okay on a client
 f. account number
5. Define a variable name for each of the items in Exercise 4.
6. Evaluate for A = 6, B = 4, and C = 2.
 a. $F = A \wedge 2/3 * B$
 b. $F = 3 * B/A \wedge 2$
 c. $F = A * C\backslash(A + C)$
 d. $F = (A + 7 - C) \text{ MOD } B$
 e. $F = (C * (B + 3 * A) + 5 * A)/C$
7. a. Joe Brown's Shoe Store expects to increase its sales by 5% during the next year. Write an equation to calculate the projected sales at the end of the year. *Hint:* The standard expression to increase a number by a percentage is

$$\text{NUMBER} * (1 + \text{PERCENTAGE})$$

 b. However, inflation will decrease his profit increase. Expected inflation is $4\frac{1}{2}\%$, and 50% of Joe's expenses are directly affected by inflation. Joe's expenses are equal to 80% of his sales. Modify the equation in Exercise 7a to calculate the projected profit, taking into consideration inflation.

8. Evaluate for A = TRUE, B = TRUE, C = FALSE, D = FALSE:
 a. R = (A OR B) AND C
 b. R = A AND B OR C
 c. R = B AND C OR A
 d. R = NOT C AND NOT D OR A
 e. R = NOT C OR NOT D AND A

9. Set up a logical expression for the following policy on using a company credit card. The card may be used if
 a. balance plus sales amount is less than maximum allowable amount
 b. last payment was less than 45 days ago
 c. credit card has not expired

10. Set up a logical expression to check data using the following criteria. The data are assumed correct when:
 a. number is < 1000
 b. number is positive
 c. number is divisible by 2

11. Develop a flowchart for the instructions to balance your checkbook. Be sure to include all steps, such as repeating the addition of each check.

12. Select one of the problem solving tools and answer the following questions:
 a. How does this tool help in the solution of a problem?
 b. Why would you use this tool?

13. Which tool do you think will be hardest for you to learn to use? Why?

WHAT'S WRONG WITH THIS?

The following problems have errors in them. Find the error, explain what is wrong, and correct the error.

1. Set up the following equations for the computer:

Mathematical Equation	Computer Equation
a. A + B = 5D	**a.** D = A + B/5
b. $\dfrac{X + Y}{4}$ = Z	**b.** (X + Y) / 4 = Z
c. 5X − 3Y = Z	**c.** Z = 5X − 3Y

2. What is the order of processing of the following equations:

Computer Equation	Order of Processing
a. D = A + B/5	**a.** D = A + B/5 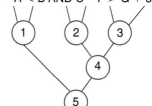
b. S = 5/D + 6* (C + B) / E	**b.** S = 5/D + 6* (C + B)/E
c. R = A < B AND C * F > G + 3	**c.** R = A < B AND C * F > G + 3

3. Write a truth table for the following equation:

R = A AND B OR NOT C

R = A AND B OR NOT C

A	B	C	R
F	F	F	F
T	T	T	T
F	F	T	F
T	F	T	T
T	F	F	F
F	T	T	F

UNIT TWO

LOGIC STRUCTURES

Chapter 4

An Introduction to Programming Structure

Overview

Pointers for Structuring a Program
Cohesion and Coupling
The Modules and Their Functions
Local and Global Variables
Parameters
Variable Names and the Data Dictionary
The Four Logic Structures

Objectives

When you have finished this chapter, you should be able to:

1. Explain the need for structured programming.
2. Explain how to design modules and functions, in terms of cohesion and coupling.
3. Explain the difference between local and global variables.
4. Explain the use of parameters.
5. List and describe the four logic structures: sequential, decision, loops, and case.

The tools described in Chapter 3 help programmers organize programs as they arise in the business world or in a textbook. The techniques explained in Chapters 4 through 8 of Unit Two provide ways to structure solutions so the computer can efficiently read and process the finished programs. That is, the techniques explained in these chapters help you write the several types of instructions that make up algorithms.

Pointers for Structuring a Program

The computer should be a tool to help people find solutions to their problems and to increase their productivity. The solutions must be developed according to methods that accomplish these goals. You can develop efficient computer solutions to problems if you heed the following pointers:

1. Use modules—break the whole into parts, each part having a particular function.
2. Use the four logic structures to assure that the solution flows smoothly from one instruction to the next, rather than jumping from one point in the solution to another.
 a. The **sequential structure** executes instructions one after another in a sequence. See Figure 4.1.
 b. The **decision structure** branches to execute one of two possible sets of instructions. See Figure 4.2.
 c. The **loop structure** executes a set of instructions many times. See Figure 4.3.
 d. The **case structure** executes one set of instructions out of several sets. (This is a special case of the decision structure. Many textbooks include the case structure with the decision structure; however, because of the unique uses of the case structure, it has been placed in a separate chapter in this text.) See Figure 4.4.
3. Eliminate the rewriting of identical processes by using modules.
4. Use techniques to improve readability, including the four logic structures, proper naming of variables, internal documentation, and proper indentation.

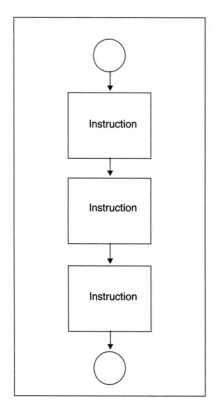

Figure 4.1 Sequential Logic Structure

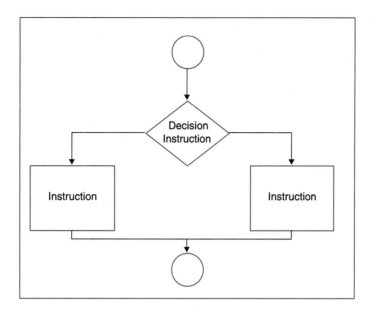

Figure 4.2 Decision Logic Structure

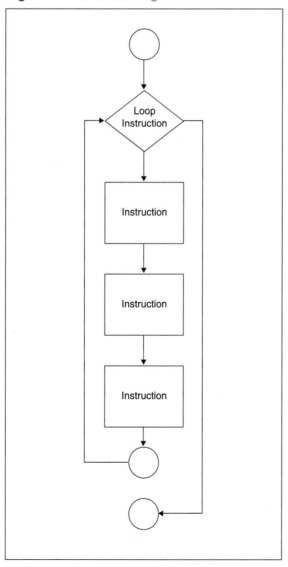

Figure 4.3 Loop Logic Structure

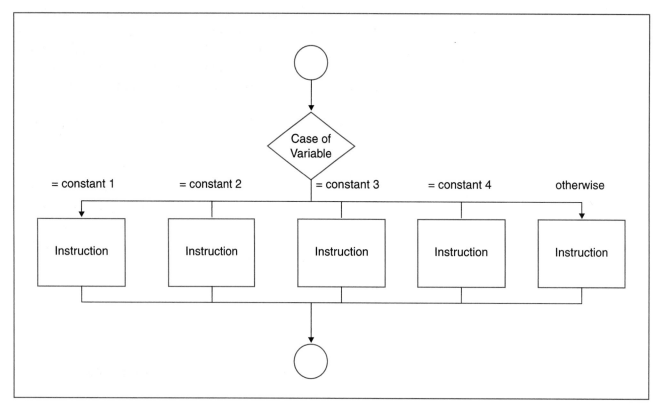

Figure 4.4 Case Logic Structure

You can structure programs more easily if you build a library of techniques that have been perfected by the experts over the years. One of the rules for programming a computer is don't reinvent the wheel. Experienced programmers are productive because, among other reasons, they have learned good programming techniques.

In addition to techniques for structuring programs, the beginning programmer needs to learn the techniques of working with data structures. Techniques used with data structures teach the student how to store data in order to access and process it as efficiently as possible. These include techniques for processing arrays, linked lists, and trees. Beginning students also need to learn techniques for programming in a graphical user interface computer world where instructions are not necessarily executed in a procedural manner. This is called object-oriented programming. These techniques are discussed in Units Three and Four in this textbook.

Cohesion and Coupling

One of the most difficult tasks in the development of a solution to a problem is to divide the solution into modules. The question arises as to what should be included in each module. Modules should be functionally independent and perform a single task. However, modules will need to be connected, primarily through the data needed to complete the specified task within the modules. These opposing concepts are called **cohesion** and **coupling.** The concepts describe conflicting principles that lead to program design compromise.

Cohesion relates to the functional independence of the module and to the performance of a single "task" within the module. The overall objective is to separate the problem into parts such that each part becomes a module that will be designed to complete an independent task, such as entering data, printing data, initializing data, and the like. Each module is independent of each other module, with a single entrance and a single exit. See Figure 4.5.

Coupling allows modules to be connected by an interface, which enables the programmer to transfer data from one module to another. Three coupling techniques are parameters, module names, and data that all modules can access, called global variables. Coupling allows for communication between modules.

Cohesion and coupling are especially important concepts to the modern programmer, who in all probability will be working with many other programmers on the same problem. Cohesion allows a programmer to write a module of a larger program and test it independently. This module can then become a "black box" in a large program. Coupling allows all programmers to use necessary variables without losing the cohesion

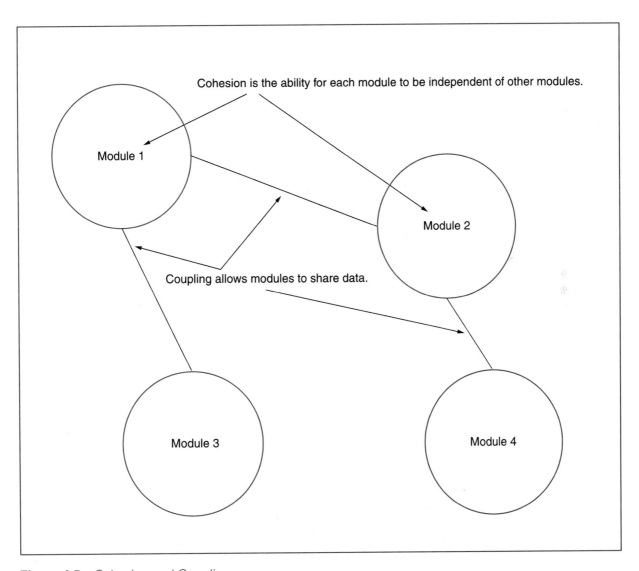

Figure 4.5 Cohesion and Coupling

of the module. For example, if Module 1 performs task 1 successfully, it may be required to provide data to Module 2. Cohesion results from having separate modules; coupling results when data are passed from Module 1 to Module 2. Another example of cohesion and coupling would be the use of the square root function. A function is another name for a module. The square root function is a set of instructions to calculate the value of the square root of a number. The independence of these instructions is *cohesion*. The function is basically a black box. The programmer really doesn't worry about what the instructions are as long as a correct value is returned. However, in order to calculate the square root, the function needs a value to calculate the square root, and a value to return to the programmer. These data links, the number and the square root of the number, are what connects or *couples* the function to the rest of the program. This is *coupling*.

The Modules and Their Functions

Just as an author would not sit down and write without thinking about the structure of the book, neither should a programmer write without thinking about the structure of the program. First, an author thinks about the subject. A programmer studies the problem. Next, the author breaks the subject matter into chapters. The programmer breaks the problem into modules, each with a specific function. Once the chapters are in the correct order then the author starts to write them, one by one. Once the modules are in the correct processing order in the interactivity chart, then the programmer writes the algorithm for each module, one by one. A good programmer tests the algorithm for each module for correctness, as thoroughly as possible, before writing the next one. It is much easier to write and test many small modules than a single large one. Not only that, but when the program needs to be modified, it is much easier to test and change small modules.

Follow these rules for designing modules when you design your own solutions:

1. Each module is an entity in itself. There is one entrance and one exit. That is, the processing starts at the top and ends at the bottom of the module. Processing does not jump out of the middle of a module to another module, nor does it jump into the middle of a module. The only way to go from one module to another module is through the exit instruction or through the use of the Process module instruction.
2. Each module has a single function, such as printing, calculating, or entering data.
3. Each module is short enough to be easily read and modified.
4. The length of a module is governed by its function and the number of instructions to be executed to perform that function.
5. A module is developed to control the order of processing. This may be where the interface of an object-oriented program is introduced or where the instructions to the computer are given as to the order of processing of the modules in a procedural-oriented program.

The types of modules needed for the solutions to most problems in this unit are the following:

1. **Control module**—This module shows the overall flow of the data through the program. All other modules are subordinate to it.

2. **Init module**—The **Initialization module** sometimes called the Preparation module, processes instructions that are executed only once during the program, and only at the beginning. These instructions include opening files, setting the beginning values of variables used in the processing, and printing report headings.

3. **Process Data modules**—These modules may be processed only once, or they may be part of a loop, which is processed more than once during the solution. There are several kinds of Process Data modules:

 a. **Calculation modules** which do arithmetic calculations, accumulating (summing to calculate totals), counting, or manipulating numerical data in some other way, such as putting numbers in numerical order. Sometimes Calculation modules manipulate string data to perform tasks such as putting a list of values in alphabetical order.

 b. **Print modules** which print output lines (the results of the processing, line by line), including headings and summaries.

 c. **Read** and **Data Validation modules,** which read or input the data into the internal memory of the computer, validate data (check the accuracy of the data), enter data from the keyboard, and so forth. Validation modules are usually separate from Read modules.

4. **Wrap-up module**—This module processes all instructions that are executed only once during the program and only at the end. These instructions include closing files and printing totals, among others.

5. Modules in an object-oriented program may include event modules such as mouse down, mouse up, key entry, etc.

Different types of modules are often combined in one solution. A module can be broken into several smaller modules to make the parts of a program more manageable. One module performing a specific function can be used by one or more other modules in which that same function is needed. Other types of modules are introduced in later chapters. The above types are the most frequently used.

Local and Global Variables

The concept of local and global variables is important in many languages. Programmers use the concept of local and global variables to allow cohesion and coupling to occur. Global variables are defined outside of the individual modules. Local variables are defined within a module. The major difference between the two is in the *scope* of the variables. The scope of the variables designates where the value of the variable can be used.

Local variables may be used only by the module itself. Other modules have no knowledge of these variables. This allows cohesion to take place. Through the use of local variables the programmer does not have to worry about variable name duplication in modules created by other programmers. If other modules need to use the value of a variable, the modules must be coupled through the use of parameters or return values.

All modules have knowledge of **global variables.** They are global to the program as they can be seen by all modules. The use of global variables defeats one of the purposes of cohesion, that is to allow for the duplication of variable names in different

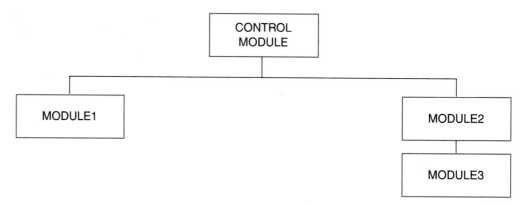

Figure 4.6 Interactivity Chart

modules. While global variables allow data coupling through the entire program, they should be used with caution. Only those variables that will be used in all modules should be designated as global variables. The programmer must be careful not to use the same variable name for a local variable and a global variable, since the hierarchy of use is to search the local variable list first, the parameter list second, and the global variable list last. The use of parameters and return values is not necessary since global variables couple the data from module to module.

Figure 4.6 shows a typical small interactivity chart. Figure 4.7 shows the scope of the variables found in this program. The variables A, B, and C are global. All modules can use and change the values of these variables. All other variables are local. Only the Control Module can see the values of X, Y, and Z. The values of D, E, and F can only be seen by Module1. Module2 can see G, H, and I. Module3 can see only X, J, and K. Notice the duplication of the variable name X. In this case, there are two memory locations named X, one that can be used by the Control Module and the other that can be used by Module3. The computer will not get confused by this duplicity, since each X is linked to its own module.

Parameters

Parameters are one of the hardest and most important concepts to understand and to use in programming. Through the use of parameters, programmers may work independently in groups, each developing a section of a very large program without the worry of duplicating variable names. In this way, a program can be developed in a relatively short amount of time. Each section is developed as a cohesive module, coupled together mainly through the use of parameters. In learning about parameters, new terminology will be introduced and discussed: the formal versus actual parameter listings, the calling module versus the called module, and call-by-value versus call-by-reference parameters. Refer to Figure 4.8 as you are reading the information about these terms.

Parameters are local variables that are passed or sent from one module to another. Parameters are another way of facilitating coupling that allows the communication of data between modules. They are placed after the module name and within parentheses,

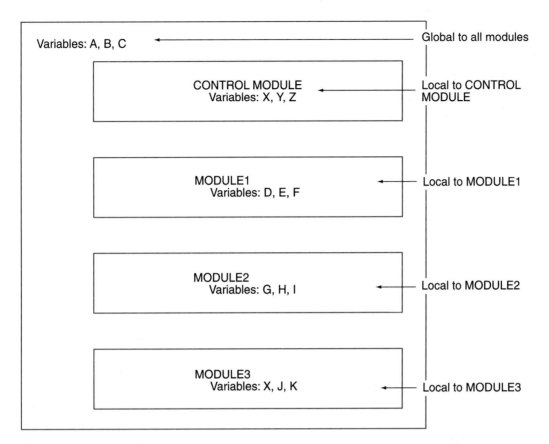

Figure 4.7 Scope of Local and Global Variables

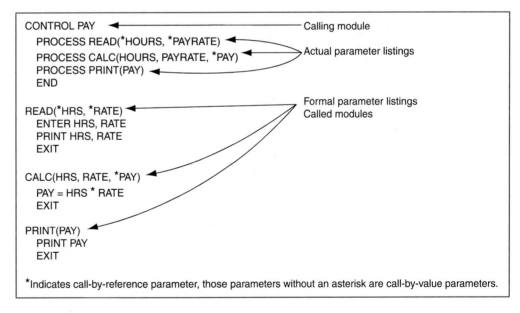

Figure 4.8 Parameter Terminology

for example: Read(A, B, C). The *calling module* is the module that processes another module. The *called module* is the module being processed. In Figure 4.8, the calling module is the Control Pay module since it processes the Read module, the Calc module, and the Print module. The Read module, the Calc module, and the Print module are the called modules since they are the ones being processed by the Control module. The *actual parameter listing* is the list of parameters that follows the module name being processed in the calling module. The *formal parameter listing* is the list of parameters that follow the module name at the beginning of the module. Notice that the variable names in the actual parameter listings may or may not be the same as those in the formal parameter listings. The values are sent and received according to the position in the listings; the first variable in the actual listing corresponds to the first variable in the formal listing, the second to the second, and so forth. In Figure 4.8, HOURS in the Control Pay module corresponds to HRS in the Read and Calc modules. Likewise, PAYRATE in the Control Pay module corresponds to RATE in the Read and Calc modules, and PAY in the Control module corresponds to PAY in the Calc and Print modules.

There are two ways to send data from one module to another through the use of parameters. The first is to send the *value* of the variable, a *call-by-value parameter*. The second is to send the *address* of the variable, a *call-by-reference parameter* that is specified by the use of an asterisk (∗) in front of the variable name in both the actual and the formal parameter listings. A variable name that has no asterisk in front of the name is assumed to be a call-by-value parameter. When a call-by-value parameter is used, the value of the variable is sent to the called module by the calling module. The called module will then make a new memory location (or address—a memory location is an address) for this variable since it has no knowledge of where the variable is stored by the calling module. Therefore, when the called module makes a change in the variable, the change will not be made in the calling module because the parameter has a different memory location in the called module than it does in the calling module (see Figure 4.9). The call-by-value parameters (referenced by circled 2s) have a memory location number in both modules. When a call-by-reference parameter (referenced by circled 1s in Figure 4.9) is used, the memory location is sent, not the value. The called module refers to the memory location to find the value. When the called module changes the value of the parameter, the calling module will see the change because they are using the same memory location—the one designated by the calling module. The call-by-reference parameters share the same memory location. When using parameters, variables might have the same location but two or more names referenced from different modules, or two memory locations may have the same name used in two different modules. Figure 4.10 shows another example of parameters.

Figure 4.11 shows the coupling diagram for Figure 4.9. A *coupling diagram* indicates how data couples modules together. A coupling diagram has a rectangle for each module and a line connecting the modules for each datum sent from one module to another. In this case, there is a rectangle for Module1 and another for Module2. Notice the connecting lines. Two double-headed arrows are drawn from the Control Pay module to the Read module, one for the hours worked and the other for the rate of pay. The double-headed arrows indicate that these parameters are call-by-reference parameters and changes can be seen by both modules. In Figure 4.11, two single-headed arrows and one double-headed arrow are drawn from the Control Pay module to the Calc module; the single-headed arrows are for the hours worked and the rate of pay and the double-headed

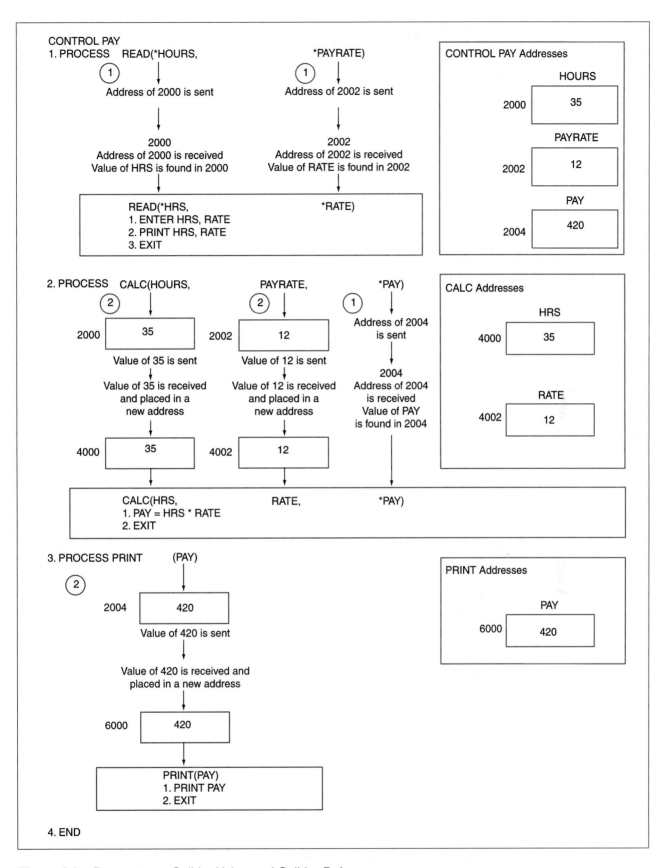

Figure 4.9 Parameters—Call-by-Value and Call-by-Reference

```
            MODULE1
              A = 3
              B = 4
              C = 5
              _____
              _____
              _____
    (1)    PROCESS MODULE2 (A, B, *C)
              _____
              _____
              _____
           EXIT

    (2)    MODULE2 (X, Y, *Z)          X AND Y ARE CALL-BY-VALUE PARAMETERS
              _____      Z IS A CALL-BY-REFERENCE PARAMETER
              _____
           X = 7
           Y = 8
           Z = 9
              _____
              _____
              _____
    (3)    EXIT
```

(1) Value of variables before PROCESS MODULE2 is executed:

A	B	C, Z	
3	4	5	← Variables for MODULE1

X	Y	
		Variables for MODULE2

(2) Value of variable at the beginning processing of MODULE2:

A	B	C, Z	
3	4	5	← Variables for MODULE1

X	Y	
3	4	Variables for MODULE2

(3) Value of variable at the end of processing of MODULE2:

A	B	C, Z	
3	4	9	← Variables for MODULE1

X	Y	
7	8	Variables for MODULE2

Figure 4.10 Parameter Example

arrow is for the amount of pay. In this case, hours worked and the rate of pay have single-headed arrows with the arrowheads into the called module, indicating that these are call-by-value parameters and changes can only be seen by the called module. This is because the called module, Calc, will be using these values and not changing them. The

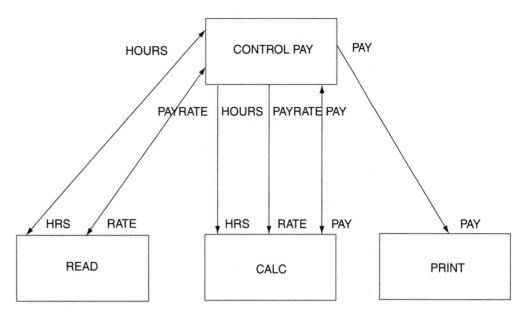

Figure 4.11 Coupling Diagram for Figures 4.8 and 4.9

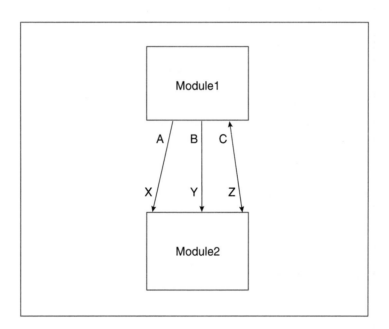

Figure 4.12 Coupling Diagram for Figure 4.10

amount of pay has a double-headed arrow, specifying a call-by-reference parameter. This is because PAY will be changed in Calc and the Control Pay module needs to know the value of the change in PAY. One double-headed arrow is drawn from the Control Pay module to the Print module, specifying that the amount of pay is a call-by-value parameter for this module since the Print module will only used the value. See Figure 4.12 for the coupling diagram for Figure 4.10.

Parameters are one of the best methods of coupling modules. The programmer does not have to worry about variable name duplication in modules that other programmers are developing. Parameters in the formal parameter listing (in the called module)

are neither local nor global variables and therefore should not be declared as local or global variables. In the event of variable name duplication, the local variable has precedence over the parameter variable, and the parameter variable has precedence over the global variable. When the program is looking for the variable name, it first looks at the local variable list, second at the formal parameter list, and last at the global variable list. If it is not found at any of the three lists, an error occurs.

The three ways to couple modules are through the use of global variables, parameters, and the module name. As we learned in Chapter 2, when functions are used within another instruction, they have a return value. The return value is the result of the function. This is accomplished through the name of the function. At the conclusion of the execution of the function, the result is placed temporarily in the name. The function must be used as part of another instruction. The value of the name of the function is no longer available once the instruction has been executed. In the case of the return value, the value is only sent out of the called module into the calling module. In a coupling diagram the arrowhead would be a single-headed arrow, but pointing to the calling module.

Many of the programs presented in this text will use parameters and will have coupling diagrams presented as part of the solution. Likewise, many of the problems at the end of the chapters will require the use of parameters. As a programmer it is very important that you learn how to use parameters.

Variable Names and the Data Dictionary

Variables must be named according to what they represent. If they represent *hours* then the variable name should be HOURS. When the variables are all local or used as parameters, variable names must be unique to the module in which they are defined. If variables are global, they must be unique to the total program. When you name a variable, that name must be used whenever you are referring to that item. You cannot use HOURS in one place and HRS in another place in one module, or again, if the variable is global, throughout the whole program. The computer references values by their names. Therefore if you change the name the computer assumes the new name represents a new item. A data dictionary is a great help to keep track of the variable usage in your program. It contains a list of all items, their variable names, their data types, the module in which they are found, pseudonyms and the module in which the pseudonym is found, and any error check that needs to be made on the variable. Figure D.5 in Appendix D is a form to use for your data dictionary. Figure 4.13 shows an example of a data dictionary. If you are using different names in different modules, you must also show the coupling diagram.

The Four Logic Structures

Up to now, this book has discussed the structure of the *problem*, that is, how to break a problem into parts, or modules, and arrange them so the data flow in a solution is logical. There are other types of structures used to design the modules themselves. These structures are called **logic structures,** and they control the logic of the data flow though the module. The instructions that make up the algorithms and the flowcharts are combinations of the four logic structures.

Data Dictionary

Item	Variable Name	Data Type	Module	Scope	Pseudonyms/Module	Error Checks
Hours worked	HOURS	Numeric—real	CONTROL PAY	Local	HRS	None
Hours worked	HRS	Numeric—real	Read/ CALC	Parameter	HOURS	Hours < 0
Payrate	PAYRATE	Numeric—real	CONTROL PAY	Local	RATE	None
Payrate	RATE	Numeric—real	Read/ CALC	Parameter	PAYRATE	Payrate < 4.00
Net pay	PAY	Numeric—real	CONTROL PAY	Local	None	None
Net pay	PAY	Numeric—real	CALC/ Print	Parameter	None	None

Figure 4.13 Example of a Data Dictionary

Chapters 5, 6, 7, and 8 in this unit are devoted to the four basic logic structures:

1. The sequential logic structure
2. The decision logic structure
3. The loop logic structure
4. The case logic structure

A brief explanation of each of these logic structures follows.

The sequential logic structure processes data one instruction after another. This logic structure is the most widely used, and it is normally used in conjunction with others. The programmer uses the decision logic structure to program the computer to make choices. The programmer uses calculations that use relational or logical operators to design these choices. The third logic structure is the loop, which enables the computer to repeat tasks. The last logic structure, the case logic structure, enables the programmer to request that the computer pick one of a set of tasks through a code or through an option using a number, a letter, or a string. When a problem is broken down in modules, the appropriate logic structures can be applied to each module. Effective use of the four logic structures results in efficient and readable solutions.

Summary

Unit Two consists of several chapters on how to structure solutions for particular computer languages and applications. You will learn four logic structures for designing programs for problem solving—plus additional techniques for record processing. These

techniques are extremely important because they can greatly increase productivity. They help you effectively organize the instructions in each of the algorithms that make up your solution. As a result, the computer can read the solution easily and process your program efficiently.

Cohesion is designing a module to be as independent of the rest of the modules as possible. Coupling refers to how data couple the modules. The three ways to couple modules are through the use of global variables, parameters, and module names. Global variables are those variables known to other modules. Local variables are those variables known to only one module. The computer searches for variables first in the local variable list, then the parameter list, and finally the global list.

Parameters are local variables that are shared with other modules. Call-by-reference parameters pass the address or location of the variable to the other module. Call-by-value parameters pass the actual value of the variable to the other module. The calling module can recognize a change made by the called module when using a call-by-reference parameter. The calling module when using a call-by-value parameter does not recognize a change made by the called module. It is very important for a programmer to learn how to effectively use parameters.

To make a computer problem manageable, break your solution into modules. For Unit Two, the types of modules you will need include Control, Init, Process Data, and Wrap-up. There are several kinds of Process Data modules, including Calculation modules, various Print modules, Read modules, and File Maintenance modules.

New Terms

cohesion

coupling

global variables

local variables

logic structures:

 sequential structure

 decision structure

 loop structure

 case structure

parameters

types of modules:

 Control

 Initialization

 Process Data

 Calculation

 Print

 Read

 Data Validation

 Wrap-up

Questions

1. Name the four logic structures.
2. What is a module? How is it used?
3. Name the major types of modules and explain their functions.
4. List the kinds of Process Data modules.
5. What is meant by the cohesion of a module?
6. What is meant by coupling modules?

7. Why are cohesion and coupling important to programmers?

8. How are modules coupled?

9. What does a coupling diagram show?

10. Define the two types of parameters. How do they differ? Under what circumstances would you use each type?

Problems

1. Create a calling module name and parameters for the following: (Remember that the asterisk before the parameter name indicates a call-by-reference parameter.)
 a. Process Print(Average, Count)
 b. Process Calc(AgeSum, Count, *Average)
 c. Process Calc(Length, Width, Height, *Volume)

2. Using the modules in problem 1, answer the following questions:
 a. Which of the parameters are call-by-value parameters?
 b. Which of the parameters are call-by-reference parameters?
 c. Are there any parameters that cannot be changed by the called module? If so, which ones. Justify your answer.
 d. Which parameters, if changed by the called module, will the calling module recognize as changed values?
 e. Which parameters have more than one location in the computer, one associated with the calling module and the other associated with the called module?
 f. Which parameters have one location in the computer, that of the calling module?

3. Create a data dictionary for the parameters in Problem 1.

Chapter 5

Problem Solving with the Sequential Logic Structure

Overview

Algorithm Instructions and Flowchart Symbols
The Sequential Logic Structure
Solution Development
 Problem Analysis
 The Interactivity (Structure) Chart
 The IPO Chart
 Coupling Diagram and Data Dictionary
 Internal and External Documentation
 The Algorithms and Flowcharts

Objective

When you have finished this chapter, you should be able to:

1. Use the sequential logic structure to develop a solution to a problem.
2. Use the proper form for instructions in an algorithm and a flowchart.
3. Use the seven problem solving tools to develop a solution to a problem.

Algorithm Instructions and Flowchart Symbols

The computer's system of communication dictates how the programmer designs algorithms and flowcharts. Recall that an algorithm is a set of instructions telling the computer how to process a module in a solution. You will write an algorithm for each module in a program. A flowchart is a visual illustration of an algorithm. You will draw a flowchart to accompany each algorithm. This chapter begins with only a few of the specific algorithm instructions and flowchart symbols you will use as a trained programmer,

those required to solve problems using the sequential structure. They are found in Table 5.1 and include the instructions to start and stop each module, to enter data, to output data, to process modules, and to perform calculations.

You write an instruction in two parts: a general instruction followed by specifications regarding what is to be processed. In Table 5.1 the general instruction is the first word in the Algorithm Instruction column, that is, START, ENTER, VARIABLE =, and

Table 5.1 Algorithm Instructions and Flowchart Symbols

Algorithm Instruction	Flowchart Symbol	Definition (What the instruction tells the computer)
START module name list of parameters (module number)		Start processing a module. The word START is optional and is not used in this text.
ENTER list of variable names		Enter data from a data block or from the keyboard into the following variable memory locations.
Variable = expression		Assign the value of the following expression to a variable location. Called the *assignment instruction*. A module may be processed as part of an expression, if there is a return value.
WRITE list of variable names		Write to the printer the values represented by the following variable names.
PRINT list of variable names		Print on the screen the values represented by the following variable names.
PROCESS module names (list of parameters)		Execute the instructions in the named module, and then return to execute the next instruction.
END/EXIT		Stop the processing of the module. Processing ceases with the End; with the Exit, processing continues with the next instruction after the PROCESS.
RETURN (variable)		Stop the processing of the module, place the value of the variable in the module name, and continue with the instruction containing the module call.

so forth. The specifications—a specific module name, a list of variable names, a specific expression, and so forth—follow the instruction. They are data and ways of processing drawn from the problem and collected for the problem, and information from the structure chart and the other charts you completed prior to the algorithms.

The START instruction is always understood, if not stated, at the beginning of an algorithm and flowchart and indicates where the module begins. The oval is the flowchart symbol for the START instruction. The instruction includes the name and number of the module. For example

<div align="center">Init (HOURS, PAYRATE) (1000)</div>

or

<div align="center">Print (HOURS, PAY) (4000)</div>

The name and number of the module also may be written at the beginning of the algorithm, which is the case throughout this text.

The ENTER instruction enables the user to enter data into memory locations with variable names. The list of variable names follows the ENTER instruction. For example

<div align="center">ENTER NAME, AGE</div>

Data are entered from the keyboard as specified in the program.

Variable = (commonly called the *assignment instruction*) enables a user to instruct the computer to store a constant, a variable, or an expression in a named variable memory location. The order of the instruction is the variable name followed by the equal sign followed by a constant, a variable name, or an expression, as in

<div align="center">PAY = HOURS * WAGE</div>

A module, called a function, may be processed in an expression if and only if the module or function has a return value. For example, the square root function is used in an assignment statement as follows:

<div align="center">A = SQRT(X) + 7</div>

If we were to look at the instructions for the square root function, we would see that the result of the square root of X would be returned in the name of the function, in this case SQRT. This value is added to 7 and then stored in A.

The WRITE instruction outputs values to the printer. The list of information to be printed follows the WRITE, as in

<div align="center">WRITE NAME, AGE</div>

The PRINT instruction is similar to the WRITE instruction except that the output is directed to the screen rather than the printer.

The PROCESS instruction tells the computer to process the instructions of a named module before continuing. The name of the module to be processed follows the PROCESS instruction. The list of parameters, enclosed in parentheses, follows the module name. For example

<div align="center">PROCESS Init (HOURS, PAYRATE)</div>

or

<div align="center">PROCESS Calc (HOURS, PAYRATE)</div>

If there are no parameters, the parentheses are not included.

The END, EXIT, or RETURN instruction specifies the completion of a module. END is used to end the Control Module and indicates that the processing of the solution is complete. EXIT is used to end a subordinate module if there is no return value, and indicates that the processing will continue in another module, the module where the PROCESS instruction originated.

The RETURN(variable) is used to place a value in the name of the module. The RETURN is used when the module is to be processed within an expression. The only way to preserve the value of the function name is either to use it in an expression or to place the value in a variable through the use of the assignment statement. The processing continues in the calling module, the module in which the instruction that contained the expression originated. The programmer writes each instruction in its entirety within the appropriate flowchart symbol. The list of instructions in Table 5.1 will be expanded in future chapters.

The Sequential Logic Structure

The most commonly used and the simplest logic structure is the sequential structure. All problems use the sequential structure, and most problems use it in conjunction with one or more of the other logic structures. To illustrate how to use this structure, this chapter develops a sample problem solution demonstrating each of the computer problem-solving steps explained in Chapter 3.

A programmer who uses the sequential logic structure is asking the computer to process a set of instructions in sequence from the top to the bottom of an algorithm. In general, the form of the algorithm looks like this:

MODULE NAME (list of parameters) (module number)

1. INSTRUCTION

2. INSTRUCTION

3. . . .

. . .

XX. END, EXIT, or RETURN (variable)

The algorithm begins with the START instruction represented by the name, the list of parameters, and number of the module. The numbered instructions following the START might be any of the instructions other than START and END/EXIT/RETURN, depending on the problem. The form of the flowchart corresponds to the form of the algorithm (see Figure 5.1). The first block is the START block. Other instruction blocks follow. These blocks can be of any type other than the flattened ellipse, which indicates START or END/EXIT/RETURN. The final block is the END, EXIT, or RETURN block. See Figure 5.2 for an example of the algorithm and the flowchart for a solution that enters a name and an age into the computer and prints them to the screen.

The name of the module is NAMEAGE and has a module number of 0000. There are no parameters. The flowchart uses the flattened ellipse to indicate that this is the start of this module. The first instruction to be executed is the ENTER instruction. The instruction is in the form of the word ENTER followed by the list of data items whose values will entered from the keyboard a data block. The flowchart uses a parallelogram to designate that this is an input instruction. The second instruction to be executed is the PRINT instruction. This instruction is in the form of the word PRINT followed by

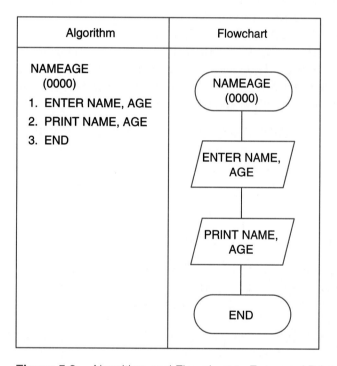

Figure 5.1 Flowchart Diagram for the Sequential Structure

Algorithm	Flowchart
NAMEAGE (0000) 1. ENTER NAME, AGE 2. PRINT NAME, AGE 3. END	

Figure 5.2 Algorithm and Flowchart to Enter and Print Two Variables

the list of data items whose values will be printed to the screen. The flowchart uses the parallelogram to indicate that this is an output instruction. The input and output instructions both use the parallelogram in the flowchart. The last instruction is the END instruction. This signals the end of the list of instructions and to stop the execution of the program. The end is placed in a flattened ellipse in the flowchart. All of the start and ending instructions use the flattened ellipse in the flowchart. Notice that the algorithm instructions are numbered starting with the first instruction after the name of the module.

Solution Development

In Chapter 3, the tools for organizing problems were introduced. Now you will use them to solve problems from the business world. These organizing tools are used in the first six steps of developing a solution.

1. The problem analysis chart—helps you define and understand the problem, develop ideas for the solution, and select the best solution.
2. The interactivity chart—breaks the solution to the problem into parts.
3. The IPO chart—helps define the input, the output, and the processing steps.
4. The coupling diagram and the data dictionary designate the data flow between modules. The data dictionary records information on the items.
5. The algorithms
6. The flowcharts

The seventh step in solving the problem is to test the solution to catch any errors in logic or calculation. Remember, the charts are aids to the efficient development of a well-written program. They will not be perfect the first time around, and will often need several revisions.

As you follow the solution to the sample sequential-structure problem that follows, take the time to think through each step. Photocopy the forms in Appendix D to complete the seven steps. The solution to the following problem will be developed. This may seem like a lot of work for this simple of a problem; however, it is easier to learn on a simple short problem than on a long difficult one. If you can understand this problem and the use of the tools involved in the solution of this problem, you will be able to solve more difficult problems as they come up.

> Problem: Mary Smith is looking for the bank that will give the most return on her money over the next five years. She has $2,000 to put into a savings account. The standard equation to calculate principal plus interest at the end of a period of time is

$$AMOUNT = P * (1 + I/M)^{\wedge}(N * M)$$

where P = PRINCIPAL (amount of money to invest, in this case $2,000)
 I = INTEREST (percentage rate the bank pays to the investor)
 N = NUMBER OF YEARS (time for which the principal is invested)
 M = COMPOUND INTERVAL (the number of times per year the interest is calculated and added to the principal)

(Refer to Table 2.5 to review the meanings of the operators used here.)

Problem Analysis

The first step in analyzing the problem is to understand what is needed and what is given, and to separate them from all of the nonessential information in the problem. Write down the input in the Given Data section and the output in the Required Results section. Record the processing that the problem demands in the section of the problem analysis chart headed Required Processing. Finally, write possible solutions for the problem and any ideas related to the solution in the Solution Alternatives section. Generate solution ideas through brainstorming and creative thinking. When you are first learning problem-solving skills, brainstorming is best accomplished in groups of two to three people. The more ideas, the easier it is to develop a good solution.

Given Data	Required Results
PRINCIPAL – $2,000 INTEREST NUMBER OF YEARS–5 COMPOUND INTERVAL (#/year)	PRINCIPAL plus INTEREST at the end of the time period
Required Processing	
AMOUNT = P * (1 + I / M)^(N * M)	
Solution Alternatives	
*1. Enter all data as variables. 2. Enter principal and interest as constants and the other data as variables. *3. Process one bank in one run. 4. Process all banks in one run. *Processes selected for the best solution.	

Figure 5.3 Problem Analysis Chart—Interest Problem

In Figure 5.3, the problem analysis chart has been filled in for the interest problem given in the previous section. The given data are the PRINCIPAL ($2,000), the INTEREST (variable), the NUMBER OF YEARS the PRINCIPAL is to be invested (5), and the COMPOUND INTERVAL (variable). The required result is the AMOUNT of PRINCIPAL plus INTEREST at the end of the time period. The required processing is the equation for the AMOUNT. Solution ideas include the following:

1. Enter all input data as variables.
2. Enter the principal and interest as constants and the other values as variables.
3. Enter the data for one bank per run.
4. Enter the data for all banks per run.

Figuring out the best solution involves weighing the ideas and alternatives in the Solution Alternatives section of the problem analysis chart. Since a solution should be as flexible as possible, it would be better to enter all the data as variables. Variables provide flexibility because their values can be changed, whereas the values of constants cannot. It is best to use variables wherever there is a possibility that the data might change, and constants only if it is known that the data won't change. Furthermore, since the user probably will not have all data from all banks at once, entering one bank per run would be best for this solution.

The Interactivity (Structure) Chart

At this point, you can develop the interactivity chart. To complete the interactivity chart you have to answer the question, How can I divide this problem into smaller, more manageable parts? In Figure 5.4, the interest problem is broken into four modules:

1. The Interest Control module, which controls the solution
2. The Read module, which enters the data

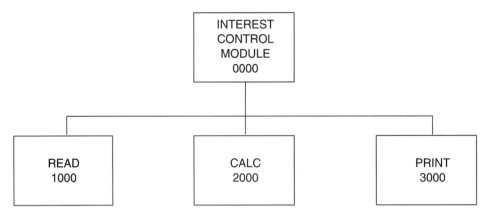

Figure 5.4 The Structure Chart—Interest Problem

3. The Calc module, which calculates the amount
4. The Print module, which prints the result

The interactivity chart shows that the Interest Control module processes the other three modules in the following order—the Read, Calc, and Print modules, respectively. These four modules are the only ones in this solution. There is no Init module since there is nothing to be initialized (no files to open, no headings to print, or the like), nor is there a Wrap-up module since there is nothing to be done at the conclusion of the processing (no files to close, no totals to print, or the like). The solution requires only that data be entered, that a calculation be performed, and that the results be printed. Each box represents a module. Each module has a function and represents two or more instructions. The modules are numbered according to their level.

The IPO Chart

Now the IPO chart can be developed from the problem analysis chart and the interactivity chart. To complete the IPO chart you have to answer the question, How can the input be processed to achieve the required output? You will provide general instructions for the solution on the IPO chart. You will not write the specific instructions to the computer until you complete the algorithms and the flowcharts.

Remember to start with the output, next specify the input, and then develop the processing method required to calculate the output from the input. In Figure 5.5, the output is the amount of the PRINCIPAL plus INTEREST after five years. The input required to produce this result includes the beginning PRINCIPAL, the INTEREST RATE (changed to hundredths after it is input), the NUMBER OF YEARS INVESTED, and the COMPOUNDED TIME INTERVAL. The processing includes entering the data (found in module 1000, the Read module), calculating the amount (found in the module 2000, the Calc module), and printing the amount (found in module 3000, the Print module). This information is found in the Processing and the Module Reference Number columns.

Coupling Diagram and Data Dictionary

Figure 5.6, the coupling diagram, and Figure 5.7, the data dictionary, are interrelated. The coupling diagram shows which variables are passed from one module to another.

Input	Processing	Module Reference Number	Output
1. BEGINNING PRINCIPAL 2. INTEREST RATE 3. NUMBER OF YEARS 4. NUMBER OF TIMES INTEREST IS COMPOUNDED YEARLY	1. Enter data (Change interest rate to hundredths) 2. Calculate ending PRINCIPAL and INTEREST AMOUNT = PRINCIPAL * (1 + INTEREST / TIME)^(YEARS * TIME) 3. Print required results	1000 2000 3000	1. Ending PRINCIPAL plus INTEREST 2. All input data

Figure 5.5　The IPO Chart—Interest Problem

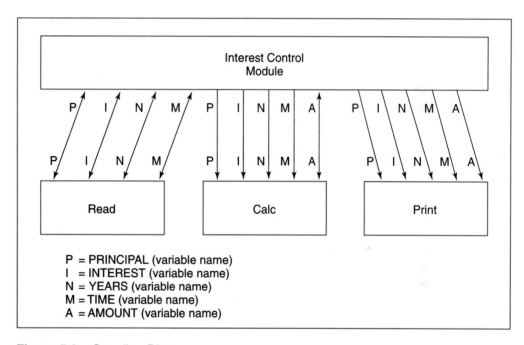

P = PRINCIPAL (variable name)
I = INTEREST (variable name)
N = YEARS (variable name)
M = TIME (variable name)
A = AMOUNT (variable name)

Figure 5.6　Coupling Diagram

The data dictionary shows which modules use each variable and the scope of each variable (whether the variable is local or global.) In this problem, all variables are local, therefore, all coupling will be done through parameters. The variables that couple the Interest Control module and the Read module are the Principal, the Interest, the Number of Years, and the Compounding Time. These are all call-by-reference parameters since the variables sent to the Read module must have values assigned to them. Since the memory locations of PRINCIPAL, INTEREST, YEARS, and TIME were sent to the Read module, both modules are using the same memory locations. Therefore, when the Read module enters values into PRINCIPAL, INTEREST, YEARS, and TIME, the Interest Control module will be able to see the new values. The same four are call-by-value parameters when coupling the Interest Control module to the Calc module. In this

Data Dictionary

Item	Variable Name	Data Type	Module(s)	Scope	Pseudonyms/ Module	Error Checks
Principal	PRINCIPAL	Numeric—real	Interest Control Read/Calc/Print	Local Parameter	None	None
Interest	INTEREST	Numeric—real	Interest Control Read/Calc/Print	Local Parameter	None	None
Number of years	YEARS	Numeric—real	Interest Control Read/Calc/Print	Local Parameter	None	None
Compounding time	TIME	Numeric—real	Interest Control/Read/ Calc/Print	Local Parameter	None	None
Amount	AMOUNT	Numeric—real	Interest Control Calc/Print	Local Parameter	None	None

Figure 5.7 Data Dictionary

case the values of PRINCIPAL, INTEREST, YEARS, and TIME are sent to the Calc module. The Calc module then assigns new memory locations for these variables. Since the Interest Control module and the Calc module have different memory locations for these variables, the Interest Control module will not be able to see any changed values and will retain the original values even if the Calc module were to change them. This is because they will not be changed, only used, in the Calc module. The amount, however, needs to be a call-by-reference parameter since the value of the AMOUNT will be calculated in the Calc module and the Interest Control module needs the new value. All five variables couple the Interest Control module to the Print module. They are call-by-value parameters since the Print module only needs to use them. The five variables will not be changed in the Print module. Usually, to determine which variables are call-by-reference parameters and which are call-by-value parameters, the programmer needs to see which variables will be changed in the called module. If the variable is changed, then it needs to be a call-by-reference parameter. If the called module is only going to use the value of the variable, then it needs to be a call-by-value parameter.

Internal and External Documentation

Documentation is useful to everyone involved with the program, so it is important that the programmer or the user of the application package write down important details about the program. The programmer's colleagues need to know what was done, how it was done, and where, in order to finish and maintain the program. The user needs to know what has been done and how, in order to use the program or application effectively, and, further, to make sure the processing is correct. Internal documentation involves noting within the instruction set what the instructions do. This is accomplished through the use of remark or comment statements. These statements appear only in the listing of the statements and nowhere else. They are for the use of the programmer only.

The end user never sees them. External documentation involves writing manuals (either printed or electronic) with detailed explanations of the program and how to use it.

You take notes for internal and external documentation during problem solving because it is easy to forget the details that you need later when you are writing the documentation. There is an Annotation section for these notes on page 500 of Figure D.4—Algorithm and Flowchart, Appendix D. You write the notes for the documentation as you develop the algorithms and the flowcharts, including any important information that is not already part of the algorithms and flowcharts. In the Interest Control module in Figure 5.8a–d the exact task of each of the Process modules (the Read, the Calc, and the Print modules) is not specified. The remarks in the Annotation section specify those tasks. For example, the Read module enters all data from the keyboard. The Annotation section is the only place you can explain such things in any depth. Only after the module has been developed completely can you tell all that belongs in the documentation, so it is best not to finish the notes on each module until then. The Annotation and Documentation sections of the Algorithm and Flowchart form are optional, but are often useful later.

Algorithm	Flowchart	Annotation	Test
Interest Control Module (0000)	INTEREST CONTROL (0000)		1. Start
1. PROCESS READ (*PRINCIPAL, *INTEREST *YEARS, *TIME)	READ	Enters all data from keyboard	2. Transfer to Read
2. PROCESS CALC (PRINCIPAL, INTEREST, YEARS, TIME, *AMOUNT)	CALC	Calculates amount	3. Transfer to Calc
3. PROCESS PRINT (PRINCIPAL, INTEREST, YEARS, TIME, AMOUNT)	PRINT	Prints data and amount	4. Transfer to Print
4. END	END		
* Specifies call-by-reference parameters			

Internal Documentation	External Documentation
1. Remark at top: Calculates principal and interest given beginning principal, interest rate, number of years, and compounded time interval 2. Include annotations	1. Same as 1 in Internal Documentation

Figure 5.8a Algorithm and Flowchart for INTEREST CONTROL Module

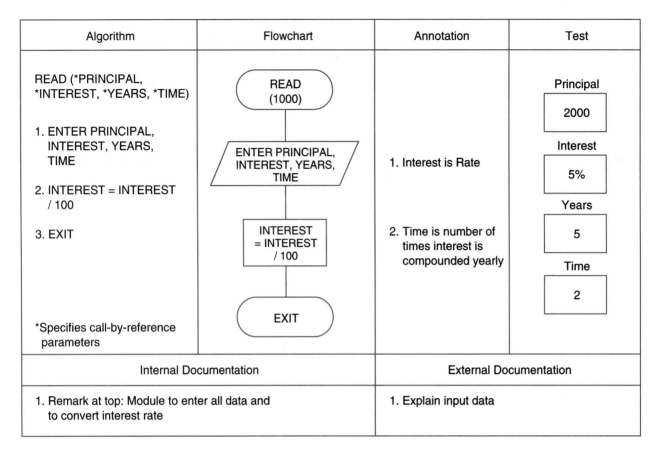

Algorithm	Flowchart	Annotation	Test
READ (*PRINCIPAL, *INTEREST, *YEARS, *TIME) 1. ENTER PRINCIPAL, INTEREST, YEARS, TIME 2. INTEREST = INTEREST / 100 3. EXIT *Specifies call-by-reference parameters	READ (1000) ENTER PRINCIPAL, INTEREST, YEARS, TIME INTEREST = INTEREST / 100 EXIT	1. Interest is Rate 2. Time is number of times interest is compounded yearly	Principal 2000 Interest 5% Years 5 Time 2
Internal Documentation		External Documentation	
1. Remark at top: Module to enter all data and to convert interest rate		1. Explain input data	

Figure 5.8b Algorithm and Flowchart for READ Module

Algorithm	Flowchart	Annotation	Test
CALC (PRINCIPAL, INTEREST, TIME, *Amount) 1. AMOUNT = PRINCIPAL * (1 + INTEREST/TIME) ^ (YEARS * TIME) 2. EXIT *Specifies call-by-reference parameters	CALC (2000) AMOUNT = PRINCIPAL * (1 + INTEREST/TIME) ^ (YEARS * TIME) EXIT	None	Amount = 2000 * (1 + .05 / 2) ^ (5 * 2) Amount = 2000 * (1 + .025) ^ 10 Amount = 2560
Internal Documentation		External Documentation	
1. Remark at top: Module to enter all data and interest		1. Specify equation	

Figure 5.8c Algorithm and Flowchart for CALC Module

Algorithm	Flowchart	Annotation	Test
PRINT (PRINCIPAL, INTEREST, YEARS, TIME, AMOUNT) 1. PRINT AMOUNT, PRINCIPAL, INTEREST, YEARS, TIME 2. EXIT	PRINT (3000) PRINT AMOUNT, PRINCIPAL, INTEREST, YEARS, TIME * EXIT	1. Print each variable on a separate line with a label.	Prints what is required

Internal Documentation		External Documentation	
1. Remark at top: Module to print required output		1. Specify output	

*Specify only what and when the information is printed, not how it is printed.

Figure 5.8d Algorithm and Flowchart for PRINT Module

The Algorithms and Flowcharts

Once the structure chart and the IPO chart are completed, you can proceed with the algorithms and the flowcharts. To complete an algorithm, you have to answer the question, How can this module complete its function? To draw the flowchart you answer the question, How does the algorithm look graphically? By using the completed algorithms and the flowcharts, you can find and correct logic errors.

Figure 5.8a–d shows the four algorithms, one for each module in the structure chart, and their corresponding flowcharts. The Interest Control module uses a sequential structure to execute the instructions, as do each of the other modules. Notice how the instructions simply flow in order from the top to the bottom of each algorithm. Remember, you must only use the algorithm instructions in Table 5.1, and each instruction must be in the form presented. The same is true for the flowcharts. The Annotation and Documentation sections of this figure have been completed to show how to use these sections in the Algorithm and Flowchart form.

The next step in developing the solution is to take a set of test data and work it through the algorithm instructions on a piece of paper or in the section provided on the Algorithm and Flowchart form. Do not skip any steps. It is important to process the data as you have instructed the computer to do it, starting at the beginning of the Interest Control module and working a set of data through the instructions in the algorithms for each module, one by one. As you process each module, write down the results. The solution is correct if the output matches the desired results. In this problem, the output should match the hand-calculated value of the amount. If the correct output is produced, then the analysis is complete. If not, you need to make corrections in the algorithm until you reach the correct output. At that point the analysis will be complete and coding into a language or application can begin. Coding is done in a language or applications course.

Summary

The instructions in every algorithm must use particular logic structures in order to increase readability for the programmer or user. The first logic structure is the sequential structure, in which the processing flows from one instruction to the next from start to end. The particular instructions used with the sequential structure include START, ENTER data, PRINT and WRITE data, assign a value to a variable (variable =), PROCESS another module, and END/EXIT/RETURN.

A solution to a problem is developed in seven steps:

1. Analyze the problem.
2. Develop the structure chart.
3. Develop the IPO chart.
4. Develop the coupling diagram and the data dictionary.
5. Develop the algorithms.
6. Develop the flowcharts.
7. Test the solution.

These steps will help you develop a solution that is efficient and easy to understand, code, and maintain.

After preparing the structure chart and the IPO chart, and while developing the algorithms and flowcharts, the programmer takes notes for the internal and external documentation that will be completed later. The internal documentation will be written within the program. The external documentation will be written in the form of manuals explaining the program.

Questions

1. What is the sequential logic structure?
2. Which flowchart symbols are used for the sequential logic structure?
3. Why is documentation important?

Problems

Complete the seven steps of solution development for the following problems:

1. A painter wants to know the amount of paint needed to paint only the walls and the interior side of the door in a room. The chosen paint covers 100 square feet per gallon. There are two windows. Test the problem with the following data:

 The room is 12 feet long, 10 feet wide, and 8 feet tall.

 The two windows are 5 by 3 feet, and 6 by 2 feet, respectively.

2. One of the jobs that Joe Roberts has been given at work is to order special paper for a report for a board meeting. The paper comes in reams of 500 sheets. He always makes five more copies than the number of people that will be there. Joe wants to know how many reams of paper he needs for a meeting. He can order only whole,

not partial, reams. Assume the required number of pages will not equal an exact number of reams. Test your solution with the following data:

The report is 140 pages long.

There will be 25 people at the meeting.

3. An instructor calculates grades by dropping the lowest homework score and the lowest test score, and adding the remaining scores together. He then calculates the final points as a percentage of the total points possible. The instructor needs to know the final points for a given student. (*Hint:* Use the minimum function, that is, the function that returns the minimum value in a list of values.) Test your solution with the following data:

Total points possible—500 points

Tom Johnson

 Homework scores—45, 46, 41, 31, 44

 Test scores—95, 84, 75, 98

4. **a.** Develop a solution that would return the fewest coins in change from a purchase of under one dollar. (Include pennies, nickels, dimes, quarters, and half dollars.)

Example:	purchase	63¢
	change	37¢
	coins	1 quarter
		1 dime
		2 pennies

 b. Develop a solution that would return the fewest bills and coins for a purchase under $50.

5. Mary Williams needs to change a Fahrenheit temperature to Celsius. Use the Fahrenheit temperature of 80 degrees to test your solution.

6. Write a solution to print the number of months (use 30 days to a month) and remaining days given the number of days between two dates.

7. A sweater is on sale for 25% off the original price. The original price is $50. Write a solution to calculate and print the sale price.

8. Bob would like to know what percentage of his income his rent is. Write a solution that would calculate and print this percentage.

9. An instructor calculates the grade percentage based on the highest score on a test. Given the highest score and one student's score, write a solution to calculate and print that student's test percentage.

10. Joe would like to build several bookcases that are different heights and widths. All will be 12 inches in depth. The bookcases will have three shelves, in addition to the bottom and the top. Write a solution to print the number of feet of 12-inch wide boards that will Joe need to complete a bookcase, given the height and width.

WHAT'S WRONG WITH THIS?

Correct the following algorithm instructions:

1. A + B = C
2. Output A, B, C
3. Input A, B, C
4. Enter a, b, and c
5. Stop
6. Goto Calc
7. A = 5 B \times C

Chapter 6

Problem Solving with Decisions

Overview

The Decision Logic Structure
Multiple IF/THEN/ELSE Instructions
Using Straight-through Logic
Using Positive Logic
Using Negative Logic
Logic Conversion
Which Decision Logic?
Decision Tables

Objectives

When you have finished this chapter, you should be able to:

1. Develop problems using the decision logic structure in conjunction with the sequential logic structure.
2. Use the seven problem solving tools when developing a solution using the decision logic structure.
3. Use nested decision instructions to develop a problem solution.
4. Distinguish the different uses of straight-through, positive, and negative nested decision logic structures.
5. Convert a positive decision logic structure to a negative decision logic structure.
6. Develop decision tables given a set of policies.
7. Develop a decision logic structure from a decision table.

The logic structures will almost always be used in some combination with each other in a program. In this chapter you will learn how to use a second logic structure, the decision structure, as a way to design instructions. The decision structure is one of the most powerful

because it is the only way that the computer can choose between two or more sets of actions. Without decisions, the computer would be nothing more than a fast calculator.

The decision logic structure is relatively easy to understand since it works much like the way humans think. Because of the complexity of some of the decisions a programmer has to design, however, it is often one of the most difficult structures to implement. The difficulty goes back to the fact that people make decisions without understanding the reasoning involved. Also, a decision can be written in many different ways, which can add to the confusion.

The Decision Logic Structure

The decision logic structure uses the **IF/THEN/ELSE instruction.** It tells the computer that IF a condition is true, THEN execute a set of instructions, or ELSE execute another set of instructions. The ELSE part is optional, as there is not always a set of instructions if the conditions are false. When there are no instructions for true, a CONTINUE statement must be used. The following algorithm instruction for the decision structure (the decision instruction) should be added to the list in Table 5.1. Notice the brackets and the indentation in the illustration of the algorithm of the decision instruction. Programmers use these features when they are writing instructions in the decision structure because they improve readability.

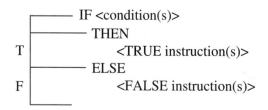

The TRUE instructions are processed when the resultant of the condition is TRUE, and the FALSE instructions are processed when the resultant of the condition is FALSE. A condition can be one of four things: a logical expression, that is, an expression that uses logical operators (AND, OR, and NOT); an expression using relational operators (greater than, less than, greater than or equal to, less than or equal to, equal to, and not equal to); a variable of the logical data type (TRUE or FALSE); or a combination of logical, relational, and mathematical operators.

Some examples of conditional expressions are as follows:

1. A < B (A and B are the same data type—either numeric, character, or string)
2. X + 5 > = Z (X and Z are numeric data)
3. E < 5 or F > 10 (E and F are numeric data)
4. DATAOK (DATAOK is a logical datum)

Logical operators are used to link more than one condition together. For example, consider a store policy requiring that to cash a check, a customer must have a driver's license, AND the check must be under $50. Two conditions are linked by the logical operator AND. (See Tables 2.5, 2.6, and 2.7 for the definitions and the hierarchy of all the operators).

The programmer can set up the conditions for a decision through the use of operands and various operators. These conditions can be used alone or linked with other

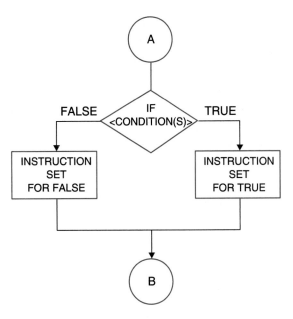

Figure 6.1 Flowchart Diagram of the Decision Structure

conditions for use in the instruction. The programmer derives the information needed to set up the condition(s) for a decision from the problem itself during problem analysis.

Figure 6.1 shows a simple flowchart illustrating the decision structure. In the figure, the resultant of the condition(s) can be TRUE or FALSE depending on the data. From the decision block (the diamond), there will be a flowline, a branch for each set of instructions: one branch for the instructions to be executed if the resultant is TRUE, and another branch for the instructions to be executed if the resultant is FALSE. There can never be more than two flowlines coming from a decision block since there are only two possibilities for the resultant of the condition(s). The TRUE branch and the FALSE branch can come from any of the lower three points of the diamond. It is best to be as consistent as possible in choosing the points of the diamond on which these branches are placed. The convention is to draw the TRUE branch on the right and the FALSE branch on the left or bottom, although it depends on the decisions to be made. This book follows that convention whenever possible. The processing flow from both the TRUE branch and the FALSE branch must connect to the rest of the processing somewhere before exiting the module. There can be no dangling flowchart blocks—blocks with no flowlines leading to another block. The processing must flow somewhere. Each module can have only one entrance and one exit. The only way to get to another module is through a PROCESS instruction, that is, by telling the computer to process the instructions in another module and then return to continue processing in the current module.

A simple decision with only one condition and one action or set of actions (one instruction or set of instructions) for TRUE and another for FALSE is relatively simple. For example, assume you are calculating pay at an hourly rate, and overtime pay (over 40 hours) at 1.5 times the hourly rate. The decision to calculate pay would be stated in the following way: IF the hours are greater than 40, THEN the pay is calculated for overtime, or ELSE the pay is calculated in the usual way (see Figure 6.2).

Decisions in which you have *multiple* conditions that lead to one action or set of actions for TRUE and another action or set of actions for FALSE are slightly more complicated than those with single conditions. In these decisions you will use logical

Algorithm	Flowchart

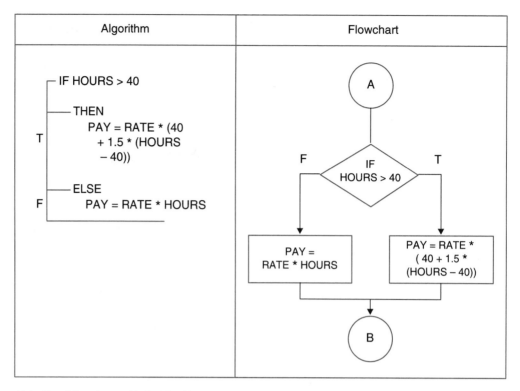

Note: For all flowcharts with decision blocks, T = TRUE and F = FALSE

Figure 6.2 Single Condition—Two Possible Actions or Sets of Actions

operators to connect the conditions. As with decisions based on a single condition, the resultant will determine whether the TRUE or the FALSE actions are taken.

The decision structure becomes more complicated as the number of conditions and/or the number of actions for a TRUE or FALSE resultant increases.

Multiple IF/THEN/ELSE Instructions

There are three types of decision logic you will use to write algorithms for solutions consisting of more than one decision. These types of decision logic include straight-through logic, positive logic, and negative logic.

Straight-through logic means that all of the decisions are processed sequentially, one after the other. There is no ELSE part of the instructions; the FALSE branch always goes to the next decision, and the TRUE branch goes to the next decision after the instructions for the TRUE branch have been processed.

On the other hand, with positive logic not all of the instructions are processed. **Positive logic** allows the flow of the processing to continue through the module instead of processing succeeding decisions, once the resultant of a decision is TRUE. Whenever the resultant is FALSE (the ELSE part of the decision), another decision in the sequence is processed until the resultant is TRUE, or there are no more decisions to process. At that time, the FALSE branch processes the remaining instructions.

Negative logic is similar to positive logic except that the flow of the processing continues through the module when the resultant of a decision is FALSE. Whenever the resultant is TRUE, another decision is processed until the resultant is FALSE, or there

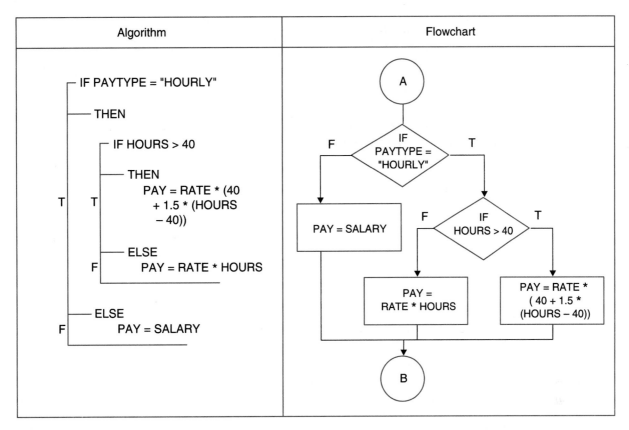

Algorithm	Flowchart

Figure 6.3 Nested IF/THEN/ELSE Instructions

are no more decisions to process. At that time, the TRUE branch processes the remaining instructions. Negative logic is the hardest to use and understand.

Some series of decisions will require a combination of two or more of these logic types. The use of each decision logic type will be further explained and illustrated later in this chapter.

In algorithms containing multiple decisions, you may have to write **nested IF/THEN/ELSE instructions.** Decisions using positive and negative logic use nested IF/THEN/ELSE instructions. Decisions using straight-through logic do not. Nested IF/THEN/ELSE instructions are sets of instructions in which each level of a decision is embedded in a level before it; like nesting baskets, one goes inside the next. You use the nested IF/THEN/ELSE structure only when one of several possible actions or sets of actions are to be processed (see Figure 6.3). In composing an IF/THEN/ELSE instruction, draw a large bracket around each IF/THEN/ELSE in the algorithm to ensure that each instruction is in the proper place. The top part of the bracket indicates the condition. The middle section of the bracket indicates the TRUE branch, or the THEN part of the decision. The bottom part of the bracket indicates the FALSE branch, or the ELSE part of the decision.

In Figure 6.3, the decision is IF PAYTYPE = "HOURLY". The decision says, Is the value of the variable PAYTYPE equal to the string constant "HOURLY"? If the two are equal, then the resultant is TRUE and the instruction set for the TRUE is processed. If the resultant is FALSE, the instruction set for the FALSE branch is processed. The TRUE branch uses a nested IF/THEN/ELSE instruction; the decision is embedded in the TRUE branch of the first decision. This decision says, Are the hours worked greater than 40? If the answer is "yes," the resultant is TRUE, and the pay is processed at the

overtime rate. When the resultant is FALSE, that is, when the hours are not greater than 40, the pay is processed at the normal rate. When the resultant of the first decision is FALSE, that is, when the PAYTYPE is not equal to "HOURLY", then the pay is processed at the salary rate. (Notice that when the PAYTYPE is not equal to "HOURLY", the embedded decision on the TRUE branch cannot be processed—an example of negative logic.) A good programmer thoroughly tests algorithms that use decision logic with many sets of test data. Every path in every algorithm is checked, and a separate set of data is used for each path. It is important to test each path so errors will be eliminated.

Using Straight-through Logic

With decisions following straight-through logic, all conditions are tested. To test a condition means to process a condition to get a TRUE or FALSE resultant. Straight-through logic is the least efficient of all types of decision logic because all the decisions must be processed. However, you must use it to solve certain problems, those that require two or more unrelated decisions, and those in which all decisions must be processed. It is often used in conjunction with the other two logic types and in data validation (checking data to make sure it is correct). It is also used with languages that have limited features in their decision instruction.

The problem illustrated in Figure 6.4 is to find the amount to charge people of varying ages for a concert ticket. When the person is under 16, the charge is $7; when the person is 65 or over, the charge is $5; all others are charged $10. The conditions are the following:

AGE	CHARGE
AGE < 16	7
AGE > = 16 and AGE < 65	10
AGE > = 65	5

Notice in Figure 6.4 that even when the age is less than 16, the other decisions must be processed.

In the algorithm and flowchart in Figure 6.4, notice that there is no ELSE in straight-through logic. If the resultant of the condition is FALSE, then the processing flow drops to the next instruction. There is no need for the ELSE since there are no instructions for the FALSE branch. If the resultant of the condition is TRUE, then the set of instructions for the TRUE part is processed, and the processing flow then drops to the next instruction. The set of conditions in Figure 6.4 is more efficiently executed by using either of the other types of logic since only one condition has to be met in order to execute the correct set of instructions.

The problem found in Figure 6.5 is to change the value of X to 0 when X becomes greater than 100, and to change the value of Y to 0 when Y becomes greater than 250. Each decision is independent of the other. The conditions are

$$X > 100 \qquad \text{X becomes 0}$$
$$Y > 250 \qquad \text{Y becomes 0}$$

Straight-through logic is required when all the decisions have to be processed, and when they are independent of each other. Based on these criteria, the only appropriate logic type for the decision in Figure 6.5 would be straight-through logic.

Algorithm	Flowchart

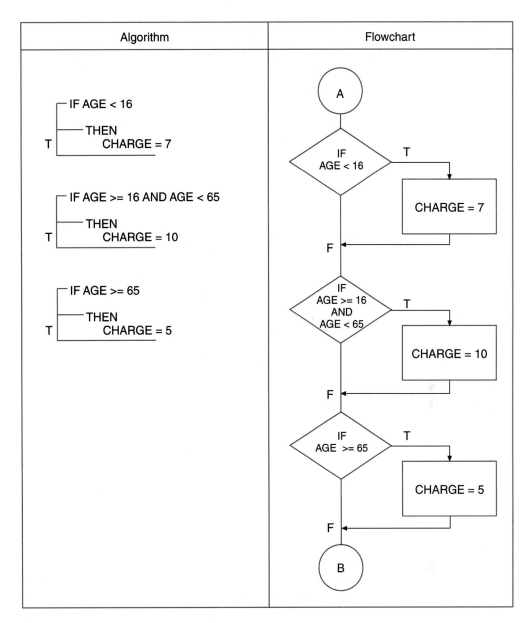

Figure 6.4 Straight-through Logic—Example 1

The following example is a similar problem, but with two sets of actions rather than one. When the age of a person is greater than 18, the person can vote; when the age is greater than 21, the person can have alcoholic beverages. Both decisions are contingent on the age of the person; however, the two actions take place at different ages. At the age of 20, a person can vote, but cannot be served alcoholic beverages. In this case, straight-through logic is required because all the decisions have to be processed, and because they are independent of each other.

Using Positive Logic

Positive logic is the easiest type for most people to use and understand because it is the way we think. Positive logic always uses nested IF/THEN/ELSE instructions. In

Algorithm	Flowchart

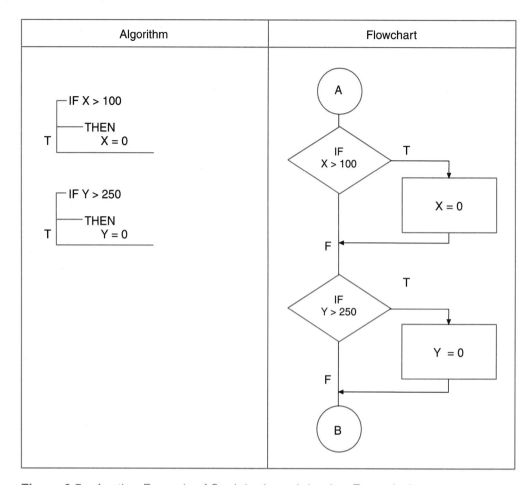

Figure 6.5 Another Example of Straight-through Logic—Example 2

general, when you use positive logic, you are telling the computer to follow a set of instructions and continue processing if the condition is TRUE; if the condition is not TRUE, then the computer processes another decision. When you use this logic, no more decisions are processed once the resultant of a condition is TRUE. Taking the same problem, and therefore, the same set of conditions as in Figure 6.4, Figure 6.6 shows how the algorithm would be structured using positive logic. Notice that there are fewer decisions processed than in Figure 6.4—two compared to three—even if all the decisions are processed. There is no need to process the third decision because the resultant of the third condition statement has to be TRUE once you have processed the second statement and found it FALSE. That is, when the age is NOT less than 16, it has to be greater than or equal to 16. Therefore, this test can be eliminated if nested IF/THEN/ELSE instructions are used. Likewise, if the age is NOT less than 65, it must be greater than or equal to 65. Therefore, the third decision in the set of conditions is not necessary.

Figure 6.7 illustrates another set of conditions using positive logic and a nested IF/THEN/ELSE structure. The problem illustrated in this figure is to calculate the commission rate for a salesperson, given the amount of sales. When the salesperson has sold less than or equal to $2,000 worth of goods, the commission is 2%. When the sales total is more than $2,000 and less than or equal to $4,000, the commission is 4%. When sales total is more than $4,000 and less than or equal to $6,000, the commission is 7%. When

Algorithm	Flowchart

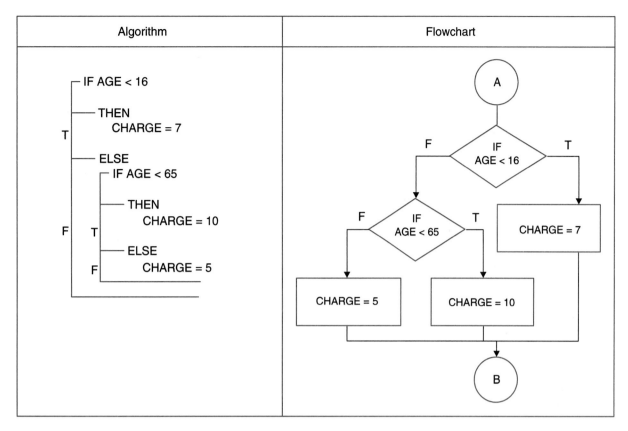

Figure 6.6 Positive Logic—Example 1

the person has sold more than $6,000, the commission is 10%. The conditions are the following:

SALES	COMMISSION
< = 2000	.02
2001–4000	.04
4001–6000	.07
>6000	.10

Notice that in Figure 6.7 there are only three decisions. The fourth decision is not necessary; when the sales are NOT < = 6000 they have to be >6000. Remember: when setting up an expression, you must have an operand on each side of the operator. SALES = 2001–4000 does not mean this is true for any sales amount between 2001 and 4000. The computer will subtract 4000 from 2001 first, then it will compare SALES to the resultant, −1999. The proper expression for this condition is SALES > 2000 and SALES < = 4000. Also, remember that if this is the second decision in a nested decision and the first decision is SALES < = 2000, the expression in the second decision will be SALES < = 4000 since SALES > 2000 will always be true. Any SALES > = 2000 will have dropped out of the nested IFs during the first decision.

Figure 6.7 is easy to read and understand because of the structure of the algorithm: Brackets and indentation are used with positive logic to show the nested levels of the decision structure. These features are important to readability.

Algorithm	Flowchart	Test

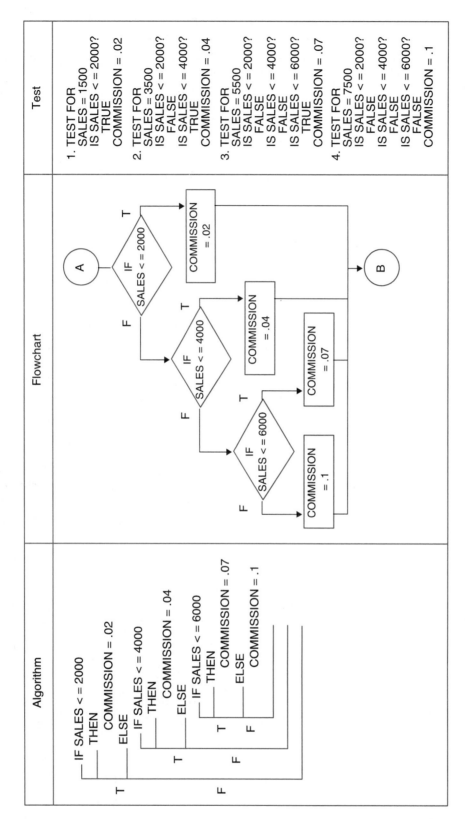

Figure 6.7 Positive Logic—Example 2

Algorithm:

IF SALES <= 2000
 THEN
 COMMISSION = .02
 ELSE
 IF SALES <= 4000
 THEN
 COMMISSION = .04
 ELSE
 IF SALES <= 6000
 THEN
 COMMISSION = .07
 ELSE
 COMMISSION = .1

Test:

1. TEST FOR
 SALES = 1500
 IS SALES <= 2000?
 TRUE
 COMMISSION = .02

2. TEST FOR
 SALES = 3500
 IS SALES <= 2000?
 FALSE
 IS SALES <= 4000?
 TRUE
 COMMISSION = .04

3. TEST FOR
 SALES = 5500
 IS SALES <= 2000?
 FALSE
 IS SALES <= 4000?
 FALSE
 IS SALES <= 6000?
 TRUE
 COMMISSION = .07

4. TEST FOR
 SALES = 7500
 IS SALES <= 2000?
 FALSE
 IS SALES <= 4000?
 FALSE
 IS SALES <= 6000?
 FALSE
 COMMISSION = .1

To test positive logic for correctness, choose data to test every path. In the Test section of the algorithm and flowchart in Figure 6.7, there are four separate sets of data tested, one for each commission. You should arrive at the correct commission (hand calculated) by working each piece of test data through the flowchart. Notice that when SALES = 1500, there is only one decision to be processed; the resultant is TRUE, the COMMISSION is replaced by .02, and the processing continues at on-page connector B. When the SALES is equal to 3500, two decisions are processed before the resultant of one of them is found to be TRUE.

There are times when it is more efficient to start an algorithm with one condition rather than another. The general rule is that the condition that will have a resultant of TRUE most of the time should be the first. This way you will have to process the fewest decisions. See Figure 6.8 for another way to set up a solution for the conditions found in the sales commission problem illustrated in Figure 6.7. Notice the changes in the operators and in the order of the decisions. If you are unable to judge the condition that is most likely to be TRUE, then it really does not matter how the logic starts.

Using Negative Logic

Negative logic is the hardest for most people to comprehend because they do not think in negative terms. In general, when you use negative logic you are telling the computer to process another decision when the resultant of the condition is TRUE; if the resultant is FALSE, then the computer processes a consequent set of instructions and then continues processing the module. It is often advantageous to use negative logic to decrease the number of tests or to improve the readability of a program.

In Figure 6.9, negative logic is used with the same set of conditions as in Figure 6.4. Notice that the actions take place on the FALSE part of the instruction unless there are no more tests. Again, the algorithm uses brackets and indentation for readability. With the indentations the algorithm makes a sideways V. If a nested IF/THEN/ELSE instruction using negative logic does not form a sideways V, then either the indentation is wrong, or some of the instructions for each TRUE or FALSE are set up incorrectly. The top section of each bracket indicates a condition or conditions. The middle section indicates the instructions processed when the resultant of the condition(s) is TRUE. This section is much larger when a decision uses negative logic than when the decision uses positive logic. The bottom section indicates the instructions to process when the resultant of the condition(s) is FALSE. This section is much smaller when a decision uses negative logic than when the decision uses positive logic. The brackets make the nested IF/THEN/ELSE instruction easier to read by partitioning the conditions, the TRUE section, and the FALSE section.

For example, in the algorithm and the flowchart in Figure 6.9, when the resultant of the first condition (AGE.516) is TRUE, another decision is processed (AGE > = 65). When the resultant of the first condition is FALSE, the charge of the ticket is $7; after this decision the processing continues through the rest of the module. When the resultant of the second condition is TRUE the charge is $5; when FALSE, the charge is $10.

Figure 6.10 shows the solution from the positive logic illustration in Figure 6.7 if we were to use negative logic. The conditions are the following:

SALES	COMMISSION
< = 2000	.02
2001–4000	.04
4001–6000	.07
>6000	.10

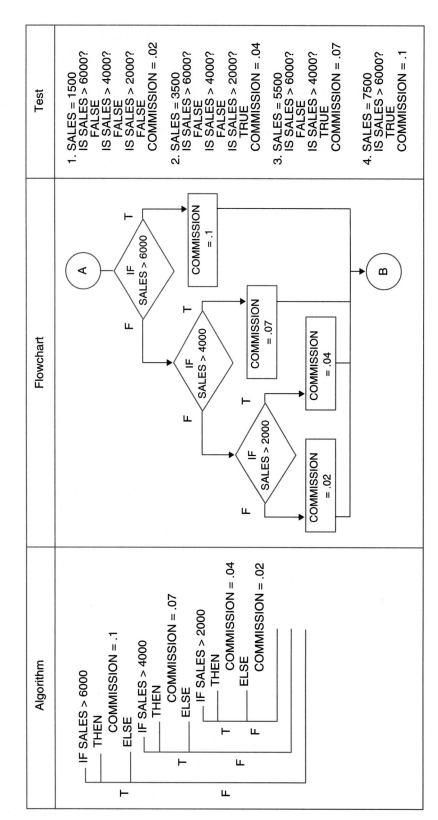

Figure 6.8 The Conditions in Figure 6.7 Set Up in a Different Way

114

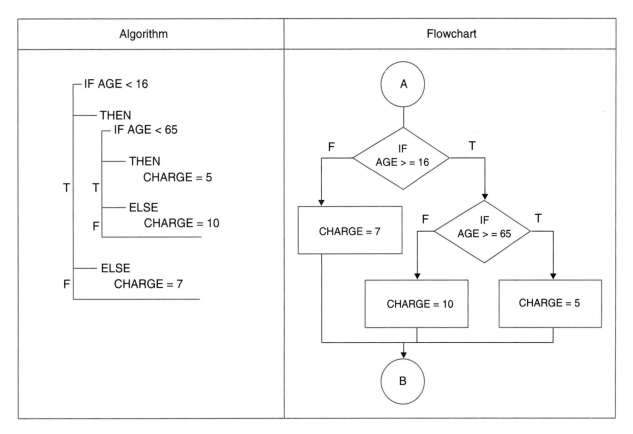

Algorithm	Flowchart

Figure 6.9 Negative Logic—Example 1

Notice how the testing takes place. Without proper testing, the problem-solving process is useless. Testing the algorithm or the flowchart reveals errors in logic, so corrections can be made before proceeding. It is far easier to make corrections during problem solving than after the coding has been done.

Figure 6.11 shows another way to organize the set of conditions in Figure 6.10 using negative logic. Here again, the rule is to have the computer make as few tests as possible.

Logic Conversion

There are times when it is necessary to change the logic from positive to negative or vice versa in order to improve the efficiency or readability of a solution. Sometimes the structure of a solution simply does not lend itself to the logic type that you've used, either because of the number and nature of the instructions to be processed, or because there are no instructions to be processed. In a decision, there must always be instructions for a TRUE section, but not always for a FALSE section. If there are no instructions for the TRUE section of a decision instruction, then it is better to convert the logic type.

To convert from positive logic to negative logic or vice versa, do the following:

1. Change all < to > =.
2. Change all < to >.
3. Change all > to < =.

Algorithm	Flowchart	Test

IF SALES > 2000
 THEN
 IF SALES > 4000
 THEN
 IF SALES > 6000
 THEN
 COMMISSION = .1
 ELSE
 COMMISSION = .07
 ELSE
 COMMISSION = .04
 ELSE
 COMMISSION = .02

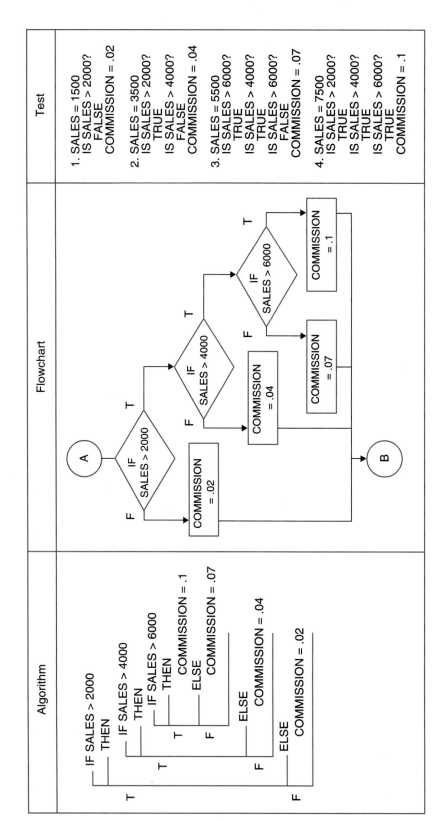

1. SALES = 1500
 IS SALES > 2000?
 FALSE
 COMMISSION = .02

2. SALES = 3500
 IS SALES > 2000?
 TRUE
 IS SALES > 4000?
 FALSE
 COMMISSION = .04

3. SALES = 5500
 IS SALES > 6000?
 TRUE
 IS SALES > 4000?
 TRUE
 IS SALES > 6000?
 FALSE
 COMMISSION = .07

4. SALES = 7500
 IS SALES > 2000?
 TRUE
 IS SALES > 4000?
 TRUE
 IS SALES > 6000?
 TRUE
 COMMISSION = .1

Figure 6.10 Negative Logic—Example 2

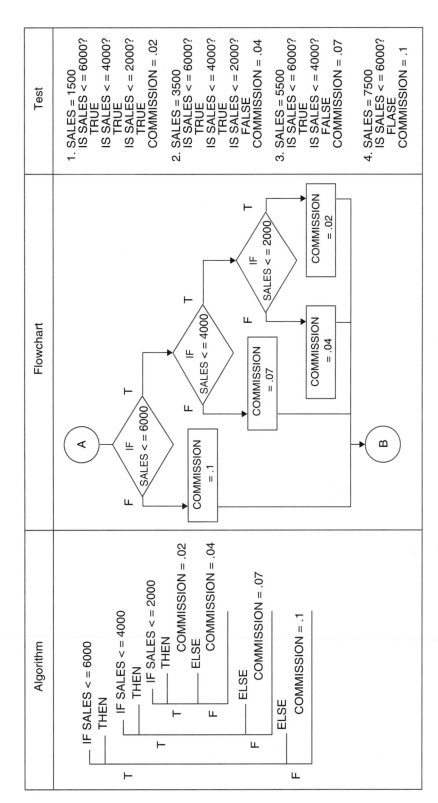

Algorithm	Flowchart	Test
IF SALES < = 6000 THEN IF SALES < = 4000 THEN IF SALES < = 2000 THEN COMMISSION = .02 ELSE COMMISSION = .04 ELSE COMMISSION = .07 ELSE COMMISSION = .1		1. SALES = 1500 IS SALES < = 6000? TRUE IS SALES < = 4000? TRUE IS SALES < = 2000? TRUE COMMISSION = .02 2. SALES = 3500 IS SALES < = 6000? TRUE IS SALES < = 4000? TRUE IS SALES < = 2000? FALSE COMMISSION = .04 3. SALES = 5500 IS SALES < = 6000? TRUE IS SALES < = 4000? FALSE COMMISSION = .07 4. SALES = 7500 IS SALES < = 6000? FLASE COMMISSION = .1

Figure 6.11 The Conditions in Figure 6.10 Set Up in a Different Way

117

4. Change all > = to <.
5. Change all = to <>.
6. Change all <> to =.
7. Interchange all of the THEN set of instructions with the corresponding ELSE set of instructions.

If you are using the convention to keep all TRUES to the right and all FALSES to the left in the flowchart, then reverse flowchart sections (see Figure 6.12). Figure 6.12 revisits the problem regarding the charge for a concert ticket based on age, illustrated in Figure 6.6 Notice that first, the relational operators are changed in the conditions. Second, the TRUE sections of the algorithms and flowcharts are exchanged with the FALSE sections.

Which Decision Logic?

How do you decide which type of decision logic to use? The general rule is to evaluate all three types for each solution and choose the one that is most readable, requires the fewest tests, and is easiest to maintain. It is easy to fall into the trap of always using the same logic type. It is also easy to fall into the habit of accepting and using a set of conditions exactly as the customer has given them. Such habits are not the most efficient ways of solving problems.

Figure 6.13 contains flowcharts showing four ways to set up the same conditions. The problem is to calculate a pay bonus given the set of conditions in the figure. From the data given, which logic type is the most efficient? Since the first and second solutions, the examples of positive logic, are easy to read, and the second solution requires the fewest tests, the second solution would be the best.

To analyze which type of decision logic would be most efficient for a particular solution, answer the following questions:

1. Which type would make the solution most readable?
2. Which type would make the solution the easiest to maintain or change?
3. Which would require the fewest tests when you don't know anything about the data?
4. Which would require the fewest tests when you're given some data?

If you find at the end of the analysis that more than one of the logic types could be equally effective, then simply choose one.

Decision Tables

You may have a problem with multiple conditions and multiple consequent actions. In such a case discovering all of the actions that correspond to particular conditions—and discovering which combination of conditions, if any, leads to meaningful actions—can be complicated. A good way to simplify this process is to draw a decision table. You would draw a decision table during problem analysis. It is an excellent tool to use as a communication vehicle with a customer or with the person for whom you are solving the problem. You can also use it to test the algorithm and the flowchart.

Algorithm	Flowchart

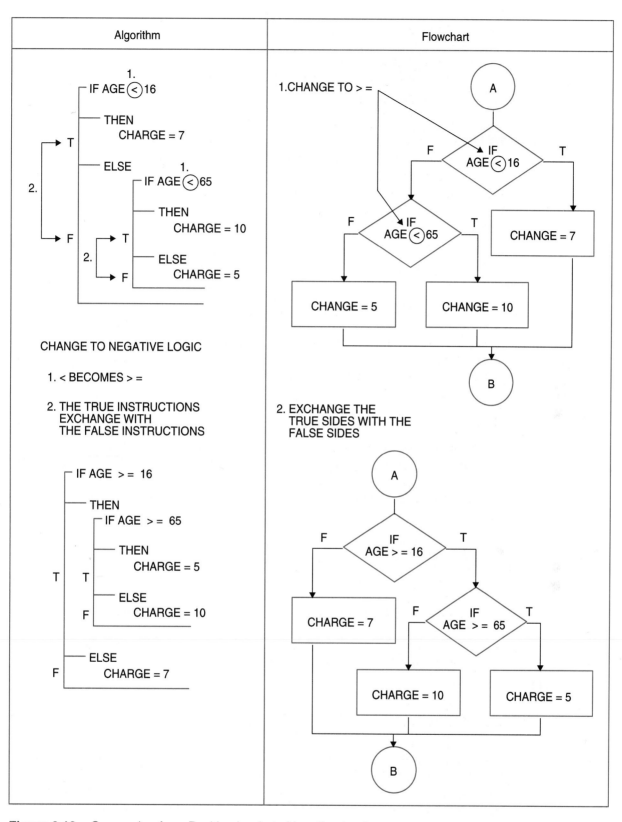

Figure 6.12 Conversion from Positive Logic to Negative Logic

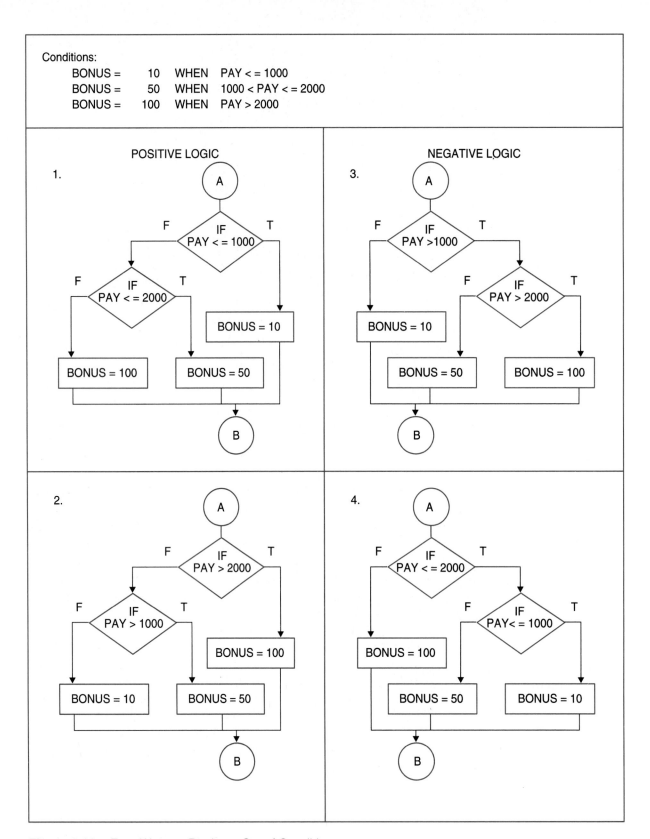

Figure 6.13 Four Ways to Design a Set of Conditions

Table 6.1 Decision Table Format

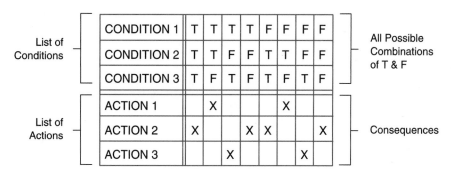

List of Conditions										All Possible Combinations of T & F
CONDITION 1	T	T	T	T	F	F	F	F		
CONDITION 2	T	T	F	F	T	T	F	F		
CONDITION 3	T	F	T	F	T	F	T	F		

List of Actions										Consequences
ACTION 1		X			X					
ACTION 2	X			X	X			X		
ACTION 3			X				X			

A **decision table** consists of four parts:

1. The conditions
2. The actions
3. The combinations of TRUE and FALSE for the conditions
4. The action to be taken or the consequences for each combination of conditions

Draw a decision table in the form of a rectangle divided into the four major sections. Each section is divided into smaller sections, as needed, according to the number of possible combinations of TRUE and FALSE (see Table 6.1). The total number of possible combinations of TRUE and FALSE for the conditions is two to the power of the number of conditions. That is, if there are two conditions, then there are $2 \wedge 2$, or 4 possible combinations of TRUE and FALSE. If there are three conditions, then there are $2 \wedge 3$, or 8 possible combinations. If there are four conditions then there are $2 \wedge 4$, or 16 possible combinations.

A flowchart must then be developed from the decision table. Figure 6.14a–c shows the flowchart development from the decision table in Table 6.1. The first step in the development of the flowchart is to draw in all decisions. Figure 6.14a shows the starting flowchart. Notice that the first decision is the first condition. The second decision is the second condition and is repeated on the true side and the false side of the first decision. The third decision is the third condition. This decision is repeated four times, once for each true and false for the second decision. This will allow all combinations of true and false values. Step two is to look at both sides of each decision, starting with decision one. Figure 6.14b shows the comparison of sides. If both sides are the same the decision is not needed. Notice that the true side of decision one is exactly the same as the false side. Therefore, this decision is not needed. The decision and the false side can be eliminated. The true side and the false side of the second decision are different and therefore, they need to stay. The same is true for the third decision. Figure 6.14c shows the final algorithm and flowchart. The four steps to develop a flowchart from the decision table are:

1. Draw all decisions in flowchart form.
2. Compare the true and false sides of each decision, starting with the first one.
3. Eliminate any decisions that have the same instructions on both the true and false sides, keeping the true consequence or action.
4. Redraw the flowchart.

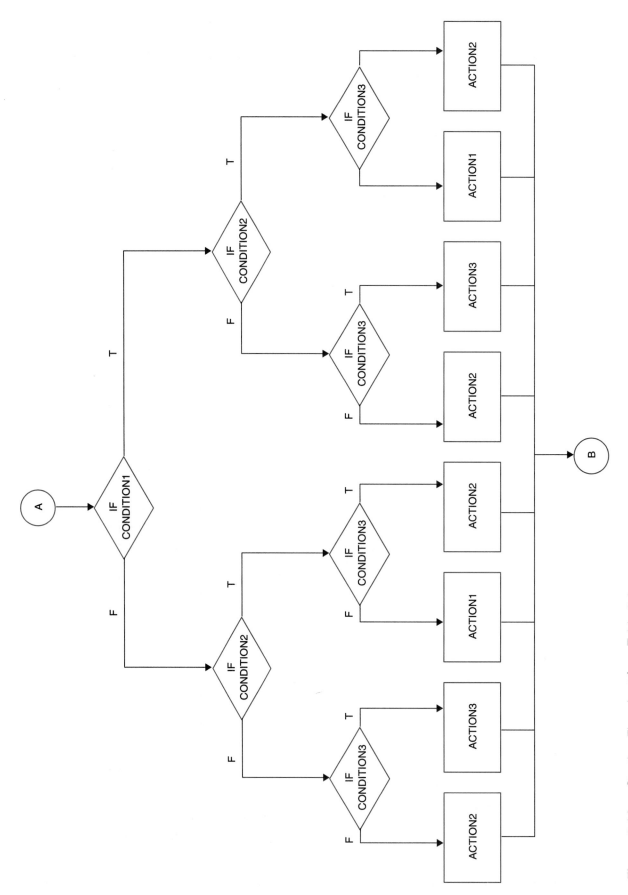

Figure 6.14a Starting Flowchart from Table 6.1

Figure 6.14b Elimination of Conditions

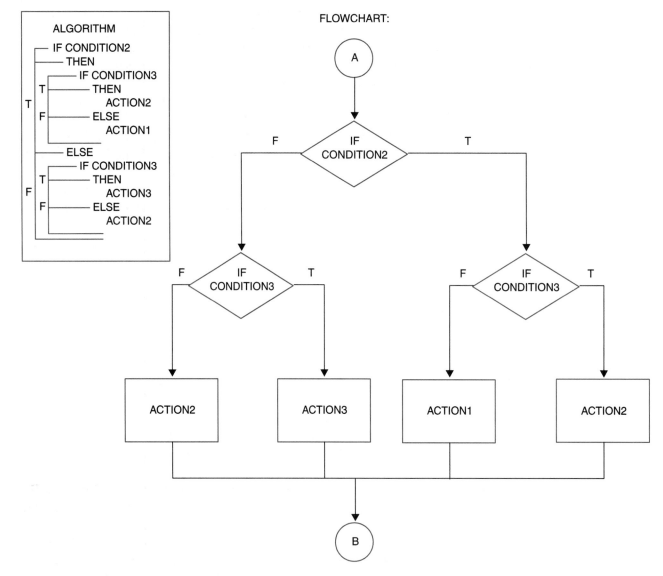

Figure 6.14c Final Flowchart

Table 6.2 Decision Table

List of Conditions	1	PURCHASE < $100	T	T	T	T	F	F	F	F
	2	LAST PAYMENT WITHIN 30 DAYS	T	T	F	F	T	T	F	F
	3	BALANCE < $1000	T	F	T	F	T	F	T	F
List of Actions	1	CREDIT OK	X		X					
	2	REFER TO CREDIT DEPARTMENT		X			X		X	
	3	CREDIT DENIED				X		X		X

The decision table is a graphical way of stating a policy for a company, such as a policy for cashing a check or for giving a bonus. Only the manager of the company can instruct the programmer regarding the company policy, so the programmer must work with the manager to complete the decision table. The manager tells the programmer the conditions and the actions, or consequences, of the decisions. The only part the programmer fills in is the upper right-hand section showing the possible combinations of TRUE and FALSE. It is important for the programmer to work with the customer to make sure all combinations have the correct set of actions. From the decision table, the programmer proceeds to write the algorithm and the flowchart to solve the problem. The major difficulty in designing the algorithm is to make decision structure as efficient as possible by weeding out all unnecessary tests and putting the conditions in the best possible order. It often takes several experiments to come up with the best structure.

Table 6.2 shows a decision table for a store policy for charging a purchase. There are three conditions:

1. The purchase is less than $100.
2. The last payment to the account was made in the last 30 days.
3. The balance of the account is less than $1,000.

Each of these three conditions could be TRUE or FALSE depending on the customer. The following actions could be taken:

1. Credit is OK, and the customer can charge the item.
2. Refer the customer to the credit department.
3. Credit is denied and the customer cannot charge the item.

These conditions and actions are supplied by the manager of the credit department. The manager also states when each of the actions will take place. The policy illustrated in Table 6.2 specifies that a customer can charge an item when all three conditions are TRUE, or when the first and third conditions are TRUE and the second one is false. The customer is referred to the credit department in the following cases: when conditions 1 and 2 are TRUE and 3 is FALSE; when 2 and 3 are TRUE and 1 is FALSE; and when 1 and 2 are FALSE and 3 is TRUE. Credit is denied in these cases: when conditions 2 and 3 are FALSE and 1 is TRUE; when 1 and 3 are FALSE and 2 is TRUE; and when all three are FALSE.

Figure 6.15 shows the development of the flowchart from Table 6.2. Figure 6.15a shows the complete flowchart with all decisions. The condition Purchase < $100 is the first decision. Last Payment within 30 Days is the second condition and is repeated twice, once each for the true and false in the first decision. The third decision is Balance < $1000. This decision is repeated four times, once for each true and false in decision two. This will give the actions for all combinations of true and false. Figure 6.15b shows the decisions which have the same true and false actions. Notice that the second decision on the false side of the first decision has identical actions on the true side and the false side. This indicates that this decision is not necessary and can be eliminated. After all, if the same action takes place whether the decision is true or false, the decision is not needed. Figure 6.15c shows the redrawn flowchart and the associated algorithm.

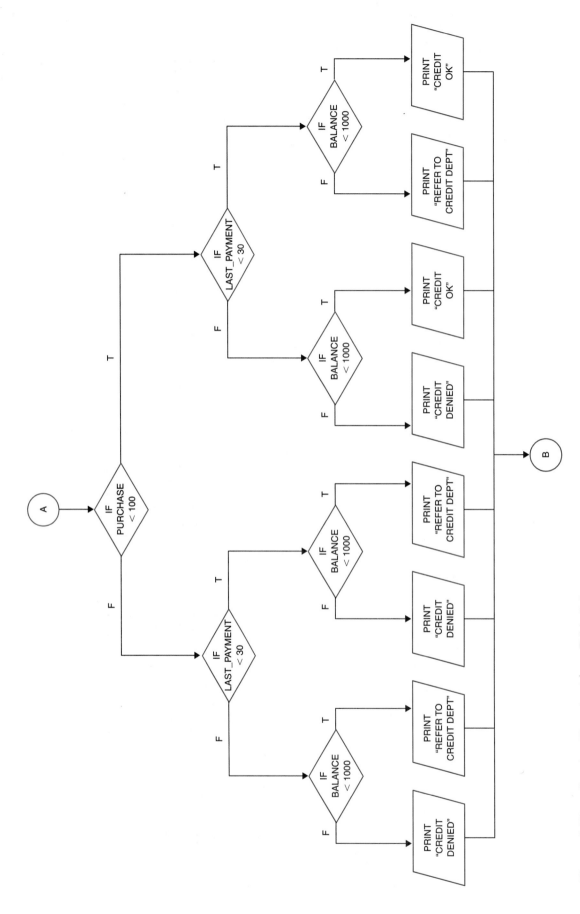

Figure 6.15a Starting Flowchart from Table 6.2

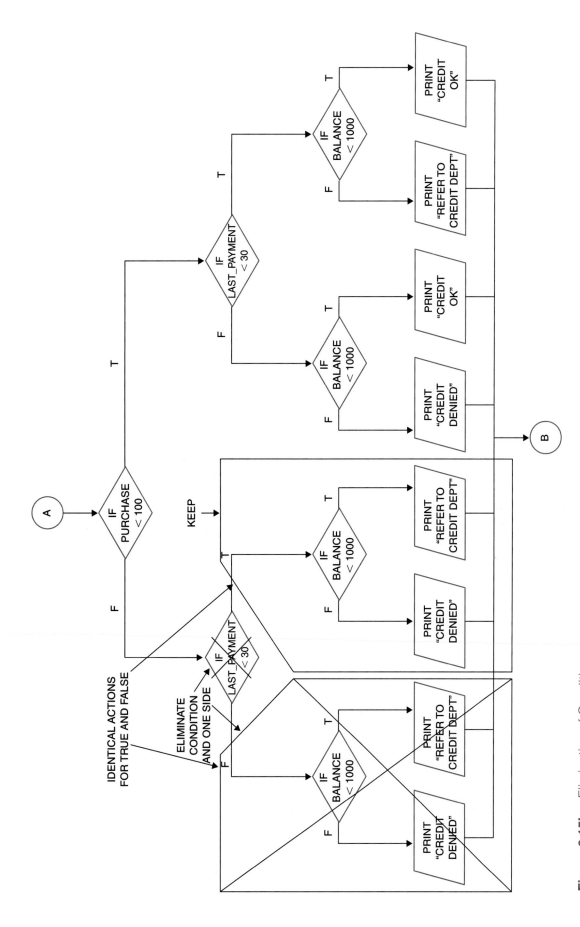

Figure 6.15b Elimination of Conditions

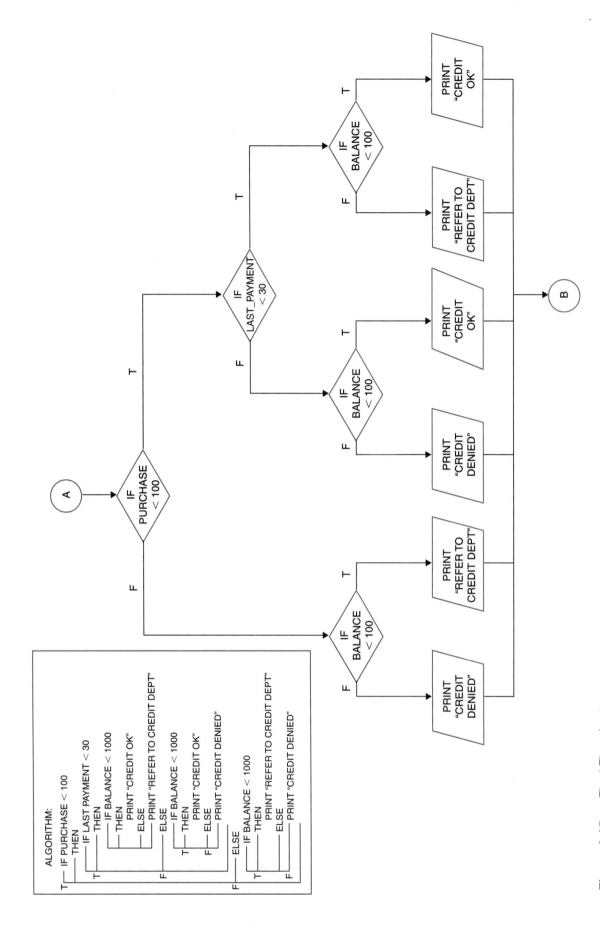

Figure 6.15c Final Flowchart

Putting It All Together

The Putting It All Together (PIAT) sections in this book are designed to show students how to pull together and use the concepts from a previous section or sections. The following PIAT demonstrates how to use the six steps of problem solving on the computer to develop a solution that uses the sequential and the decision logic structures. The problem is the following:

The Fantastic Floral Company sells to wholesale and retail buyers. The wholesale buyer needs a resale number in order to buy at no tax and to receive discounts. The retail buyer pays 6% tax. The following are the discounts to the wholesale buyer:

AMOUNT < $100	DISCOUNT = 2%
AMOUNT > = $100 and < $500	DISCOUNT = 5%
AMOUNT > = $500	DISCOUNT = 10%

Given an amount of purchase, how much will the customer owe the company? (See Figure 6.16a–h.)

Given Data	Required Results
Wholesale buyers: • Must have resale no. • Amount of purchase discount < 100 2% >= 100 & < 500 5% >= 500 10% Retail buyers: • No discount Amount of purchase (AP) 6% Tax	Amount owed
Processing Required	**Solution Alternatives**
Retail = 1.06 * AP Wholesale = (1–D) * AP	*• Use positive logic for discount • Use negative logic for discount • Use straight-through logic for discount *This solution was chosen because it results in the fewest number of tests and it is easily read.

Figure 6.16a Fantastic Floral Company—Problem Analysis Chart

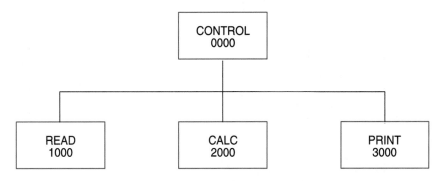

Figure 6.16b Fantastic Floral Company—Structure Chart

Input	Processing	Module Reference Number	Output
1. Amount of purchase 2. Resale number (If 0 then a retail customer)	1. ENTER input data 2. CALC discount 3. CALC amount owed 4. PRINT output information	1000 2000 2000 3000	1. Amount of purchase 2. Resale # 3. Amount owed

Figure 6.16c Fantastic Floral Company—IPO Chart

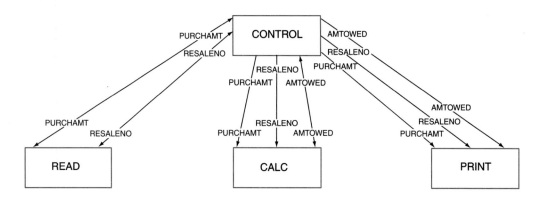

Item	Variable Name	Data Type	Module	Scope	Pseudonyms	Error Cks.
Purchase Amt	PURCHAMT	Numeric	All	Local	None	None
Resale Number	RESALENO	Numeric	All	Local	None	None
Amount Owed	AMTOWED	Numeric	Control, Calc, Print	Local	None	None

Figure 6.16d Coupling Diagram and Data Dictionary

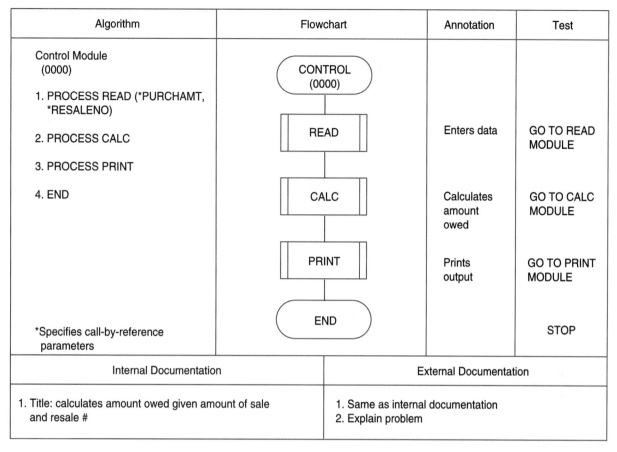

Figure 6.16e The Algorithms and Flowcharts—CONTROL Module

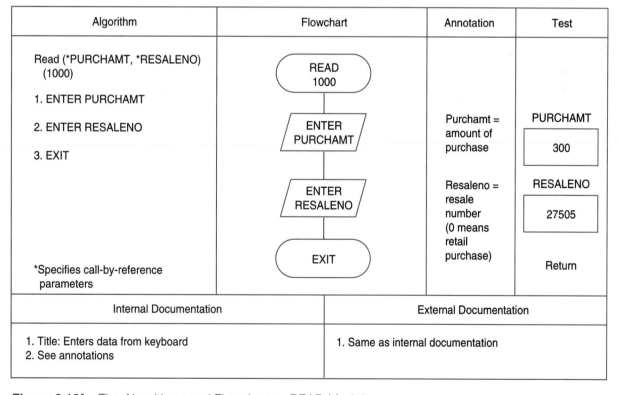

Figure 6.16f The Algorithms and Flowcharts—READ Module

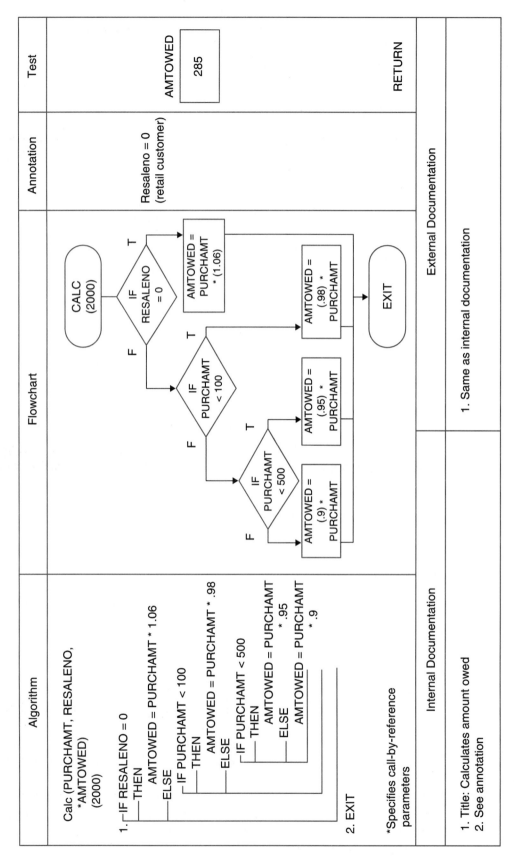

Algorithm	Flowchart	Annotation	Test
Calc (PURCHAMT, RESALENO, *AMTOWED) (2000)			AMTOWED
1. IF RESALENO = 0	CALC (2000)	Resaleno = 0 (retail customer)	285
THEN AMTOWED = PURCHAMT * 1.06 ELSE	IF RESALENO = 0		
IF PURCHAMT < 100 THEN AMTOWED = PURCHAMT * .98 ELSE	AMTOWED = PURCHAMT * (1.06)		
IF PURCHAMT < 500 THEN AMTOWED = PURCHAMT * .95 ELSE AMTOWED = PURCHAMT * .9	IF PURCHAMT < 100		
2. EXIT	IF PURCHAMT < 500		
*Specifies call-by-reference parameters	AMTOWED = (.98) * PURCHAMT		
	AMTOWED = (.95) * PURCHAMT		
	AMTOWED = (.9) * PURCHAMT		
	EXIT		RETURN
Internal Documentation		External Documentation	
1. Title: Calculates amount owed 2. See annotation		1. Same as internal documentation	

Figure 6.16g The Algorithms and Flowcharts—CALC Module

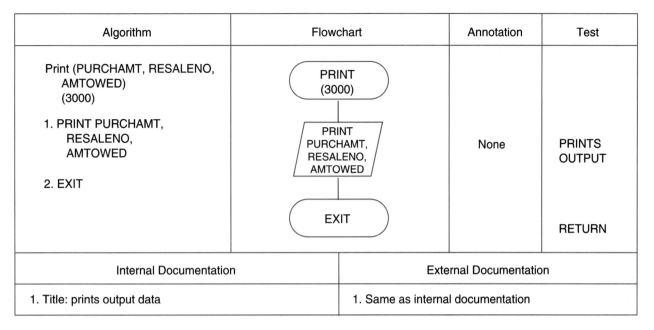

Algorithm	Flowchart	Annotation	Test
Print (PURCHAMT, RESALENO, AMTOWED) (3000) 1. PRINT PURCHAMT, RESALENO, AMTOWED 2. EXIT	PRINT (3000) → PRINT PURCHAMT, RESALENO, AMTOWED → EXIT	None	PRINTS OUTPUT RETURN

Internal Documentation	External Documentation
1. Title: prints output data	1. Same as internal documentation

Figure 6.16h The Algorithms and Flowcharts—PRINT Module

Summary

The decision logic structure is the second basic logic structure used to design algorithms, the first being the sequential logic structure. You now know the following types of instructions for communicating with the computer:

- START
- STOP
- PRINT
- assignment
- IF/THEN/ELSE
- PROCESS module
- END/EXIT
- RETURN
- CONTINUE

The instruction added in this chapter was IF/THEN/ELSE, the decision instruction. This instruction is powerful because it enables the computer to make decisions. It is an instruction that is easy to understand but more difficult to put to work.

There are three decision logic types: straight-through, positive, and negative logic. The programmer evaluates the conditions that arise in the problem to decide which logic type to use.

When you are using the decision structure, a decision table is a tool to help you develop the best set of decision instructions. You can use it to find the consequences following from every condition in a decision, so you can design the most efficient algorithms to solve the problem.

New Terms

decision table

IF/THEN/ELSE instruction

negative logic

nested IF/THEN/ELSE instruction

positive logic

straight-through logic

Questions

1. What is the decision logic structure?
2. What are the three types of decision logic?
3. Without specific data, outline an algorithm and a flowchart for each of the three types of decision logic.
4. What factors should determine which decision logic type to use for a specific problem?
5. How do you use a decision table?

Problems

1. Set up four different algorithms and flowcharts for calculating a student's letter grade given the following (do not use straight-through logic):

$$90–100 = A$$
$$80–89 = B$$
$$70–79 = C$$
$$60–69 = D$$
$$\text{below } 60 = F$$

2. Which of the four solutions from Problem 1 would be the best? Justify your answer.

3. Using first positive and then negative logic, write the algorithms and draw the flowcharts for the following set of conditions:

$$R = 50 \text{ for } S < = 1000$$
$$R = 100 \text{ for } S = 1001–4000$$
$$R = 250 \text{ for } S = 4001–8000$$
$$R = 75 \text{ for } S > 8000$$

4. Change the positive logic algorithm and flowchart from Problem 3 to reflect the following:

$$R = 100 \text{ for } S = 1000–4000$$
$$R = 250 \text{ for } S = 4001–8000$$
$$\text{otherwise } R = 75$$

5. a. Write the Calculation module to choose the largest number from a set of three numbers, A, B, and C. Write the algorithm and draw the flowchart.
 b. Modify the logic in the calculation module to print A, B, and C in the correct order.

6. Complete the seven problem-solving steps for the following problem

 Eureka Water Company charges the homeowner one of the following:

 a. A flat rate of $15.00 for usage up to and including 1,000 cubic feet.
 b. $.0175 per cubic foot for usage over 1,000 cubic feet and up to and including 2,000 cubic feet.
 c. $.02 per cubic foot for usage over 2,000 cubic feet and up to and including 3,000 cubic feet.
 d. A flat rate of $70 for usage over 3,000 cubic feet.
 Test your flowchart with actual data.

7. A company that issues check-cashing cards uses an algorithm to create card numbers. The algorithm adds the digits of a four-digit number, and then adds a fifth digit of 0 or 1 to make the sum of the digits even. The last digit in the number is called the *check digit*. Complete the seven problem-solving steps to develop a solution which accepts a four-digit number into one variable, adds the check digit, and prints the original number and the new number. Test your flowchart with the following data: Original number = 4737 (47371) and 4631 (46310).

 Hint: You may use any or all of the functions:

 $$\text{INTEGER (X)—Integer function}$$
 $$\text{STRING (X)—Numeric to string}$$
 $$\text{VALUE (A)—String to numeric}$$

 and the principle of concatenation of strings.

 [*Note:* The INTEGER (X) gives the whole number value of the real number X. When X is 546.43, the INTEGER (X) is 546; when X is 23.899 the INTEGER (X) is 23. The STRING (X) and VALUE (A) are conversion functions. The resultant of the function STRING (X) is the string value of the numeric X. The resultant of VALUE (A) is the numeric value of the string A. Concatenation is the combining of strings by placing the first string in front of the second one. For example, the resultant of "4" + "5" would be "45".]

8. Create a decision table, flowchart, and algorithm for Problem 11 in Chapter 2.

9. An admission charge for The Little Rep Theater varies according to the age of the person. Develop a solution to print the ticket charge given the age of the person. The charges are as follows:
 a. Over 55: $10.00
 b. 21–54: $15.00
 c. 13–20: $10.00
 d. 3–12: $5.00
 e. Under 3: Free

10. A customer needs a specific amount of paper. The charges on the paper are $0.10 for $0.10 for single sheets, $0.055 for amounts in multiples of 100 sheets, $0.04 in multiples of 500 sheets and $0.03 in multiples of 1000 sheets. Develop a solution to calculate the type and number of package(s) for the least amount of money the customer should buy, given the minimum amount of sheets the customer needs. For example, if the customer needs 380 sheets, the amount she would pay when buying in multiples of 100 would be $22.00. However if the customer bought 500 sheets the cost would be $20.00. It would be cost effective for the customer to buy a package of 500 sheets.

WHAT'S WRONG WITH THIS?

1. **a.** Is the following nested IF/THEN/ELSE statement valid? Why or why not?

 IF X < Y
 THEN
 IF Y < Z
 THEN
 X = X + 1
 ELSE
 Y = Y + 1

 b. Which IF does the ELSE go with? Explain your answer.

 c. How would you change the statement to clarify the statement? (*Hint:* Use a CONTINUE statement.)

 d. Discuss your answer with your classmates and your instructor.

2. Correct the following algorithmic instructions:

 a. IF A < B + C THEN
 ELSE
 A = 5

 b. IF LEN(S) < "5" THEN
 PRINT S

 c. IF WIDTH = LENGTH THEN
 PRINT "OBJECT IS A SQUARE."
 OTHERWISE
 PRINT "OBJECT IS NOT A SQUARE."

Chapter 7

Problem Solving with Loops

Overview

The Loop Logic Structure
Incrementing
Accumulating
WHILE/WHILE-END
REPEAT/UNTIL
Automatic-Counter Loop
Nested Loops
Indicators
Algorithm Instructions and Flowchart Symbols
Recursion

Objectives

When you have finished this chapter, you should be able to:

1. Develop problems using the loop logic structure in conjunction with the decision and sequential logic structures.
2. Use the seven problem-solving tools when developing a solution using the loop logic structure.
3. Use counters and accumulators in a problem solution.
4. Use nested loop instructions to develop a problem solution.
5. Distinguish the different uses of three types of loop logic structures.
6. Use recursion in a simple problem.

The Loop Logic Structure

A third logic structure for designing decisions is the loop structure. The loop logic structure is the repeating structure. Most problems in business involve doing the same task over and over for different sets of data, so this structure is extremely important. Through the loop structure, a program can process many payroll records or inventory items, or put a mailing list in alphabetical or zip code order. This structure is harder to understand than the decision structure, but easier to use. The main difficulty in using it as part of a solution is identifying what instructions should be repeated. Besides being used to repeat instructions in a solution, the loop structure is also used to return the processing to an earlier point in the solution. When it is used in this way, it is a replacement for the GOTO statement. (The GOTO is an instruction that tells the computer to transfer to another instruction in the solution instead of processing the next instruction in sequence.) Replacing the GOTO statement in this way increases the readability of the program.

There are three types of loop structures, and each of them can be written in another way as a set of instructions using the IF/THEN/ELSE structure and the GOTO instruction. However, the IF/THEN/ELSE is a decision logic structure and generally should not be used to facilitate a loop. The loop structure is easier to read and to maintain than the IF/THEN/ELSE and GOTO combination. Since you already understand the decision logic structure, the decision equivalence to each loop structure will be illustrated to aid your understanding of loops.

The first type of loop structure is the WHILE/WHILE-END loop, which repeats instructions while a condition is TRUE and stops repeating when a condition arises that is not TRUE. The second type is the REPEAT/UNTIL loop, which repeats instructions while a condition is FALSE or until a condition is TRUE. The third is the automatic-counter loop in which a variable is set equal to a given number and increases in equal given increments until it is greater than an ending number. Each of the three loop structures has a specific use according to the language and/or the problem.

The algorithm and the flowchart differ with each type of loop structure. It is extremely important to indent the instructions in algorithms using the loop structure—as it is with the decision structure—to improve readability. Also, nested loops are common, and in these solutions indentation with bracketing allows you to easily identify each loop.

Several standard types of tasks are accomplished through the use of the loop structure. Two of these are counting (also called incrementing and decrementing), and accumulating (also called calculating a sum or a total). In both tasks a number is added or subtracted from a variable and the result is stored back into the same variable. The basic difference between counting and accumulating is in the value that is added or subtracted. When counting, the value is a constant; when accumulating, the value is a variable. In each case the resultant variable is assigned the value of zero before starting the loop. This is called *initializing* the variable.

Incrementing

The task of **incrementing,** or counting, as we said, is done by adding a constant, such as 1 or 2, to the value of a variable. To write the instruction to increment a variable, you use an assignment statement. For example

$$COUNTER = COUNTER + 1 \text{ or } C = C + 1$$

when counting by ones. An assignment statement allows the variable on the left side of the equal sign to be assigned the value of, or to be replaced by, the resultant of the expression on the right side of the equal sign. The counter or incrementor instruction is a special version of the assignment instruction. This instruction enables the programmer to count the number of items, people, temperatures, and so on, as part of the solution to a problem.

Notice the structure of the assignment instruction for incrementing. The same variable name is on both sides of the equal sign; the amount the variable is to be incremented follows the plus sign. This instruction takes the value of the counter, adds one to it, and then replaces the old value of the counter with the new. The increment can be one, two, or any other constant, including negative numbers if you want to decrement rather than increment. In this example COUNTER or C must be initialized to zero before starting the loop.

Accumulating

Another task that a program must often perform is **accumulating,** or summing, a group of numbers. The process of accumulating is similar to incrementing, except a variable instead of a constant is added to another variable, which holds the value of the sum or total. The instruction for accumulating is the following:

$$SUM = SUM + VARIABLE \text{ or } S = S + V$$

For example, the instruction to find total sales would be

$$TOTALSALES = TOTALSALES + SALES$$

Like the instruction to increment, the instruction to accumulate is a special version of the assignment instruction. As in incrementing, the same variable name appears on both sides of the equal sign (TOTALSALES) but in accumulating, a variable rather than a constant is on the right-hand side of the plus sign (SALES in this case). In other words, with an accumulator you are adding a variable (the item you are accumulating) to the value of another variable, which holds the sum or total. In these examples SUM and TOTALSALES must be initialized to zero before starting the loop.

Calculating the product of a series of numbers is similar to finding the sum of a series of numbers with two exceptions. First, the plus sign is replaced by the multiplication sign (∗). Second, the product variable must be initialized to 1 instead of 0. For example

$$PRODUCT = 1$$
$$PRODUCT = PRODUCT * NUMBER$$

Finding the product of a series of numbers is used very little. However, it will be used in the discussion on recursion at the end of this chapter.

WHILE/WHILE-END

The first of the three types of loop structures is the **WHILE/WHILE-END** structure. This type of loop tells the computer that while the condition is TRUE, repeat all

instructions between the WHILE and the WHILE-END. The form of the algorithm is the following:

```
  ┌─  WHILE<CONDITION(S)>
  │      INSTRUCTION
  │      INSTRUCTION
  │      .
  │      .
  └─  WHILE-END
```

When you design WHILE/WHILE-END loops, use brackets and indentation to improve the readability of the algorithms. In the illustration, notice how the bracket connects the beginning (WHILE) and the end (WHILE-END) of the loop, and how the instructions are indented within the brackets to make the algorithm easily readable. The general form of the WHILE/WHILE-END flowchart is shown in Figure 7.1.

The symbol used for the WHILE part of the instruction is the diamond, the symbol for a decision. Since this flowchart shows only part of an algorithm, on-page connectors are used. The completed algorithm would have one part that would connect to A and another part that would connect to B.

At the beginning of the WHILE/WHILE-END loop, the program processes the condition in order to decide whether or not to execute the loop instructions, as illustrated in the diamond at the top of the flowchart. When the resultant of the condition is FALSE at the start, the instructions within the loop will not be executed. When the resultant of the condition is TRUE, the complete set of instructions will be executed, and then the

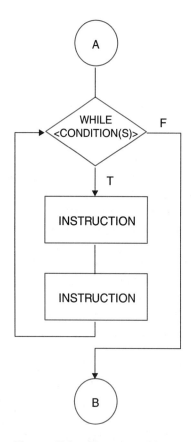

Figure 7.1 Flowchart Diagram of WHILE/WHILE-END

Algorithm	Flowchart

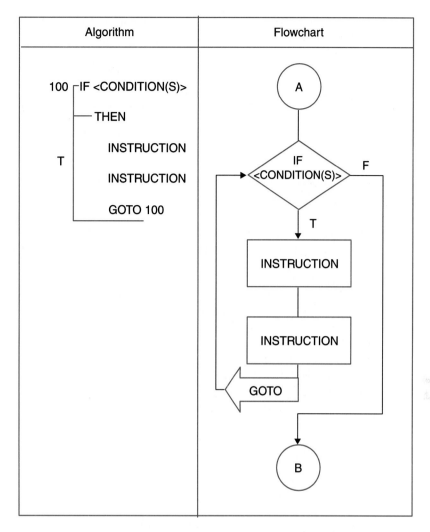

Figure 7.2 Decision Equivalent to WHILE/WHILE-END

computer will go back to the beginning of the loop and process the condition again. The loop will repeat until the resultant of the condition is FALSE, at which time the processing will continue with the instruction following the WHILE-END.

Figure 7.2 shows the decision-structure equivalent to the WHILE/WHILE-END loop structure. Notice that the decision structure requires a separate GOTO instruction to return to the top of the loop. Also notice that when the condition is FALSE, the line of execution transfers to the instruction after the loop instructions.

Use the WHILE/WHILE-END structure when you do not know the number of times the instructions will be repeated, or if there are cases when the instructions in the loop should not be processed. In those cases, the resultant of the condition is FALSE at the start of the loop, so the loop is not processed. Examples of such cases are when youare entering information on clients and you don't know the number of clients; when you are finding the total amount of sales for the day and you don't know the number of sales; or when you are calculating test averages for each student in a class. In the case of calculating test averages, the loop instructions should not be executed for students who for some reason have no test grades. To calculate the average you would accumulate and count the test scores and then divide the total score by the number of tests. Since the number of grades would be zero for the student who didn't take any tests, the computer

would try to divide by zero, and an error would result. The WHILE/WHILE-END loop would enable the processing to pass over the student and go on to the next student.

Putting It All Together

Problem: Create the algorithm and the flowchart to find the average age of all the students in a class.

Figure 7.3 demonstrates the use of a WHILE/WHILE-END structure to calculate the average age of a group of people. The first instruction initializes the value of the SUM (the total of all ages) to zero. The second instruction initializes the value of the counter variable to zero. Next the first age is entered. An age must be entered before the loop so that the condition has data to use to get a resultant. This is called a **primer READ** because it gives the WHILE/WHILE-END loop a valid value for the variable AGE in order for the conditions to be true the first time through the loop. If there is no primer READ, the value of AGE is unknown. The next read is at the end of the loop for two reasons. First, the first value of AGE must be included in the calculations. Second, the test for the trip value (the value of age that makes the condition false and signals the end of the list of students) must be just before the condition, so the trip value will not be calculated as part of the average age. The WHILE begins the loop and processes the condition AGE < > 0. When the resultant is TRUE, the instructions in the loop are processed; when the resultant is FALSE, the processing continues with the calculation of the average. The first loop instruction accumulates the ages by adding each age to the total, followed by an instruction that counts the people by adding one to the counter variable. The final instruction in the loop enters the next age. The processing then returns to process another condition. The last age should be zero. This zero is called a *trip value*, or a *flag*. It allows the value of AGE to control when to stop the looping process and continue with the rest of the solution. When the looping process stops, the processing continues at the calculation of the average age, and finally, the printing of the average age.

The data for this problem can be entered from the keyboard; the user enters a 0 when there are no more ages to enter.

REPEAT/UNTIL

The second type of loop structure is the **REPEAT/UNTIL** structure. This type of loop tells the computer to repeat the set of instructions between the REPEAT and the UNTIL, until a condition is TRUE. There are two major differences between this instruction and the WHILE/WHILE-END. First, in the WHILE/WHILE-END loop, the program continues to loop as long as the resultant of the condition is TRUE; in the REPEAT/UNTIL loop, the program stops the loop process when the resultant of the condition is TRUE. Second, in the WHILE/WHILE-END loop, the condition is processed at the beginning; in the REPEAT/UNTIL loop, the condition is processed at the end. When the condition is processed at the end, the instructions in the loop are processed entirely at least once, whether it is necessary or not.

It is important that you be able to distinguish between the loop structures so you can use them appropriately. Both the setup and the uses for the two loop structures are different. When you are using the WHILE/WHILE-END, you must initialize the data so that the resultant of the condition is TRUE the first time through the loop. Otherwise, the loop instructions will never be processed. With the REPEAT/UNTIL, you can set the

Algorithm	Flowchart

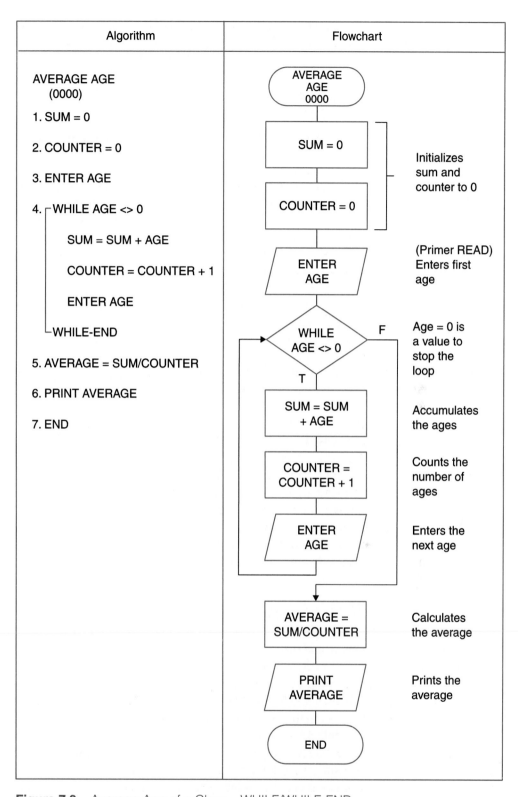

AVERAGE AGE
 (0000)

1. SUM = 0

2. COUNTER = 0

3. ENTER AGE

4. ⌐WHILE AGE <> 0

 SUM = SUM + AGE

 COUNTER = COUNTER + 1

 ENTER AGE

 └WHILE-END

5. AVERAGE = SUM/COUNTER

6. PRINT AVERAGE

7. END

Figure 7.3 Average Age of a Class—WHILE/WHILE-END

operands of the conditions anywhere within the loop since the condition is processed at the end. If there is any reason that the loop instructions should not be processed the first time, then the WHILE/WHILE-END must be used because the REPEAT/UNTIL always processes the instructions at least once.

The format of the REPEAT/UNTIL algorithm is the following:

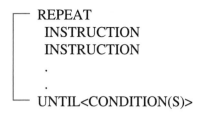

Here again, use brackets and indentation to improve readability. Figure 7.4 shows the form of the flowchart. When you are designing a REPEAT/UNTIL flowchart, write the word REPEAT above the first instruction of the loop.

Figure 7.5 shows the equivalent decision structure. Unlike the WHILE/WHILE-END decision equivalent, in which the processing of the condition is always at the beginning of the instructions to be repeated, with the REPEAT/UNTIL decision equivalence, the processing of the condition always takes place at the end of the instructions to be repeated. Also notice that there is no evidence of a loop until the condition is reached. The loop structure provides a solution that is easier to read and understand.

Use the REPEAT/UNTIL structure when you do not know the number of times the instructions will be executed, when you know they will be executed at least once, or when you do not know the condition for repeating until after the instructions in the loop are processed. Usually, the problem will dictate which loop to use.

Putting It All Together

Using the REPEAT/UNTIL loop structure, this section illustrates the same problem as in the Putting It All Together section for the WHILE/WHILE-END loop.

Figure 7.6 demonstrates the use of the REPEAT/UNTIL structure to calculate the average age of a group of people. The first instruction initializes the value of the SUM (the total of all ages) to zero. The second instruction initializes the value of the counter variable to zero. Next the first age is entered. An age must be entered so that there is a value for AGE to use in the calculations the first time through. The data must be entered just before processing the condition, or the trip values will be used in the last calculations in the solution. This is the *primer READ*. However, unlike the WHILE/WHILE-END loop structure, when a zero is entered at this place in the processing, the instructions within the loop will be executed at least once, depending on what is entered at the second ENTER instruction. For this reason, the WHILE/WHILE-END loop structure is often the preferred choice. The REPEAT begins the loop. The first instruction in the loop accumulates the ages by adding each age to the total, followed by an instruction that counts the people by adding 1 to the counter variable. The last instruction in the loop enters the next age. The next instruction is the loop-end instruction. This instruction processes the condition AGE50. When the resultant is FALSE the instructions in the loop are processed; when the resultant is TRUE the processing continues with the calculation of the average. The last age is a zero, the trip value or dummy value. It allows the value of AGE to control when to stop the loop process and continue with the rest of the solution. When the looping process stops, the processing continues at the calculation of the average age, and finally, the printing of the average age.

The programmer or the language dictates which loop structure to use. Sometimes the WHILE/WHILE-END and the REPEAT/UNTIL could be used with equal efficiency. The choice is then a matter of the preference of the programmer.

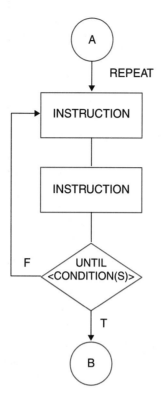

Figure 7.4 Flowchart Diagram of REPEAT/UNTIL

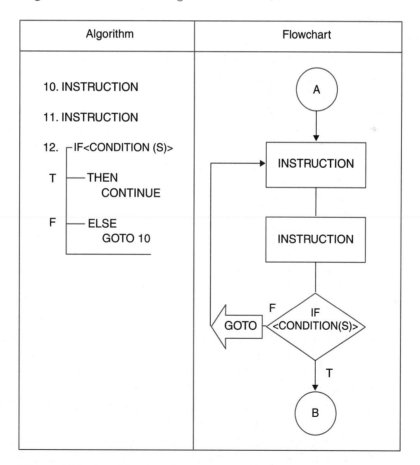

Figure 7.5 Decision Equivalent to REPEAT/UNTIL

Algorithm	Flowchart
AVERAGE AGE (0000) 1. SUM = 0 2. COUNTER = 0 3. ENTER AGE 4. ┌REPEAT SUM = SUM + AGE COUNTER = COUNTER +1 ENTER AGE └UNTIL AGE = 0 5. AVERAGE = SUM/COUNTER 6. PRINT AVERAGE 7. END	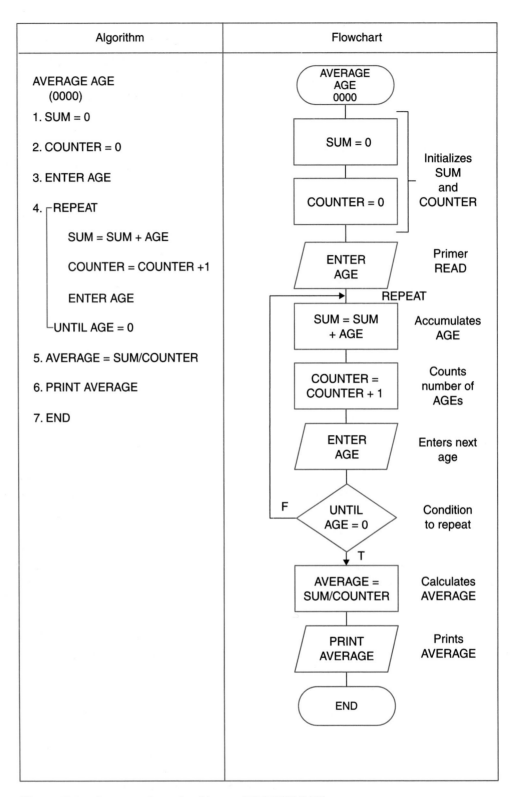

Figure 7.6 Average Age of a Class—REPEAT/UNTIL

Automatic-Counter Loop

The third type of loop structure is the **automatic-counter loop.** This type of loop increments or decrements a variable each time the loop is repeated. To design an automatic-counter loop, the programmer uses a variable as a counter that starts counting at a specified number and increments the variable each time the loop is processed. The amount to be incremented is specified by the instruction. The set of instructions within the loop repeats until the counter is greater than an ending number. The beginning value, the ending value, and the increment value may be constants, variables, or expressions (calculations). They should not be changed during the processing of the instructions in the loop. They may be changed after the loop is completed. The test for whether or not to process the loop instructions in an automatic-counter loop will be found at the beginning or at the end of the loop depending on the language or the version of the language.

The form of the algorithm for the automatic-counter loop is the following:

```
┌─  LOOP: COUNTER = BEGIN TO END STEP S
│     INSTRUCTION
│     INSTRUCTION
│       .
│       .
│       .
└─  LOOP-END: COUNTER
```

Here again, a bracket connects the beginning (LOOP) and the end (LOOP-END) of the algorithm, and the instructions are indented to improve readability, just as in the other loop structures.

The flowchart in Figure 7.7 shows the automatic-counter loop structure. C is the variable used for the counter, BEGIN is the beginning value, END is the ending value, and STEP is the value by which the counter is to be incremented. To decrement, STEP needs to be a negative number and BEGIN must be greater than END.

Figure 7.8 shows the decision equivalents to the automatic-counter loop structure. There are two solutions illustrated since there are two places the decision can be processed: at the beginning of the loop or at the end of the loop. Which decision equivalent would be used depends on the language or the version of the language. You will not find both versions in the same language or language version.

There are rules that the computer follows when processing the automatic-counter loop. The rules are simple, and it is important that the user understands and follows them when testing a solution.

When incrementing the counter, remember the following rules:

1. When the computer executes the LOOP instruction, it sets the counter equal to the beginning number (see a, a_1, and a_2 in Figures 7.7 and 7.8).
2. When the computer executes the LOOP-END, it increments the counter. Since the counter is usually incremented before the condition is processed, this book assumes the incrementing of the counter at LOOP-END (see b, b_1, and b_2 in Figures 7.7 and 7.8).
3. When the counter is less than or equal to the ending number, the processing continues at the instruction that follows the LOOP instruction. When the counter is greater than the ending number, the processing continues at the instruction that follows the LOOP-END instruction (see c, c_1, and c_2 in Figures 7.7 and 7.8).

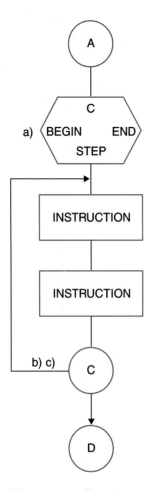

Figure 7.7 Flowchart of Automatic-Counter Loop

There are two differences in these rules when decrementing the counter. First, the counter is decremented instead of incremented at the end of the loop. Second, when the counter is greater than or equal to the ending value, the loop continues, instead of when it is less than or equal to the ending value when incrementing. It is important to understand these rules in order to use this structure correctly.

Use the automatic-counter loop structure when you know from the start the number of times the loop will be executed, for example, when you know the exact number of people in the group for the average age problem. Do not use a decision to exit from this loop. A decision cannot be used to exit a loop since to do so would require a GOTO instruction. GOTO instructions tend to decrease readability, so they are used as little as possible in structured programming. If a decision instruction is necessary, then consider using one of the other types of loop structures for your solution. Counter loops are overused because they are easy to code. A good programmer is careful not to fall into this trap. It is important to choose the best loop structure for the problem.

Putting It All Together

This section illustrates the same problem, using the automatic-counter structure, that the previous two Putting It All Together sections did, using the other two types of loop

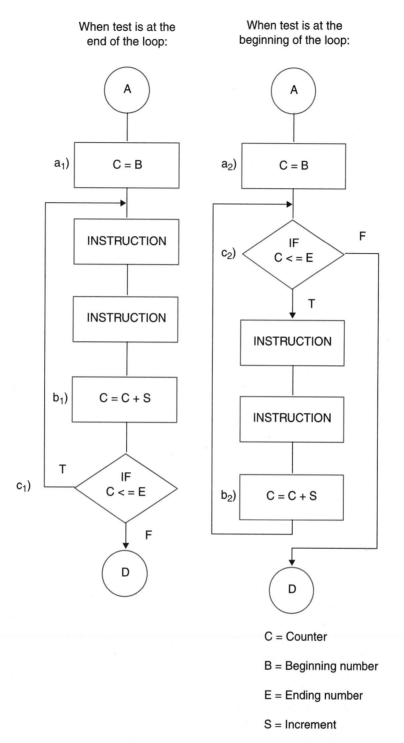

When test is at the end of the loop:

When test is at the beginning of the loop:

$a_1)$ C = B

$a_2)$ C = B

INSTRUCTION

INSTRUCTION

$b_1)$ C = C + S

$c_1)$ IF C < = E

INSTRUCTION

INSTRUCTION

$b_2)$ C = C + S

$c_2)$ IF C < = E

C = Counter

B = Beginning number

E = Ending number

S = Increment

Figure 7.8 Decision Equivalents to Automatic-Counter Loop

instructions. The solution calculates the average age of a group of people. The programmer would use the automatic-counter loop structure for this solution only if the number of students in the class—in this case, 12—is known.

In the automatic-counter loop in Figure 7.9, as in the other two loop structures, the first two instructions initialize the variables SUM and COUNTER. The next instruction begins the loop. The A is the counter variable. The counter begins at 1 and is incremented

Algorithm	Flowchart
AVERAGE AGE 1. SUM = 0 2. COUNTER = 0 3. ┌ LOOP: J = 1 TO 12 │ SUM = SUM + AGE │ COUNTER = COUNTER + 1 └ LOOP-END: J 4. AVERAGE = SUM/COUNTER 5. PRINT COUNTER, AVERAGE 6. END	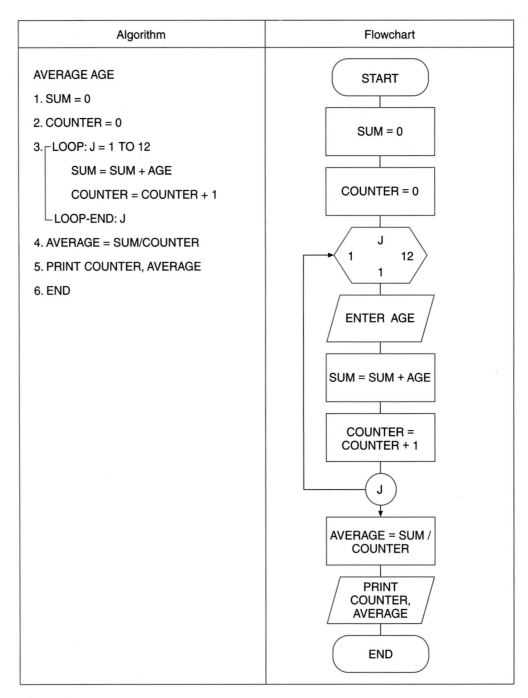

Figure 7.9 Example of an Algorithm and Flowchart for a Problem Using an Automatic-Counter Loop

by 1 until the counter is greater than 12, so the loop will be repeated 12 times. Notice that there is no increment value. When the increment value is left out, the computer assumes the value to be 1. When the counter increases to a value greater than 12, the average age is calculated and printed. Also notice the absence of the primer READ. The primer READ is not necessary because there is no trip value needed since the number of students is known. Whenever the number of items to be counted is known, the automatic-counter loop is preferred. If that number is not known, then one of the other loop structures is preferred.

Nested Loops

Loops can be nested like decisions can. Each loop must be nested inside the loop just outside it. The general rules regarding loops, such as where the condition is processed and how indentation and brackets are used, hold true for nested loops as well as single loops. The inner loops do not have to be the same types of loop structures as the outer loops; that is, a WHILE/WHILE-END may be nested inside a REPEAT/UNTIL loop, or vice versa.

Examples of nested loops are found in Figure 7.10. Notice that the nested loop in each flowchart results in the same output. The output for all four nested loops is

R (*outer loop counter*)	C (*inner loop counter*)
1	1
1	2
1	3
2	1
2	2
2	3

In the first flowchart in Figure 7.10, the nested loops are both WHILE/WHILE-END structures. The inner loop will be processed as long as the value of C is less than 3. The inner loop increments C and, later, prints the values of R and C. During the processing of the inner loop the value of R does not change. When C becomes equal to 3, the inner loop has completed the processing, and the processing continues with the outer loop. The outer loop increments R, sets C equal to 0, and then processes the inner loop. When R becomes equal to 2, processing continues at B. This whole process is essentially what happens in the other three examples of nested loops. The outer loop increments R, sets C equal to 0 or 1, and processes the inner C loop. The inner loop increments C and prints the values of R and C.

Indicators

Indicators are logical variables that a programmer sets within a program to change the processing path or to control when the processing of a loop should end. They are sometimes called *flags, switches*, or *trip values*. The user has no knowledge of these indicators during processing. These indicators are used to detect changes in the output line, in the processing of data, and so forth. An error indicator designates that an error has occurred in the input or the output. An end-of-data indicator designates that there are no more data to be entered.

The main thing to remember about indicators is that they are set internally to change the processing. An indicator was used in the average-age problem in Figures 7.3 and 7.6 (the zero value of AGE) to stop the looping when there were no more ages to be entered. An indicator can be a variable of logical data type or a value that the variable can never equal. For example, an indicator for AGE might be 0 or 500 or −300. An indicator for an error could be a variable with the value TRUE when an error has occurred, and FALSE when there have been no errors.

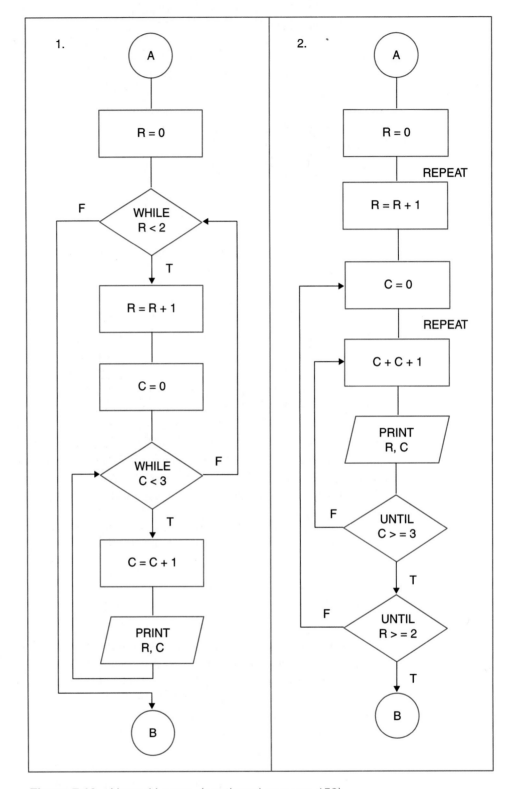

Figure 7.10 Nested Loops *(continued on page 153)*

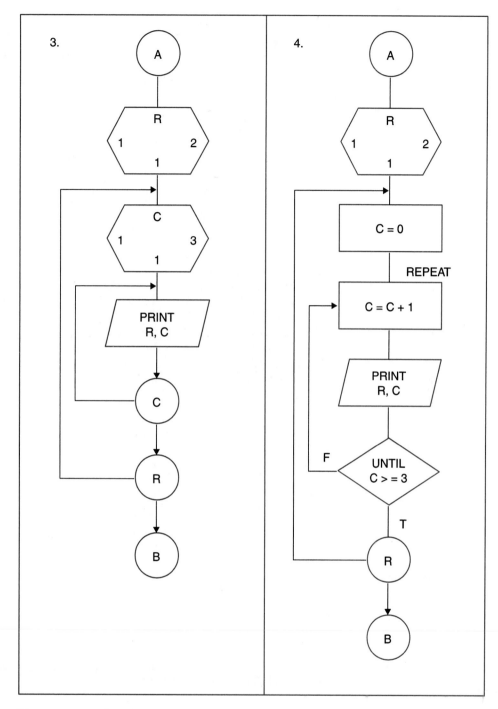

Figure 7.10 *(continued from page 152)*

Algorithm Instructions and Flowchart Symbols

Table 7.1 is an update of the Algorithm Instructions and Flowchart Symbols table, last seen in Chapter 5. The new instructions and logic structures for combinations of instructions shown on pages 104, 138, 139, 140, 145, and 148 have been added to this table.

- Decision Logic Structure IF/THEN/ELSE
- Instructions
 Incrementing COUNTER = COUNTER + 1
 Accumulating SUM = SUM + VARIABLE
- Loop Logic Structures WHILE/WHILE-END
 REPEAT/UNTIL
 LOOP: COUNTER = BEGIN TO END
 STEP S
 LOOP-END: COUNTER

Table 7.1 Algorithm Instructions and Flowchart Symbols *(continued on page 155)*

Algorithm Instruction	Flowchart Symbol	Definition (What the instruction tells the computer)
START module name (module number)		Start processing a module.
ENTER list of variable names		Enter data from a data block or from the keyboard into the following variable memory locations.
variable = expression		Assign the value of the following expression to a variable location. Called the *assignment instruction*. A module may be processed as part of an expression, if there is a return value.
WRITE list of variable names		Write to the printer the values represented by the following variable names.
PRINT list of variable names		Print on the screen the values represented by the following variable names.
PROCESS module name (list of parameters)		Execute the instructions in the named module, and then return to execute the next instruction.

Table 7.1 *(continued from page 154)*

Algorithm Instruction	Flowchart Symbol	Definition (What the instruction tells the computer)
END/EXIT	(rounded rectangle)	Stop the processing of the module. Processing ceases with the End; with the Exit, processing continues with the next instruction after the PROCESS.
RETURN (variable)	(rounded rectangle)	Stop the processing of the module, place the value of the variable in the module name, and continuing with the instruction continuing the module call.
IF/THEN/ELSE	(diamond)	When the condition is TRUE, use one set of instructions; when the condition is FALSE, use another set of instructions.
CONTINUE	(rectangle) or no box	A placeholder. Does *no* processing and continues to next instruction.
COUNTER = COUNTER + 1 (C = C + 1)	(rectangle)	Increments the counter C.
SUM = SUM + VARIABLE (S = S + V)	(rectangle)	Accumulates the values of the variable V.
WHILE/WHILE-END	(diamond)	Repeat a set of instructions as long as a condition is TRUE.
REPEAT/UNTIL	(diamond)	Repeat a set of instructions as long as a condition is FALSE.
LOPP: C = BEGIN TO END STEP S LOPP-END: C	(hexagon with C, A, B, S)	Repeat a set of instructions using a counter C that starts at A(BEGIN) and increments by S(STEP) and C is greater than B(END).

Recursion

Another type of loop structure is **recursion.** Recursion occurs when a module or a function calls itself. The condition that ends the loop must be within the module. Recursive procedures can be replaced by conventional loop structures; however, recursive procedures are usually faster. Figure 7.11 is an example of a recursive function that calculates a factorial number such as 5! (5! = 5 * 4 * 3 * 2 * 1). Notice that FACTORIAL (N) continues to call itself until N = 1. There must be a way to stop the recursion. See Figure 7.12 for the execution of this problem where N = 4.

Summary

The third basic logic structure is the loop logic structure. By using this structure a programmer can design solutions that repeat processes.

There are three common tasks associated with the loop logic structure. The first is counting, or incrementing, to determine a total number of items. The second is summing, or accumulating, to find totals. The third is processing arrays, which are presented in Chapter 9.

The loop logic structure includes three types of loops. The WHILE/WHILE-END loop repeats a set of instructions as long as the resultant of a condition is TRUE. The condition is processed before the first instruction of the loop. The second type of loop structure is the REPEAT/UNTIL loop. This loop repeats the instructions until a condition is TRUE. The condition in this case is processed after the last instruction in the loop. The third loop structure is the automatic-counter loop. In this case the instructions are repeated a given number of times, directed by a loop counter, which starts at a beginning number and is increased by a given increment until the counter is greater than an ending number. The condition can be processed at the beginning or the end of an automatic-counter loop depending on the language. For a programmer, the type of loop to use is determined by the problem and the language or application.

Indicators are used with loop structures to change the course of processing and to control the ending of a loop. Indicators are written in the instructions within the solution and normally are not entered externally by the user.

New Terms

accumulating	primer READ
automatic-counter loop	recursion
incrementing	REPEAT/UNTIL
indicators	WHILE/WHILE-END

Questions

1. Write the instruction to increment a variable by one.
2. Write the instruction to accumulate a variable A.
3. Name the three types of loop structures.
4. State the major differences in the three types of loop structures.

Algorithm	Flowchart

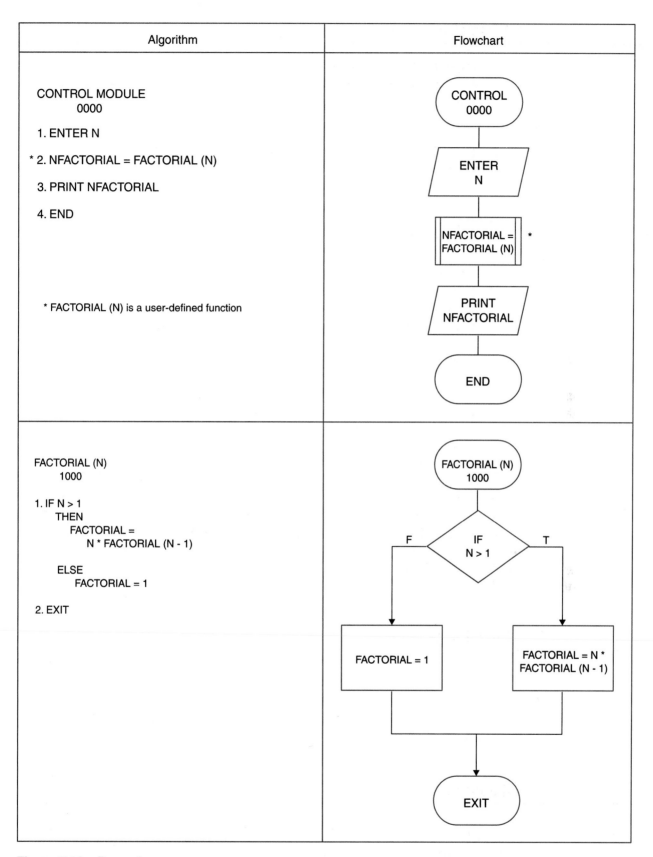

Figure 7.11 Recursion

The figure text rotated within the illustration reads:

This figure shows how recursion works during execution. N = 4 in this processing and therefore variables will be replaced by the actual numbers.

1. ENTER N [4]
2. NFACTORIAL = FACTORIAL(4) (FACTORIAL has a value of 24 upon return)

FACTORIAL(4)
IF 4 > 1
 THEN
 FACTORIAL = 4 * FACTORIAL(3) (FACTORIAL has a value of 6 upon return)

FACTORIAL(3)
IF 3 > 1
 THEN
 FACTORIAL = 3 * FACTORIAL(2) (FACTORIAL has a value of 2 upon return)

FACTORIAL(2)
IF 2 > 1
 THEN
 FACTORIAL = 2 * FACTORIAL(1) (FACTORIAL has a value of 1 upon return)

FACTORIAL(1)
IF 1 > 1 (This condition is now false and does not call FACTORIAL)
 THEN
 FACTORIAL = 1 * FACTORIAL(0)
 ELSE
 FACTORIAL = 1
 RETURN (1)

 ELSE
 FACTORIAL = 1
 RETURN (2)

 ELSE
 FACTORIAL = 1
 RETURN (6)

 ELSE
 FACTORIAL = 1
 RETURN (24)

3. PRINT NFACTORIAL (The value of NFACTORIAL is now 24.)
4. END

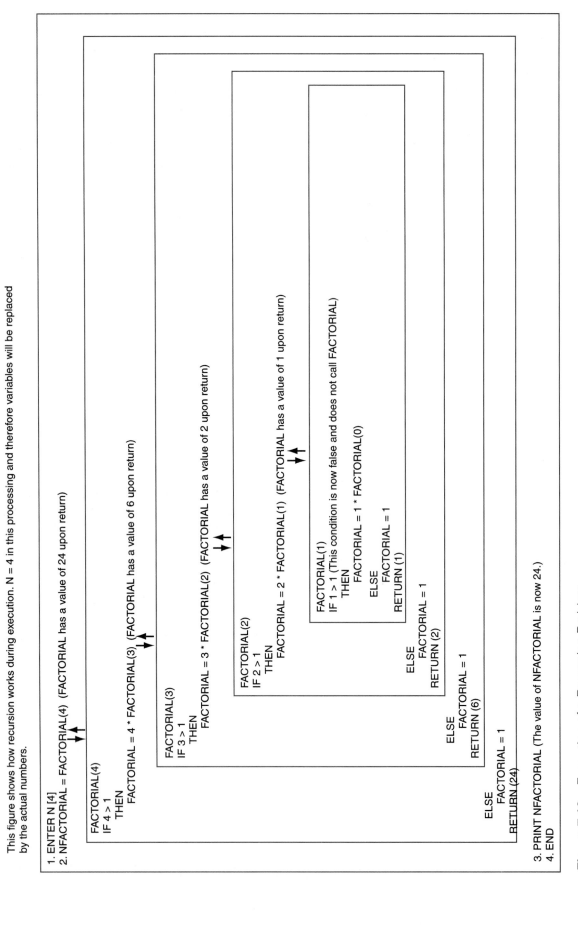

Figure 7.12 Execution of a Recursion Problem

158

Problems

Complete the seven steps of solution development and a data dictionary for the following problems:

1. The Last Stop Boutique is having a five-day sale. Each day, starting on Monday, the price will drop 10% of the previous day's price. For example, if the original price of a product is $20.00, the sale price on Monday would be $18.00 (10% less than the original price). On Tuesday the sale price would be $16.20 (10% less than Monday). On Wednesday the sale price would be $14.58; on Thursday the sale price would be $13.12; and on Friday the sale price would be $11.81. Develop a solution that will calculate the price of an item for each of the five days, given the original price. Test the solution for an item costing $10.00.

2. Make changes in the solution to Problem 1 to *set* the price at wholesale when the sale price drops below the wholesale price. Test the solution with the following prices:

Retail Prices:	$20.00, $25.00
Wholesale Prices:	$10.00, $20.00

3. Develop a solution that will calculate the average temperature, given a set of temperatures. The number of temperatures may differ from time to time. (Use a trip value to stop the processing of the loop.) Test the solution with the following 10 temperatures:

78°	90°
85°	80°
87°	83°
75°	90°
86°	70°

4. Mr. Brown has given a test to his class. He would like to have the average score for the class, and the highest and lowest scores. Develop a solution to calculate and print out these values. (Use a trip value to stop the processing of the loop.)

5. Mary Smith, a student, has borrowed $3,000 to help pay her college expenses. After setting up a budget, $85 was the maximum monthly payment she could afford to make on the loan. Develop a solution to calculate and print the interest, the principal, and the balance on the loan per month. Other information she would like to know is the number of years and months it will take to pay the loan back and the total interest she will pay during that period. The interest rate is 1% per month on the unpaid balance. Keep in mind these formulas:

$$\text{interest} = \text{balance} * \text{interest rate}$$
$$\text{payment} = \text{balance} - \text{interest}$$
$$\text{new balance} = \text{balance} - \text{payment}$$

6. Develop a solution to flip a coin a given amount of times and then print the number of heads and the number of tails. The equation to toss the coin is

$$\text{COIN} = \text{INT}(\text{RANDOM} * 2) + 1$$

(See page 23 for explanation of this formula.) When COIN = 1 the toss is heads, and when COIN = 2 the toss is tails. RANDOM returns a number between 0 and 1, including 0 but not 1. Therefore, when RANDOM is less than .5, COIN will equal 1: and when RANDOM is greater than or equal to .5 and less than 1, COIN will equal 2.

7. Develop a solution to calculate a student's grade average for one semester. The letter grades should be entered and the grade average printed out. An A is equivalent to 4 grade points, a B is 3 grade points; a C is 2 grade points, a D is 1 grade point, and an F is zero grade points.

8. Write a solution to tell the user whether a number is a palindrome. (A palindrome number is a number that is the same written both forward and backwards, such as 81318.)

9. Write a solution to tell the user whether a string is a palindrome. (A palindrome string is a list of characters that spell the same word(s) forward or backwards, such as wow or radar.)

10. Write a solution to tell the user whether a number is a prime number. (Hint: A prime number can be divided only by one and itself. There are no other factors.)

11. Write a solution to print out all of the factors of a number. Specify to the user whether the number is a prime number, that is, if the only factors are one and itself.

12. Using the recursion technique, develop a solution to calculate the power of a number, given the number and the exponent.

WHAT'S WRONG WITH THIS?

1. Correct the following set of instructions:

```
COUNTER = 1
WHILE COUNTER < 5
    ENTER SCORE
    TOTAL = TOTAL + SCORE
    COUNTER = COUNTER + 1
WHILE-END
```

2. Correct the following set of instructions:

```
COUNTER = 1
REPEAT
ENTER SCORE
TOTAL = TOTAL + SCORE
COUNTER = COUNTER + 1
UNTIL X > 5
```

3. Correct the following set of instructions:

```
SUM = 0.0
LOOP: COUNTER = 10 TO 100 STEP 15
    SUM = COUNTER
LOOP-END: COUNTER
PRINT "SUM OF THE NUMBERS =", SUM
END
```

Chapter 8

Problem Solving with the Case Logic Structure

Overview

The Case Logic Structure
Codes

Objectives

When you have finished this chapter, you should be able to:

1. Develop problems using the case logic structure in conjunction with the loop, decision, and sequential logic structure.
2. Use the seven problem-solving tools when developing a solution using the case logic structure.

Sometimes a solution to a problem requires the computer to choose one action from a set of actions. This kind of solution can be designed through the decision logic structure, but the case logic structure is more efficient for this purpose. The case logic structure is a special variation of the decision structure, although it is different enough to present here separately. This structure is used in many business-related problems, such as those dealing with types of accounts in a bank or types of insurance available to an employee.

The Case Logic Structure

The case logic structure is made up of several or many sets of instructions, only one of which will be selected by the user and executed by the computer. Through the use of the case logic structure, a programmer can enable a user to enter the value of a variable from the keyboard or from a file to select one of several options in a list. Then the instruction following the END-OF-CASE instruction is executed. (The END-OF-CASE

instruction is the last instruction in the case logic structure.) The case logic structure does not enable the program to loop back to select another option. If the solution requires looping back, then the programmer has to use the loop logic structure. The form of the case logic structure is the following:

CASE OF VARIABLE
 = constant1:
 actions for VARIABLE = constant 1
 = constant2:
 actions for VARIABLE = constant 2
 = constant3:
 actions for VARIABLE = constant 3

 .

 .

 OTHERWISE:
 actions for VARIABLE = anything else
END-OF-CASE

Notice how the indentation improves the readability of the algorithm.

When the case logic structure is processed, the computer compares the value of the variable against each of the constants. When it finds a constant that is equal to the value of the variable, it executes the corresponding set of instructions. Then the computer skips to the instruction after the END-OF-CASE. The variable must be an integer or a character. For example, take an inventory problem in which the name of the vendor for each item is entered as a variable in the computer in the form of a code (S = Small's Auto Parts, W = Wilson's Water Pumps, J = Johnson's Ignition Systems, and so on). The computer selects the vendor's name, when the user enters, as the value of the variable, the value of the code (S, W, J, and so on). The variable with the value of the vendor code would be used in the CASE OF VARIABLE part of the logic structure. S, W, J, and so on, would be the constants. Under each constant would be the instruction to be executed once the vendor is selected. The last section is the OTHERWISE section of the structure. This section has the set of instructions executed when the variable has a value of anything but the named constants in the list of options. Inclusion of the OTHERWISE option will allow for all possible responses and may be used as a response for validation of data; however, it is optional. The END-OF-CASE tells the computer where the case logic structure ends.

Other examples of the appropriate use of the case logic structure would be a solution for the computer to choose the deduction for medical insurance or to choose the retirement plan for an employee when there are several different plans. The values of the constants, the codes for the insurance or retirement plans, and the set of instructions corresponding to each constant would be part of the input data and processing given in the problem.

The number of possible options, or sets of instructions, in the case logic structure is limited only by the language or the application package in use. The options may be in any order, but it is best to arrange them according to how frequently they're used, starting with the most frequently used option. For a given problem, the programmer would rely on the customer to find out which options would be used most often. When

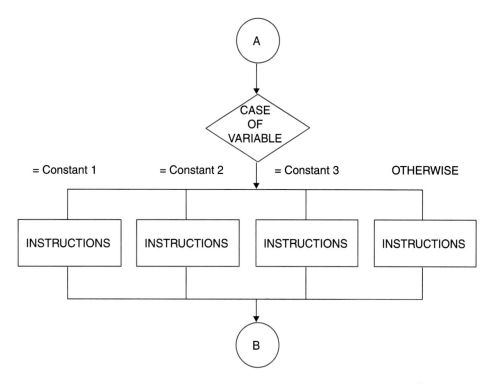

Figure 8.1 Flowchart Diagram for the Case Logic Structure

the options are ordered this way, the computer will have to check all possible options fewer times before it finds the one selected by the user. If no one knows which options would be used more often, then the options are arranged in the most logical order. That arrangement might be alphabetical, numeric, the order given by the customer, or some other order. The OTHERWISE section is always last.

Figure 8.1 shows the flowchart of the case structure. The decision equivalent would be a set of nested IF/THEN/ELSE instructions using positive logic (see Figure 8.2). In the case logic structure, the relative operator is always *equals* (=). The case logic structure is really a special form of the decision logic structure; it allows the computer to select one set of instructions from many sets through the use of the relative operator, equals. Compare Figures 8.1 and 8.2 to see how much easier it is to design a solution requiring the computer to choose one option out of many by using the case logic structure rather than the decision logic structure.

Codes

Codes are characters, character strings, numbers, or some combination of these types of data that a programmer uses to name the options, the constants, in a case structure. They are abbreviations for each of the options. They direct the processing path just as indicators do by specifying the choice that the user prefers—and, therefore, that the computer will make, such as a particular employee medical plan out of many plans: B for Best Medical Insurance Company, D for Direct Medical Insurance Company, and the like.

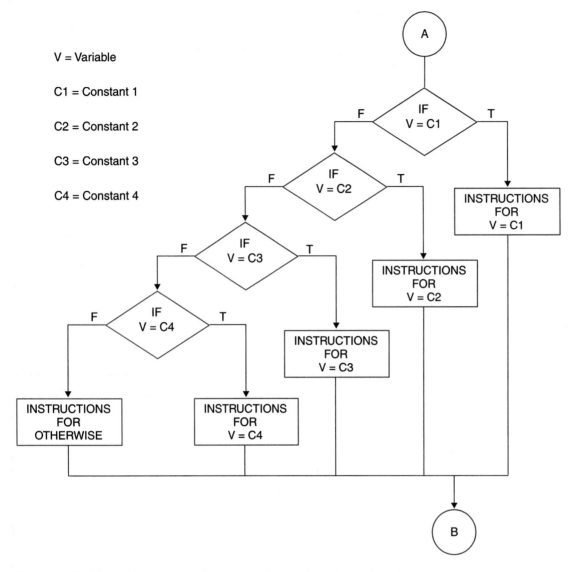

Figure 8.2 Decision-Structure Equivalent to the Case Logic Structure

The major differences between indicators and codes are the following:

1. Codes are data to be entered by the user. Indicators are internal signals to change the processing path.
2. A code can have a value of many different types. The value of an indicator can be logical data, TRUE or FALSE, or it can have an implausible value for the variable it represents in the solution.

A code is used to indicate choices such as the type of insurance an employee is buying, the sex of an individual, or the vendor in an inventory problem. When using codes the OTHERWISE option indicates an error. Without the use of codes, solutions requiring a choice from among several options would be longer, harder to read, and more difficult to maintain.

Putting It All Together

This PIAT illustrates the use of codes in a solution using the case logic structure.

> Problem: A company has four different medical plans. The programmer has given each plan a code corresponding to the beginning initial of the company: Plan 1 = F, Plan 2 = B, Plan 3 = K, Plan 4 = E.

> The company pays for all of Plan 1. The individual has to pay for part of the others. The payroll deduction for Plan 2 = 4.65, for Plan 3 = 7.85, and for Plan 4 = 5.50. Any other codes are considered in error. Write the algorithm and draw the flowchart for a module to determine the payroll deduction.

> Figure 8.3 shows the module to calculate the insurance deduction. The variable name for the medical plan code is MEDCODE. MEDCODE can have only the value of F, B, K, or E. Any other value is unacceptable, and therefore, an error. When MEDCODE has the value of F, then MEDDEDUCTION, the medical deduction, is 0; when it has the value of B, MEDDEDUCTION is 4.65; when it has the value of K, MEDDEDUCTION is 7.85; when it has the value of E, MEDDEDUCTION is 5.50. If MEDCODE has any other value, then an error module is processed. An error module is simply the module that prints out an error message.

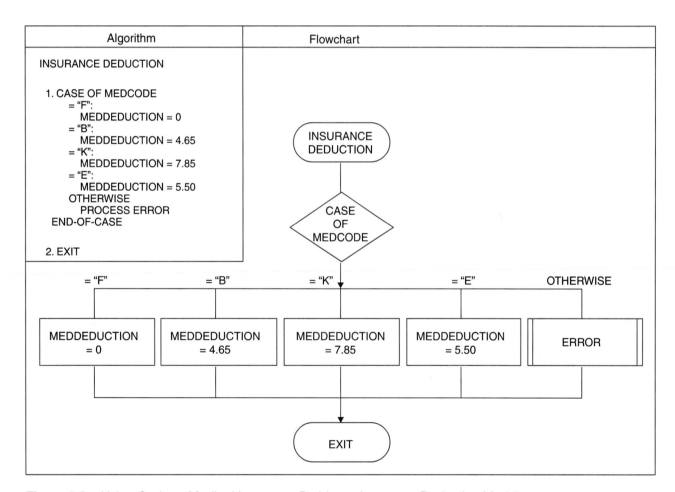

Figure 8.3 Using Codes—Medical Insurance Problem—Insurance Deduction Module

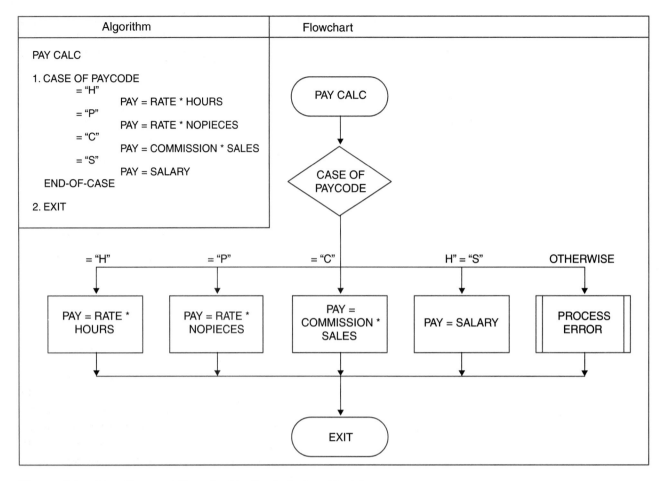

Algorithm	Flowchart

PAY CALC

1. CASE OF PAYCODE
 = "H"
 PAY = RATE * HOURS
 = "P"
 PAY = RATE * NOPIECES
 = "C"
 PAY = COMMISSION * SALES
 = "S"
 PAY = SALARY
 END-OF-CASE

2. EXIT

Figure 8.4 Algorithm and Flowchart to Illustrate Pay Module

Another Putting It All Together

This PIAT illustrates the use of a pay-type code to calculate an employee's pay. The codes are as follows:

H = Hourly	pay = rate * hours
P = Piece work	pay = rate * number of pieces
C = Commission	pay = commission * sales
S = Salary	pay = salary

The module to calculate an employee's pay is illustrated in Figure 8.4. Notice that there is an instruction box for each pay-type which explains how to calculate each employee's pay. The OTHERWISE instruction processes an error module to catch any keyboard entry errors. If there is an error, the processing returns to an instruction that re-enters the employee's data.

Summary

The case logic structure is the fourth basic logic structure. It enables the computer to select one set of actions from many by selecting an integer or a character from a list of coded choices designed by a programmer. Codes are one of the tools of the case logic

structure. The case structure replaces a multiple nested IF/THEN/ELSE instruction that uses positive logic and the relative operator *equals*. By using the case structure rather than the nested IF/THEN/ELSE alternative, a programmer can improve the readability and reduce the complexity of the program.

New Terms

codes

Questions

1. Why would the case logic structure be used instead of the decision structure?
2. What are the differences between an indicator and a code?

Problems

1. Develop the other modules to complete the pay-type problem in the second PIAT. Allow the user to enter as many employees as needed. Develop a trip value to end the execution of the program. Remember that when an error occurs in the pay code, the execution for that employee is ignored and the employee's information must be re-entered. The information to enter includes the employee's name, social security number, pay-type code, and pay information (commission, hours, or number of pieces—do these have to be three entries?).

2. Modify problem 1 so that only the pay-type code is re-entered when an error in the code is entered.

3. Develop the algorithm and flowchart for a module that calculates the retirement deduction rate according to the following codes:

$$R = 10\%$$
$$P = 8\%$$
$$G = 5\%$$

4. Develop a solution to change Roman numerals into their decimal equivalents. Consider all possibilities of Roman numerals—no letters, only one letter, two letters, and so forth. Assume that the Roman numerals are in correct order. One Roman numeral character is entered at a time. The Roman numeral is entered from left to right. Roman numeral equivalents are as follows:

$$I = 1$$
$$V = 5$$
$$X = 10$$
$$L = 50$$
$$C = 100$$
$$D = 500$$
$$M = 1000$$

Examples of input numbers are IV, XIX, MMCDXL, and CCLV.

5. A student wants to know his grade average for the semester. The grades are given in letter grades with numeric equivalents. Develop a solution to calculate a grade average given the letter grades. (Remember, the grade average is figured per unit of credit, not per course). An A = 4.0, B = 3.0, C = 2.0, D = 1.0, F = 0.0. Test the solution with the following data:

History	B	3 units
Economics	A	3 units
P.E.	A	1 unit
Chemistry	C	4 units
Art	B	3 units

(Use a trip value to stop the processing of the loop and a case structure to find the grade points.)

6. A bank has five different types of savings accounts. The interest rates on these accounts change often. The bank manager would like to have the computer print out information about the account type of her choice. Develop a solution to solve the manager's problems.

7. Mr. Johnson would like to know how many A's, B's, C's, D's, and F's his students received on a test. He has 200 students who took the test. He would like to enter the student number and the number grade for the test for each student. Develop the solution to print out each student's student number, number grade, letter grade, and the total number of A's, B's, C's, D's, and F's. His grading scale is 90–100 is an A, 78–89 is a B, 65–77 is a C, 50–64 is a D, and below 50 is an F.

8. How would you modify problem 7 to process any number of students?

WHAT'S WRONG WITH THIS?

Correct the following sets of algorithm segments:

1. ENTER L

 CASE OF L

 < 5:

 PROCESS M5

 <10:

 PROCESS M10

 OTHERWISE

 PROCESS M_OTHERWISE

 END-OF-CASE

2. ENTER V

 CASE OF V

 = CAR1

 PROCESS M_CAR1

 = CAR2

 PROCESS M_CAR2

 = CAR3

 PROCESS M_CAR3

 END-OF-CASE

Unit Two

Supplementary Exercises

Develop a complete solution and develop a data dictionary for the following problems:

1. Draw a solid diamond on the screen using a single character. The size of the diamond should be entered from the keyboard:

 Example: SIZE = 5 CHARACTER = "*"

   ```
       *
      ***
     *****
      ***
       *
   ```

2. Draw the same diamond except make it hollow:

 Example: SIZE = 5 CHARACTER = "*"

   ```
       *
      **
     *  *
      **
       *
   ```

3. Modify Problem 4 in Chapter 5 so that the program stops when the remaining change becomes zero.

4. Reverse the digits of a number such that 1234 becomes 4321.

5. Calculate the ending balance of your checking account, given the beginning balance, the checks written, the deposits made, and the bank charges.

6. Create your own dice game. Print directions and rules. Have the user play against the computer.

7. Print all prime factors of a number entered from the keyboard.

8. Print out four cards dealt randomly from a 52-card deck. Differentiate between hearts, spades, clubs, and diamonds.

9. Calculate the average rainfall over a two-week period.

10. The counselor of a camp would like to know the average age of those attending a camp, and who the oldest and the youngest are. The number of people attending a camp can vary from 50 to 100.

UNIT THREE

DATA STRUCTURES

Chapter 9

Processing Arrays

Overview

Objectives

When you have finished this chapter, you should be able to:

1. Develop problem solutions for the computer using one-dimensional and two-dimensional arrays.
2. Develop problem solutions requiring table look-ups, sequential and binary searches, frequency distributions, and cross-tabulations.

Arrays

Data are stored in the computer in a logical way in order to find values as quickly as possible. The way data are stored is called the structure of the data, hence the term *data structures*. The only data structure you have used so far in this text is the single variable. Using this data structure a single value is stored in a single named location. This is the simplest of all data structures. Another type of data structure is the array.

It is often advantageous for a user to store several values for the same variable in the internal memory of the computer because it decreases processing time. This multiple storage means there has to be more than one memory location in the computer for each variable name. When more than one memory location is designated for a single variable, it is called an **array.** Sometimes these variables are called *subscripted variables*. This term comes from mathematics, where subscripts are used to designate multiple values. Arrays are also called *tables,* a term from business, where data are often put into tables. For the purposes of this book, *array* will be used because it is universal and is becoming the most common term. The other two terms may be used in other sources.

Arrays are used in the problem-solving process. A programmer who has many values of data of the same kind—many ages, temperatures, names, or the like—can store them in the computer in the form of an array, which makes them easier to read and use. The programmer can also use the values in one array to point to values in another array. This is the way the computer tallies the number of times a value appears in an array. Without arrays this task would be very difficult.

A programmer uses arrays when more than one value of a variable is used in the solution—for example, when calculating the percentage of each store's sales out of the total company's sales or when calculating each student's test grade based on the highest and lowest score. The store sales or the student grade would be the variable with more than one value—a value for each store or for each student. If the user does not need to use the data more than once in the solution, then there is no need for the programmer to use arrays. Their overuse is a common mistake for beginning programmers.

The programmer tells the computer how many memory locations to save for an array through a special instruction or within the definitions of the data types. In many languages this is called *dimensioning the variable*. The number of memory locations to be saved is equal to or greater than the number of locations used in the solution. If it is less than the number of locations used, an error will result. For problem-solving purposes, this number is part of the annotation of the flowchart.

Most languages use **static arrays.** This means that once the computer is told how many locations to save, that number cannot be changed unless the instruction is changed. In most languages, the number in the instruction is a constant. However, some languages allow the number to be a named constant, which is set once, and only once, during the execution of the solution. In this case, the number of memory locations saved for an array can change from execution to execution, but it cannot change during an execution. This type of array is still considered static.

Some languages allow **dynamic arrays.** When using dynamic arrays, the programmer designates the number of array locations as a variable, which can be expanded or contracted during the execution of the solution. Dynamic arrays are more flexible and use less memory space than static arrays, but are usually more time-consuming during processing.

If the number of the designated memory locations is a constant, then the programmer has to know how big the array will be, at its largest, while writing the solution. When that number is a variable but cannot be changed during execution, then the size of the array is

part of the input data. This kind of static array allows more flexibility in the solution and in the use of memory than the static array in which the number of memory locations is a constant. The most flexible arrays are dynamic. The solution itself determines the size of a dynamic array, and thus, the programmer does not have to predict the size of the array.

Each array memory location is called an *element* and is given a number or numbers corresponding to the position of the location in the array. This number is a reference number relative to the location of the first value of the array. The name of the element has two parts. The first part is the variable name; the second part is the reference number, also called the *index number*, an *element number* or subscript in parentheses.

Because computers are zero-based, for counting purposes, many programming languages are also zero-based. This means that the first array element is numbered *zero*, and not *one*; it is a **base-zero system.** In this way, the computer adds the element number to the memory location number of the beginning box to locate the needed value. If the language uses the **base-one system,** then to identify a specific data memory location, add the element number minus one to the beginning number (see Figure 9.1). Base one is easier for the programmer to understand since the first element is the first box, the second element is the second box, and so on. When using zero-based arrays, the zero element is the first box, the first element is the second box, and so on. Often programmers will avoid zero elements in an array because they make it more difficult to keep track of element numbers. In this chapter, base one will be assumed.

You do not have to use an array in any special order. The data used as the actual values of the elements must have the same data type as you have identified for the variable name of the array. By using the assignment instruction, you can assign the value of a constant, a variable, or an expression to an element.

Figure 9.1 Base-Zero Versus Base-One Arrays

One-Dimensional Arrays

The simplest array structure is the **one-dimensional array.** This array consists of one column of memory locations. In Figure 9.2, there is a one-dimensional array named AGE. In mathematics, the reference name for each of the equation elements would be age_1, age_2, age_3, and so forth. Array elements are given similar reference names. Since the computer has trouble with subscripts (because it is hard to roll the screen up a little to print subscripts), the reference names are written as AGE(1), AGE(2), AGE(3), and so forth. The number in the parentheses is a reference number only and can be a constant, a variable, or an expression. AGE(5) would be the fifth element of the AGE array. In most languages the reference numbers must be of the integer data type, although a few computer languages allow any data type for the index or reference value. Since this practice is not common, and since the use of each of the data types is very similar to the use of integers as element numbers, this book will use only integers for the index.

Parallel arrays are two or more arrays in which values in the same elements relate to each other. For example, in Figure 9.3 there are three arrays: The first array contains area numbers, the second array contains temperatures for day 1, and the third, temperatures for day 2. The data in the first elements in the second and third arrays pertain to the area in the first array; the data in the second elements in the second and third arrays pertain to the area in the second element of the first array, and so forth. They are called parallel arrays because the same element number in all arrays has data pertaining to the same subject.

Entering Data into an Array

When you enter data into an array, or *load an array*, you use a loop. If you know the number of elements, then use the automatic-counter loop. If you do not know the number of elements, then use an indicator code with the REPEAT/UNTIL or the WHILE/WHILE-END loop. Use a counter to keep track of how many elements are entered into the array.

	Array AGE	Variable Reference Name	
1	32	AGE(1)	The number in the parentheses refers to the box number in the array, the element number.
2	54	AGE(2)	
3	25	AGE(3)	
4	36	AGE(4)	
5	45	AGE(5)	
6	20	AGE(6)	
7	28	AGE(7)	
8	50	AGE(8)	
9	42	AGE(9)	

Figure 9.2 One-Dimensional Array

AREANO		DAY1		DAY2	
1	5	1	82	1	80
2	9	2	75	2	72
3	15	3	80	3	82
4	35	4	90	4	87
5	50	5	95	5	92
6	73	6	78	6	80
7	88	7	85	7	83

Figure 9.3 Parallel Arrays

Figure 9.4 shows the loop process for entering the data into an array. In Figure 9.4a, the number of elements in the array is known to be the value of N. The counter (R) of the loop allows the computer to increase the element number by 1 each time a piece of data is entered into a memory location. The computer can find a particular memory

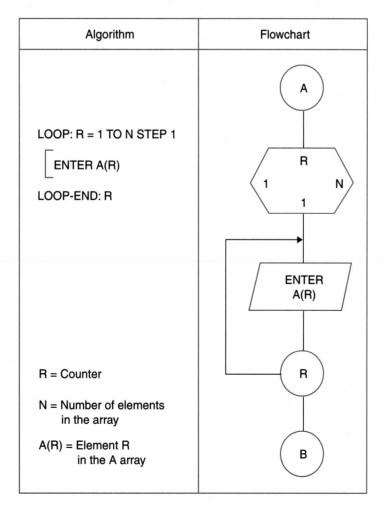

Figure 9.4a Loading an Array Using the Automatic-Counter Loop

location for an element in the array by first finding the set of locations for the array A, and then using the reference number, which has the value of R. Figure 9.4b shows the use of the REPEAT/UNTIL loop structure. Here, the elements are counted to find the total number of elements (N) in the array. The number N is then used for the number of elements through the rest of the solution. Figure 9.4c shows the use of the WHILE/WHILE-END loop structure. Here again, the elements are counted to find the total number of elements. Notice that in the REPEAT/UNTIL and the WHILE/WHILE-END loops, the indicator element number (indicating the end of the array) is not included in the total number of elements since it is not needed. The number of elements is known and, therefore, the indicator is not needed to signal the end of the array.

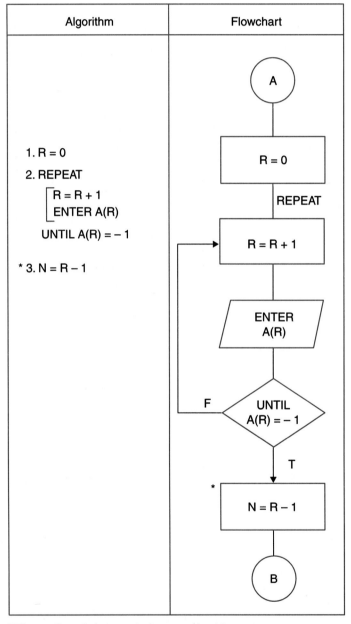

Algorithm	Flowchart

1. R = 0
2. REPEAT
 [R = R + 1
 [ENTER A(R)
 UNTIL A(R) = − 1

* 3. N = R − 1

* The number of elements in the array should
 not include the trip value of −1.

Figure 9.4b Loading an Array Using the REPEAT/UNTIL Loop

Algorithm	Flowchart
1. R = 1 2. ENTER A(R) 3. WHILE A(R) <> −1 ⌈ R = R + 1 ⌊ ENTER A(R) WHILE - END * 4. N = R − 1	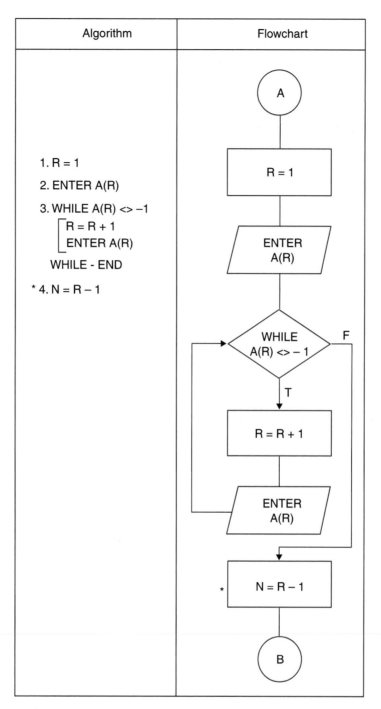

* The number of elements in the array should
 not include the trip value of −1.

Figure 9.4c Loading an Array Using the WHILE/WHILE-END Loop

Printing an Array

After the array is loaded, it can be used in other calculations or printed. You write the algorithm to print an array by using the automatic-counter loop since the number of elements in the array is now known (see Figure 9.5). The array will be printed one element at a time down the page. To print the array across the page, you must use the WRITE or

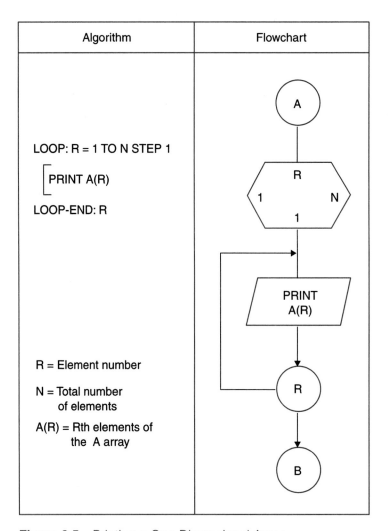

Algorithm	Flowchart

LOOP: R = 1 TO N STEP 1

PRINT A(R)

LOOP-END: R

R = Element number

N = Total number
 of elements

A(R) = Rth elements of
 the A array

Figure 9.5 Printing a One-Dimensional Array

PRINT instruction without a carriage-return. With parallel arrays, the corresponding elements will be printed in columns on the same line. The headings to the columns have to be printed before the processing starts into the automatic-counter loop.

Accumulating the Elements of an Array

Often you need the sum of the elements in an array. Use the accumulation instruction to total the sum (see Figure 9.6). Initially, set the variable containing the value of the sum to zero. The accumulation instruction (SUM = SUM + A(R)) tells the computer to add the value of the element A(R) to the old value of the sum (SUM) and to store the result in the sum memory location (SUM). The value of the element number R changes as the loop counter progresses from one element to the other during summing.

Observe the testing of the flowchart in Figure 9.6 through the use of the various variable memory locations. The first time the loop is processed, the value of R is 1, and the value of A(R), or A(1), is 2. This value is added to the previous value of the SUM, which is zero since the value of SUM is zero at the start. The resultant is 2, which is

Algorithm	Flowchart

Figure 9.6 Accumulating the Elements of a One-Dimensional Array

placed into the SUM memory location. The second time the loop is processed, R is incremented to 2 and the value of A(R), or A(2), is 4. This value is added to the previous value of SUM, which is 2, and placed into the SUM memory location. SUM now has the value of 6. The third time the loop is processed, R is incremented to 3, and A(R), which has the value of 6, is added to SUM and placed into the SUM memory location. The value of SUM is now 12. This sequence is repeated each time the loop is processed until the value of R has increased to a number greater than the value of N. The final value of SUM is 42.

Putting It All Together

This PIAT illustrates how to enter, use, and print one-dimensional arrays.

Problem: The president of the Too Little Variety Stores would like to know each store's percentage of the total sales of the corporation. There are 15 stores altogether.

Figure 9.7 shows the complete solution to this problem. Parts d–g show the algorithms and flowcharts for each module. The Control module processes each of the modules, Init, Read, Calc, and Print. The SUM is set equal to zero in the Init module. The

Given Data	Required Results
1. 15 stores 2. Sales per store	Percent of total sales per store
Processing Required	Solution Alternatives
Percent (store) = sales (store)/total sales	Must use arrays

Figure 9.7a Problem Analysis Chart

Interactivity (Structure) Chart

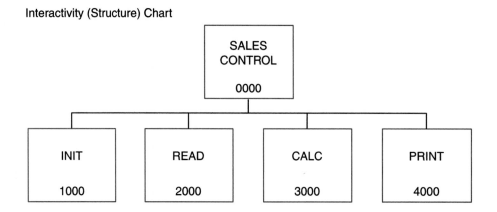

IPO Chart

Input	Processing	Module Reference Number	Output
1. Sales per store	1. Enter data 2. Set sum to 0 3. Accumulate sales 4. Calculate percent 5. Print output	2000 1000 3000 3000 4000	For each store; 1. Store number 2. Sales 3. % of total sales

Figure 9.7b Interactivity (Structure) Chart and IPO Chart

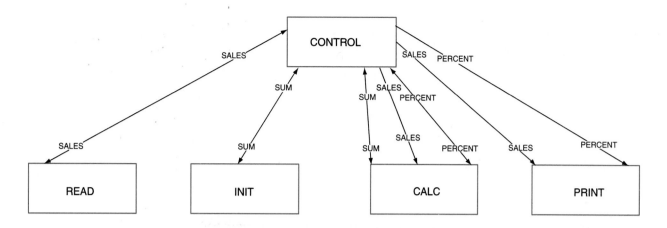

Item	Variable Name	Data Type	Module	Scope	Pseudonyms	Error Cks.
Sum of Sales	SUM	Numeric	Control, Init, Calc	Local	None	None
Sales Array	SALES	Numeric	All	Local	None	None
Sales Percentage	PERCENT	Numeric	Control, Calc, Print	Local	None	None
Element No.	R	Numeric	Read, Calc, Print	Local	None	None

Figure 9.7c Coupling Diagram and Data Dictionary

Read module enters the values of the sales per store into the SALES array. The value of R becomes the store number. In the Calc module, the sum of the sales for all 15 stores is calculated through the use of a loop and the accumulation instruction. The next loop calculates the value of the percentage for each store and places that value in the PERCENT array. The SALES array and the PERCENT array are parallel arrays since each element corresponds to the same store. The values of the store number, the SALES, and the PERCENT, are printed, element by element, in the Print module.

Two-Dimensional Arrays

A **two-dimensional array** is a block of memory locations associated with a single memory variable name and designated by row and column numbers. Each element number is written as A (Row #, Column #). The row number is always first, and the column number, second. As in one-dimensional arrays, the row and column numbers can be constants, variables, or expressions, and they are of the integer data type.

Algorithm	Flowchart	Testing

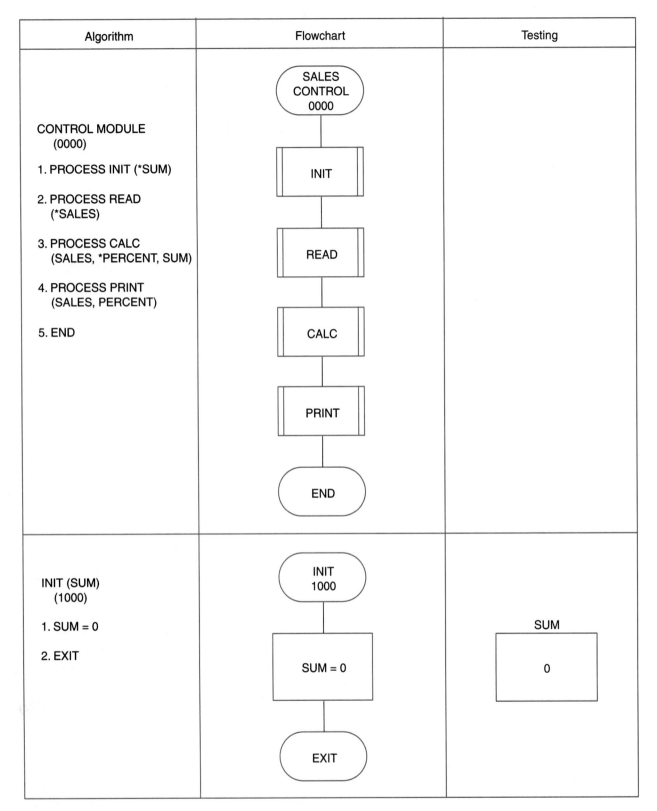

CONTROL MODULE
 (0000)

1. PROCESS INIT (*SUM)

2. PROCESS READ
 (*SALES)

3. PROCESS CALC
 (SALES, *PERCENT, SUM)

4. PROCESS PRINT
 (SALES, PERCENT)

5. END

SALES
CONTROL
0000

INIT

READ

CALC

PRINT

END

INIT (SUM)
 (1000)

1. SUM = 0

2. EXIT

INIT
1000

SUM = 0

EXIT

SUM

0

Figure 9.7d Algorithms and Flowcharts for the CONTROL and INIT Modules

Algorithm	Flowchart	Testing

READ (*SALES)
(2000)

1. LOOP: R = 1 TO 15

　ENTER SALES(R)

　LOOP-END: R

2. END

Flowchart:
READ 2000

R
1 ＼ 15
1

ENTER SALES (R)

R

EXIT

Testing:

R
1̶ 2̶ 3̶ 4̶ 5̶ 6̶ 7̶
8̶ 9̶ 1̶0̶ 1̶1̶ 1̶2̶1̶3̶
1̶4̶ 15

SALES

1	200
2	300
3	150
4	400
5	275
6	125
7	175
8	450
9	475
10	350
11	375
12	100
13	325
14	225
15	500

Figure 9.7e　Algorithm and Flowchart for the READ Module

Whatever processing takes place, the row and column numbers always remain in the same positions in the parentheses. As with one-dimensional arrays, you can change the values in a two-dimensional array by using the assignment instruction.

Two-dimensional arrays are used for tables of values. They can replace parallel arrays if the data are of the same data type in all of the arrays. In Figure 9.8, the two-dimensional array A is shown. This figure shows the general form of a two-dimensional array, and of reference names (including the variable name and the row and column numbers), for each location. For example, A(3,2) is found at the intersection of row 3 and column 2.

Loading a Two-Dimensional Array

You load a two-dimensional array with nested loops (see Figure 9.9). The data are normally loaded row by row. When you load the data row by row, the outer loop represents the row, and the inner loop represents the column. This order of the loops allows the row number to remain constant while the column number varies. In Figure 9.9, compare the data block to the array block. In the flowchart, notice that the first four numbers in the data block are read into the first row of the array, the second four into the second row, and the last four into the third row. The array is now loaded with the numbers 1 to 12.

Algorithm	Flowchart	Testing

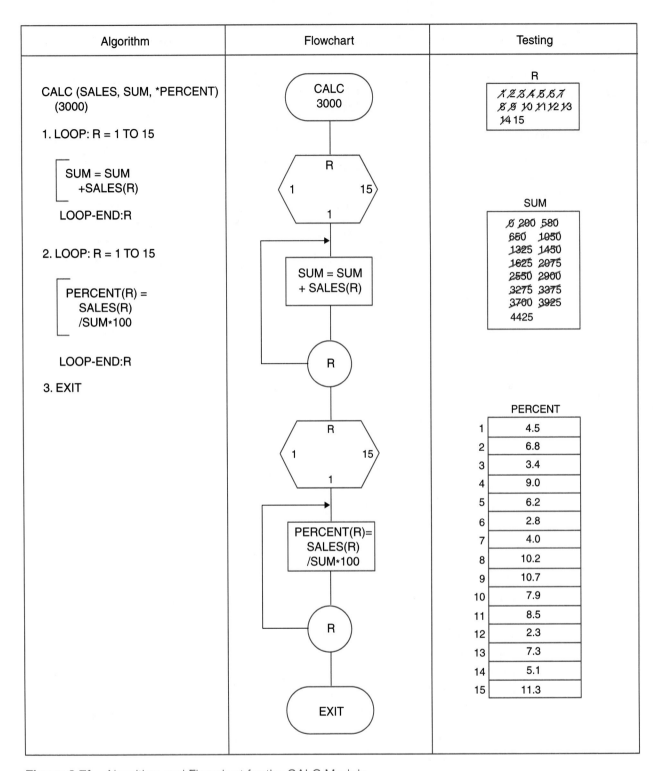

CALC (SALES, SUM, *PERCENT)
(3000)

1. LOOP: R = 1 TO 15

 ⎡ SUM = SUM
 ⎢ +SALES(R)

 LOOP-END:R

2. LOOP: R = 1 TO 15

 ⎡ PERCENT(R) =
 ⎢ SALES(R)
 ⎢ /SUM*100

 LOOP-END:R

3. EXIT

Figure 9.7f *Algorithm and Flowchart for the CALC Module*

Algorithm	Flowchart	Testing

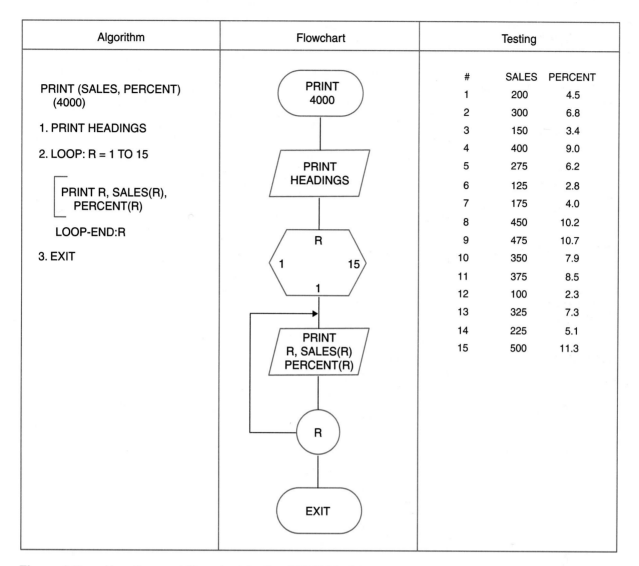

Algorithm column:

PRINT (SALES, PERCENT)
 (4000)

1. PRINT HEADINGS

2. LOOP: R = 1 TO 15

 ⌐ PRINT R, SALES(R),
 PERCENT(R)

 LOOP-END:R

3. EXIT

Flowchart column:

PRINT
4000

PRINT
HEADINGS

R
1 15
1

PRINT
R, SALES(R)
PERCENT(R)

R

EXIT

Testing column:

#	SALES	PERCENT
1	200	4.5
2	300	6.8
3	150	3.4
4	400	9.0
5	275	6.2
6	125	2.8
7	175	4.0
8	450	10.2
9	475	10.7
10	350	7.9
11	375	8.5
12	100	2.3
13	325	7.3
14	225	5.1
15	500	11.3

Figure 9.7g Algorithm and Flowchart for the PRINT Module

A

COLUMN ⟶

ROW	1	2	3	4	...	
1	A(1,1)	A(1,2)	A(1,3)	A(1,4)	...	A(1,C)
2	A(2,1)	A(2,2)	A(2,3)	A(2,4)	...	A(2,C)
3	A(3,1)	A(3,2)	A(3,3)	A(3,4)	...	A(3,C)
.
R	A(R,1)	A(R,2)	A(R,3)	A(R,4)	...	A(R,C)

Figure 9.8 General Form of a Two-Dimensional Array

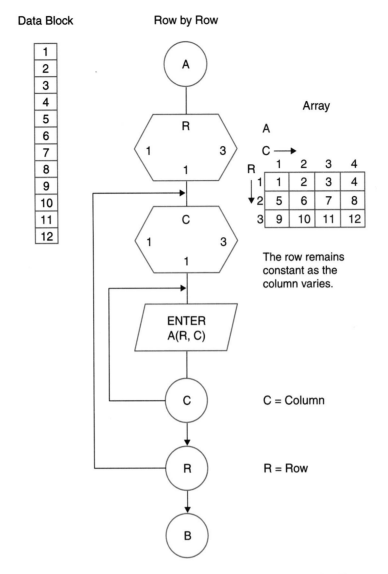

Data Block | Row by Row

A

R
1 3
 1

C
1 3
 1

ENTER
A(R, C)

C C = Column

R R = Row

B

Array

A

C →
 1 2 3 4
R
 1 1 2 3 4
↓ 2 5 6 7 8
 3 9 10 11 12

The row remains
constant as the
column varies.

Figure 9.9 Loading a Two-Dimensional Array, Row by Row and Column by Column

Printing a Two-Dimensional Array

After the array is loaded, it can be used in calculations or it can be printed. As in loading, use nested loops to print a two-dimensional array. Normally, the array is printed row by row. For example, in Figure 9.10 the outer loop (R loop) is the row loop and the inner loop (C loop) is the column loop. Notice that the flowchart is designed so that the element is printed without a carriage return. The carriage return is executed after a row is printed, that is, after the end of the column loop. Place the CARRIAGE RE-TURN instruction between the two LOOP-END instructions. The column headings are printed before either loop is executed. Each row heading is printed between the two begin loop instructions.

Follow the execution of the loop in Figure 9.10 to see how the array is printed. Notice that the column headings are printed first. Then the row loop (R loop) prints row 1 and continues to print through row NR, the total number of rows. Within each row, the row heading must be printed first. Then, through the use of the C loop, each element of the row, designated by the column number (C), is printed without a cursor or carriage

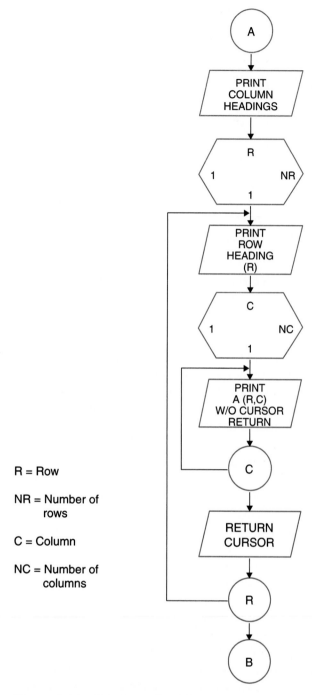

R = Row

NR = Number of rows

C = Column

NC = Number of columns

Figure 9.10 Printing a Two-Dimensional Array

return. Once the elements of the row are printed, the cursor or carriage has to be returned before the next row can be printed.

Accumulating the Rows and Columns of a Two-Dimensional Array

In many solutions you need a total of rows, columns, and the grand total of all elements in a two-dimensional array, and as in the one-dimensional array, you use the assignment

A

C →	1	2	3	4	5
R ↓ 1	1	2	3	4	10
2	5	6	7	8	26
3	9	10	11	12	42
4	15	18	21	24	78

- Column 5 holds the sum of each of the rows.
- Row 4 holds the sum of each of the columns.
- A(4,5) holds the grand total.

Figure 9.11 Accumulating a Two-Dimensional Array

statement to do so. In the solution, you can use a row following the last row of the array for the sums of the columns, and a column to the right of the last column for the sums of the rows (see Figure 9.11). For the grand total, use the intersection of the last row and the last column. This placement of row, column, and grand totals facilitates printing them in the printout of the array.

In the array in Figure 9.11, the fourth row is used for the sum of each column, and the fifth column is used for the sum of each row. The element in the lower right-hand corner (A(4,5)) holds the grand total. In the solution, the sum of both the rows and columns can be done in the same nested loop. It does not make any difference if the sum of a single row is completed before the start of the sum of the next row, or if the sums of all rows are done at the same time, column by column. Notice in Figure 9.11 that the final sum is the same if you accumulate a whole row at one time or all of the rows at the same time. Either way, the sum of the first row is 10, the sum of the second row is 26, and the sum of the third row is 42. The same is true of the columns.

In Figure 9.12, the algorithm and the flowchart are shown for summing the rows and columns of the array in Figure 9.11. Notice the differences between the assignment instructions for summing the rows, the columns, and the grand total. The only difference is in the reference numbers of the array elements containing the sums. The grand total is the sum of the last column since the sum of the rows is complete at the end of the inner loop.

In the instruction to sum the rows in Figure 9.12, the element numbers are represented by the variable R for the row number, and the expression NC + 1 for the column number containing the sum of the rows. In the instruction to sum the columns, the element numbers are represented by the expression NR + 1 for the row containing the sum of the columns, and the variable C for the column number. The element numbers of the box containing the grand total, at the intersection of the sums of the rows and the columns, are NR + 1 and NC + 1. When the rows are summed in the outer loop, they are summed one row at a time; the columns are summed when all the rows have been processed. The summation of the grand total is placed between the two LOOP-ENDs because it is the summation of the sums of the rows. The elements to be accumulated for the grand total are the row sums in A(R,NC + 1).

Algorithm	Flowchart

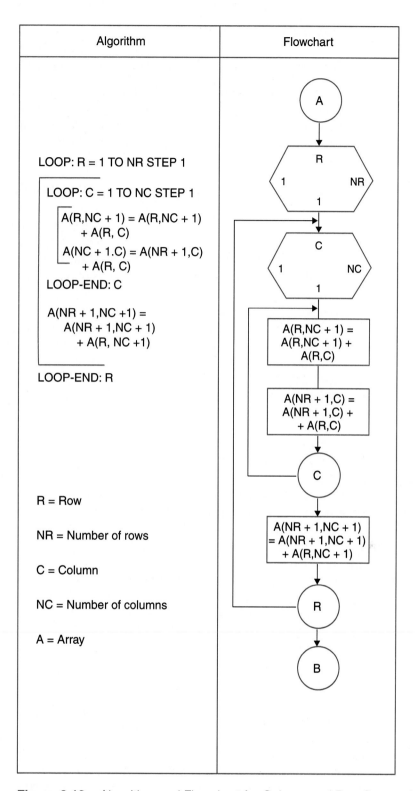

LOOP: R = 1 TO NR STEP 1

 LOOP: C = 1 TO NC STEP 1

 A(R,NC + 1) = A(R,NC + 1)
 + A(R, C)

 A(NC + 1.C) = A(NR + 1,C)
 + A(R, C)

 LOOP-END: C

 A(NR + 1,NC +1) =
 A(NR + 1,NC + 1)
 + A(R, NC +1)

LOOP-END: R

R = Row

NR = Number of rows

C = Column

NC = Number of columns

A = Array

Figure 9.12 Algorithm and Flowchart for Column and Row Summation

Putting It All Together

This PIAT illustrates how to enter, use, and print a two-dimensional array.

Problem: The president of the Too Little Variety Stores has collected sales data on the 15 stores in the corporation for seven days. He would like to know the total sales for each store for the week, the total sales for the company for each of the seven days, and the total sales for the company for the week (see Figure 9.13a–i).

In this problem the data can be stored in a two-dimensional sales array; the sum of the store sales per week can be stored in a one-dimensional array; the sum of the store sales per day can be stored in another one-dimensional array; and the grand total for the week can be stored in a single variable. An alternative way to store the data would be to place the totals in an additional row and column in the two-dimensional sales array. The latter would be the better alternative since there is only one array to keep track of and to print. This makes the processing easier and more readable.

The problem is divided into five modules: the Control, Init, Read, Calc, and Print modules. The Control module uses the sequential logic structure to process the modules. The Init module sets the elements of the array to zero. The Read module enters the sales data into the SALES array, row by row. The Calc module sums the rows to give the totals for each store, the columns to give the totals for each day, and the grand total for the total sales for the company for the week. The Print module prints the SALES array including the totals.

Given Data	Required Results
1. 15 stores 2. Sales for all stores for seven days	1. Total sales for each store for the week 2. Total sales for the company for each day 3. Total sales for the company for the week
Processing Required	Solution Alternatives
1. Summing store sales 2. Summing day's sales 3. Summing all sales	1. Use two-dimensional array for data, one dimensional array for store totals and day's totals, and a single variable for grand total 2. Use a two-dimensional array to store all values

Figure 9.13a Problem Analysis Chart for Array Problem

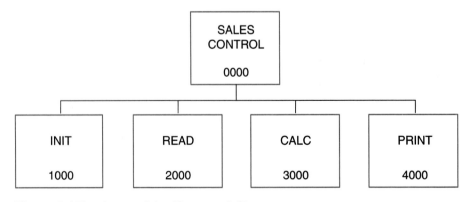

Figure 9.13b Interactivity (Structure) Chart

Input	Processing	Module Reference Number	Output
1. Sales/day/store 2. 15 stores	1. Enter data 2. Set array = 0 3. Calculate totals 4. Print output	2000 1000 3000 4000	1. Total sales for each store for the week 2. Total sales for each day for company 3. Total sales for week for company

Figure 9.13c IPO Chart for Array Problem

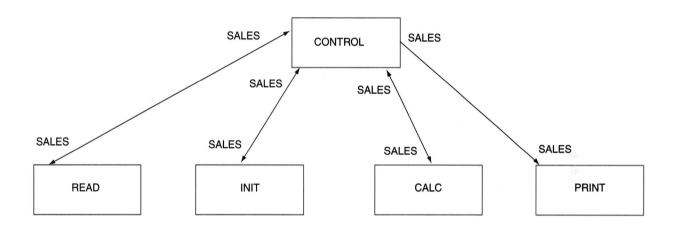

Item	Variable Name	Data Type	Module	Scope	Pseudonyms	Error Cks.
Sales Array	SALES	Numeric	All	Local	None	None
Row No.	R	Numeric	Read, Calc, Print	Local	None	None
Column No.	C	Numeric	Read, Calc, Print	Local	None	None

Figure 9.13d Coupling Diagram and Data Dictionary

Algorithm	Flowchart

CONTROL MODULE
(0000)

1. PROCESS INIT

2. PROCESS READ

3. PROCESS CALC

4. PROCESS PRINT

5. END

Figure 9.13e Algorithms and Flowcharts for CONTROL Module

Algorithm	Flowchart
INIT MODULE (1000) 1. LOOP: R = 1 TO 16 LOOP: C = 1 TO 8 SALES(R,C) = 0 LOOP-END: C LOOP-END: R 2. EXIT	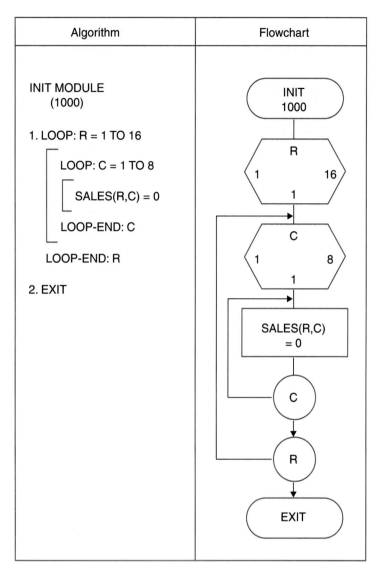

Figure 9.13f Algorithm and Flowchart for INIT Module for Array Problem

Algorithm	Flowchart

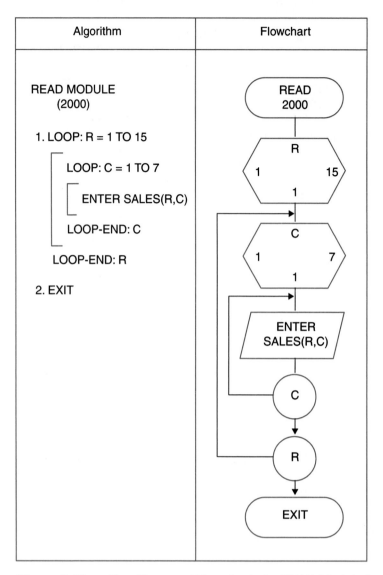

READ MODULE
(2000)

1. LOOP: R = 1 TO 15

 LOOP: C = 1 TO 7

 ENTER SALES(R,C)

 LOOP-END: C

 LOOP-END: R

2. EXIT

Figure 9.13g Algorithms and Flowcharts for READ Module for Array Problem

Algorithm	Flowchart	Annotation

CALC MODULE
 (3000)

1. LOOP: R = 1 TO 15

 LOOP: C = 1 TO 7

 SALES(R,8) = SALES(R,8)
 + SALES(R,C)

 SALES(16,C) = SALES(16,C)
 + SALES(R,C)

 LOOP-END: C

 SALES(16,8) = SALES(16,8)
 + SALES(R,8)

 LOOP-END: R

2. EXIT

* Column 8 contains the sum of each
 of the rows.
** Row 16 contains the sum of each of
 the columns.
Element row 16, column 8, contains
 the grand total.

CALC
3000

R
1 15
1
Row loop

C
1 7
1
Column loop

SALES(R,8) =
SALES(R,8) +
SALES(R,C)
Sum of each row*

SALES(16,C) =
SALES(16,C) +
SALES(R,C)
Sum of each column**

C

SALES(16,8) =
SALES(16,8) +
SALES(R,8)
Grand total
 of all elements

R

EXIT

Figure 9.13h Algorithms and Flowcharts for CALC Module for Array Problem

Algorithm	Flowchart

PRINT MODULE
 (4000)

1. PRINT COLUMN
 HEADINGS

2. LOOP: R = 1 TO 16

 PRINT ROW HEADINGS(R)
 W/O {CR}
 LOOP: C = 1 TO 8

 PRINT SALES(R,C)
 W/O {CR}

 LOOP-END: C

 PRINT {CR}

 LOOP-END: R

3. EXIT

Figure 9.13i Algorithms and Flowcharts for PRINT Module for Array Problem

Multidimensional Arrays

In some cases there is a need for arrays with a third or even a fourth dimension. These arrays are called **multidimensional arrays.** You will use these multidimensional arrays relatively infrequently. They can facilitate an understanding of the data, improve the readability of algorithms, and facilitate processing since the data can be processed through the use of three or more nested loops. For example, with input data on temperatures referenced by day, city, county, and state, *day* would be the first dimension, *city* would be the second dimension, *county* would be the third dimension, and *state* would be the fourth dimension of the array. In any case, any temperature could be found as long as the day, the city, the county, and the state are known. A multidimensional array allows the programmer to use one array for all the data.

Figure 9.14 shows a block of memory locations and their reference names for a three-dimensional array. The recommended variable names for the indexes of a three-dimensional array are R or ROW for row, C or COLUMN for column, and D or DEPTH for depth. For a fourth dimension, V or VOLUME for volume is used. These arrays are processed much like two-dimensional arrays, and, therefore, are not discussed further.

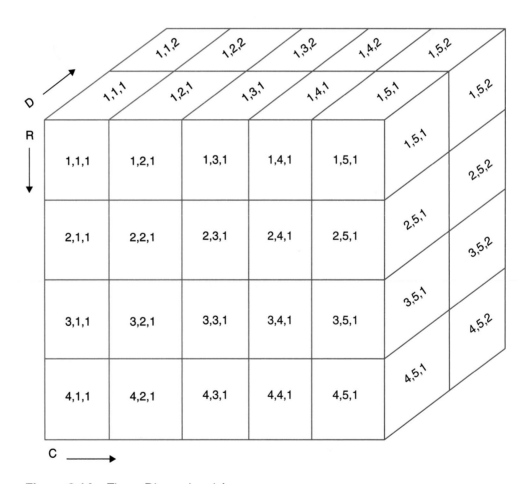

Figure 9.14 Three-Dimensional Array

Table Look-up Technique

A common application for arrays is using a value to look up another value in a table. A one-dimensional array would be used if the element number can be utilized as the given value (see Figure 9.15). A two-dimensional array with two columns would be used if the element number cannot be utilized. In this case the first column would contain the values to compare with the given value, and the second column would contain the values to be located. Figure 9.15 illustrates a one-dimensional array and the algorithm and flowchart for a table look-up problem. In this problem we need to find the number of days in a month given the month number. In this case we can utilize the element number as the month number. For example, given month number 3, the number of days in the month would be 31.

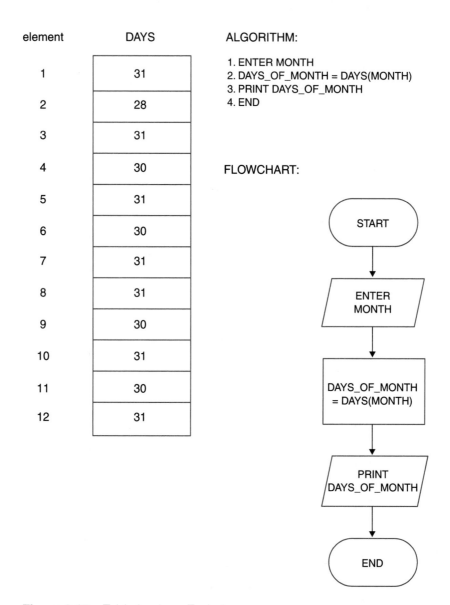

Figure 9.15 Table Look-up Technique

Sequential Search

There are instances when you do not know the element number, but you do know the value of the element. In this case a search must be made on the array to find the correct element. The easiest way to search an array is called a **sequential search.** Figure 9.16 shows the algorithm and the flowchart of a sequential search. The methodology is simple. Given the search value of the element you want to find, element number one is tested to see if it matches the search variable. If it does, the flow of the program drops out of the loop with the desired element number. If it doesn't match, the element number is incremented and loops back to test again.

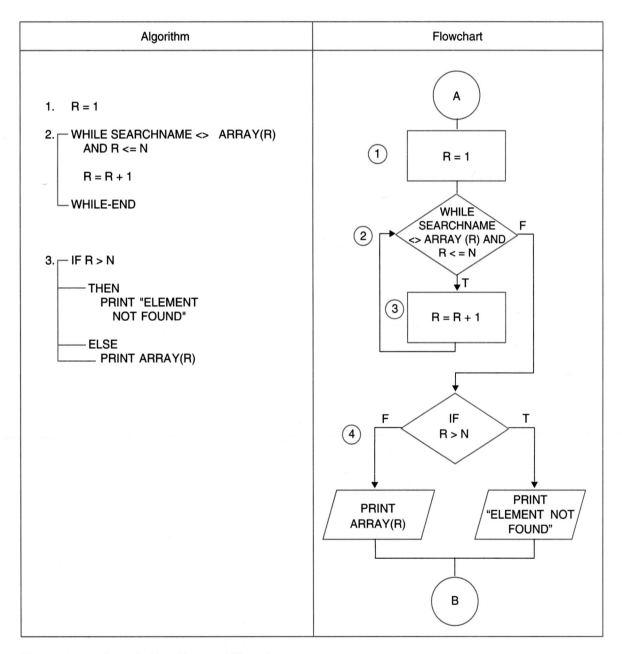

Algorithm	Flowchart

1. R = 1

2. ┌ WHILE SEARCHNAME <> ARRAY(R)
 │ AND R <= N
 │
 │ R = R + 1
 │
 └ WHILE-END

3. ┌ IF R > N
 │
 ├── THEN
 │ PRINT "ELEMENT
 │ NOT FOUND"
 │
 ├── ELSE
 └──── PRINT ARRAY(R)

Figure 9.16 Search Algorithm and Flowchart

Notes on Figure 9.16

The numbers of the notes correspond to the circled numbers in Figure 9.16.

1. The element number is set to 1.
2. A loop is set up to test the search value against the value of the present element of the array. This test also includes a test to make sure the element number is less than or equal to the number of elements in the array. If the element number is greater than the number of elements, then the value was not found.
3. The element number is incremented.
4. A test is made on the element number. If the element number is greater than the number of elements in the array, an error message is printed; if not, the value of the element is printed.

Binary Search

A faster method of finding the desired element of an array is the **binary search.** Figure 9.17a shows the algorithm and flowchart of the binary search. The binary search technique involves comparing the mid-element of all or part of the array. If it compares, the flow of the program drops out of the loop with the desired element number. If it does not compare, the program checks to see if the value is lower or higher than the middle value. The boundaries are then reset to select a new section of the array in which the search value should be found. The midpoint of this new section is then checked. The program continues to divide the remaining elements by half until the desired element is found. At that time the program drops out of the loop with the element number of the desired element. Even with 1,000 elements, this method will take less than 10 comparisons. Figure 9.17b shows an example of a binary search using the algorithm and flowchart in Figure 9.17a.

Notes on Figure 9.17a

The numbers of the notes correspond to the circled numbers in Figure 9.17a.

1. The lower boundary is set at 1 (lowest element number).
2. The upper boundary is set at the maximum number of elements in the array.
3. The flag is set at zero. This flag is set at 1 when the element is found and 2 when the element could not be found. The flag is zero until one of the two preceding conditions is met.
4. The loop is started and will continue as long as FLAG = 0.
5. The mid-element number is calculated. This always rounds down because the INT function is used.
6. This is the test to see if the mid-element is equal to the value of the desired element. If it is, FLAG is set equal to 1, and the flow goes back to the WHILE statement. The condition at the WHILE statement is now false so the flow drops out of the loop. If the mid-element is not equal to the desired value, then the flow of the program will need to loop again to test another element against the search value.
7. The next test is made so the upper or lower boundaries can be reset. This will designate the new section of the array. This is accomplished by testing the value of the element against the search value. If the search value is greater than the value of the mid-element number, then the lower boundary is set to one more than the midpoint. This means that the desired element is

Algorithm	Flowchart

Binary Search

LB = Lower bound
UB = Upper bound
N = # IN LIST
R = Current record
 position
S = Search value

1. LB = 1

2. UB = N

3 FLAG = 0

4. ─ WHILE FLAG = 0
 R = INT ((LB + UB) / 2)
 ─ IF A(R) = S
 ─ THEN
 FLAG = 1
 ─ ELSE
 ─ IF S > A(R)
 ─ THEN
 LB = R + 1
 ─ ELSE
 UB = R - 1
 ─ IF LB > UB
 ─ THEN
 FLAG = 2
 └─WHILE-END

5. ─ IF FLAG = 2
 ─ THEN
 PRINT "RECORD
 NOT FOUND"
 ─ ELSE
 PRINT
 RECORD R

6. EXIT

Figure 9.17a The Binary Search

Order of comparisons to find the element number of 44:
1. Element 7 (UB = 13, LB = 1, Element Number = 7)
2. Element 10 (UB = 13, LB = 8, Element Number = 10)
3. Element 12 (UB) = 13, LB = 11, Element Number = 12)
Number 44 is found in three comparisons, compared to 12 with the sequential search.

Figure 9.17b Example of a Binary Search

in the upper half of this section. If not, the upper boundary is set to one less than the midpoint. This means the desired element is in the lower half of the section.

8. This test is to make sure there are still elements in the section to be tested. If the lower boundary is less than the upper boundary, the flag is set to 2, indicating that the element could not be found.

9. The flow then returns to the WHILE statement and continues or drops out of the loop according to the value of FLAG.

10. After the flow of the program drops out of the loop, the flag is checked. When the FLAG is 1, the element (or elements if using parallel arrays or multidimensional arrays) information is printed. If FLAG is equal to 2, an error message is printed.

The Pointer Technique

Another way to use arrays is to specify the value of an element in one array as the element number in another array. This use of array is called the **pointer technique** because the value of the element in the first array points to the element in the second. The pointer technique is used in frequency distributions. A *frequency distribution* is a tally of one type of value in an array, such as how many students in a school are in each class or how

many of a company's customers live in each zip code area. The result of a frequency distribution is a one-dimensional array that contains the tally for the value of each element number. The pointer technique is also used for calculating statistics from a questionnaire by cross-tabulating items, such as how many students are in each major and each class. The result of this cross-tabulation is a two-dimensional array containing the tally for the combination of the majors and the classes; the majors would be the rows and the classes would be the columns.

Frequency Distribution

In the development of the solution for a frequency distribution, the arrays must be defined. First, you define the array from which the frequency distribution is to be calculated. This array contains data that are coded using the numbers from 1 to the number of different values in the array. For example, when doing a frequency distribution on grades in a class, the grades must be coded numerically, such as, 1 = A, 2 = B, 3 = C, 4 = D, 5 = F. Or if the frequency distribution deals with the number of freshmen, sophomores, juniors, seniors, and graduate students in a college, each class would be given a number from 1 to 5. The number of values of the code then becomes the number of elements in the frequency distribution array.

Figure 9.18a–d illustrates the solution to the following problem: Using a grading scale of 1 to 10, design a solution to find the number of students out of a class of 30 who received each grade (1 through 10) on a test. The GRADE array contains 30 elements, one element for each student, which have the values of 1, 2, 3, 4, 5, 6, 7, 8, 9, or 10, depending on the student's grade. The F (frequency distribution) array contains 10 elements, one for each different grade: The first element contains the number of students who received a 1, the second element contains the number of students who received a 2, the third element contains the number of students who received a 3, and so forth (see Figure 9.18d).

Figure 9.18a–c demonstrates the use of pointers for a frequency distribution. Notice in the Calc module that the value of the GRADE array element is the element number of the frequency distribution array. The GRADE array element value points to the location of the element in the frequency distribution array that needs to be incremented, an example of the pointer technique. Follow the notes for Figure 9.18a for an explanation of each major step in the solution to the frequency distribution problem. Figure 9.18d shows an example of data and outcomes of this problem.

Notes for Figure 9.18a–c

The following notes explain the flowcharts for a frequency distribution. The numbers of the notes correspond to the circled numbers in Figure 9.18a–c.

1. The INIT module initializes the variables to zero.
2. The READ module enters the data.
3. The CALC module calculates the frequency distribution.
4. The PRINT module prints the frequency distribution.
5. The frequency distribution array (F) is 10 elements long since there are 10 grades. Each element in this array is set to zero since each element is a counter.
6. There are 30 elements in the GRADE array since there are 30 students. This loop enters the grades into the GRADE array. There is one element per student.

7. In this loop the ending number is the number of students, since that will determine the frequency distribution.
8. Through the pointer technique, the value of the GRADE(R) points to the element number of the frequency distribution array to indicate which element will be incremented. The P variable is the pointer variable. It has the same value as the grade of student number R.
9. The P element of the frequency array is incremented. At the conclusion of the loop, the frequency array will contain the values of the number of students receiving each of the 10 grades.
10. This loop executes 10 times since there are 10 elements in the frequency array.
11. This loop prints the number of the element (R), which is equal to the grade and the value of the R element of the frequency array (F), which is the number of students receiving the grade (R).

Figure 9.18a Algorithm and Flowchart of CONTROL Module for a Frequency Distribution

Algorithm	Flowchart

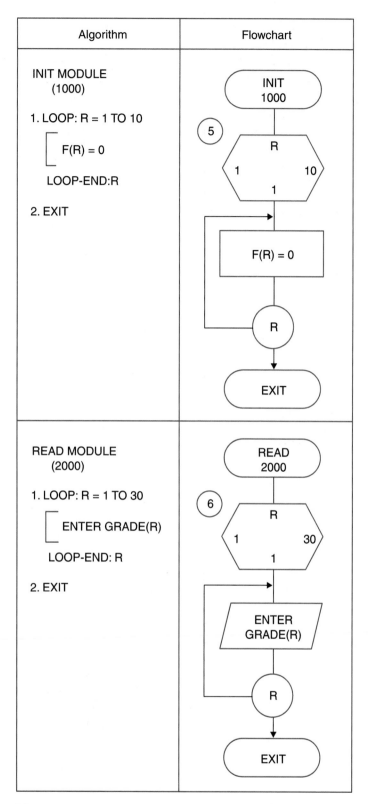

INIT MODULE
(1000)

1. LOOP: R = 1 TO 10

F(R) = 0

LOOP-END:R

2. EXIT

READ MODULE
(2000)

1. LOOP: R = 1 TO 30

ENTER GRADE(R)

LOOP-END: R

2. EXIT

Figure 9.18b Algorithms and Flowcharts of INIT/READ Modules for a Frequency Distribution

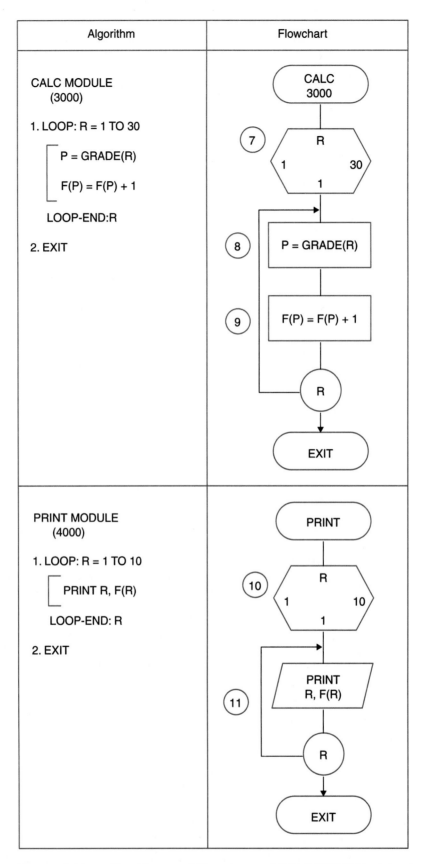

Figure 9.18c Algorithms and Flowcharts of CALC/PRINT Modules for a Frequency Distribution

GRADE		F	
1	10	1	1
2	8	2	1
3	5	3	2
4	10	4	1
5	9	5	1
6	10	6	3
7	6	7	4
8	1	8	5
9	3	9	6
10	8	10	6
11	7		
12	7		
13	10		
14	9		
15	8		
16	9		
17	9		
18	6		
19	7		
20	4		
21	3		
22	2		
23	7		
24	6		
25	8		
26	9		
27	10		
28	9		
29	8		
30	10		

The *element* number of the frequency distribution array is the number of points correct and the *value* of the element is the number of students that received that grade. Therefore, the frequency distribution array F shows that there is 1 student who received 1 point, 1 that received 2 points, 2 that received 3 points, 1 that received 4 points, 1 that received 5 points, 3 that received 6 points, 4 that received 7 points, 5 that received 8 points, 6 that received 9 points and 6 that received 10 points.

Figure 9.18d Example of a Frequency Distribution

Cross-Tabulation

Figure 9.19a–f demonstrates the use of pointers for a cross-tabulation. Notice that in this case, one value points to the row and another value points to the column in the cross-tabulation array. The data for cross-tabulation problems come from questionnaires. A cross-tabulation shows the number of people who gave the same responses to two questions from the questionnaire. A frequency distribution shows the number of people who gave the same answer to one question. For example, in a frequency distribution, a set of test scores would be separated into the number of students who received each grade. In a cross-tabulation, a set of test scores could be separated into the number of females that received each test grade and the number of males that received each test grade, or the number in each class that received each test grade. A cross-tabulation allows the user to have more detail in the analysis of the data.

Figure 9.19a–f shows the algorithms and flowcharts for a cross-tabulation for the following problem: A questionnaire has been sent out by a company that conducts polls. The company needs a program to calculate the cross-tabulation for any two items on the questionnaire. The questionnaire has five items. Items 1 and 2 are *age* and *education* (5 possible answers). Items 3 through 5 are *yes* or *no* questions. There were 150 questionnaires returned. The following would be a sample output given option 1 = sex and option 2 = question 1.

Algorithm	Flowchart
CROSSTAB CONTROL (0000) 1. PROCESS INIT 2. PROCESS ENTERDATA 3. PROCESS CHOOSEOPTIONS 4. WHILE C1 > 0 PROCESS CALC PROCESS PRINT PROCESS INIT PROCESS CHOOSEOPTIONS WHILE-END 5. END	

Figure 9.19a Algorithms and Flowcharts of CONTROL Module for Cross-Tabulation Solution

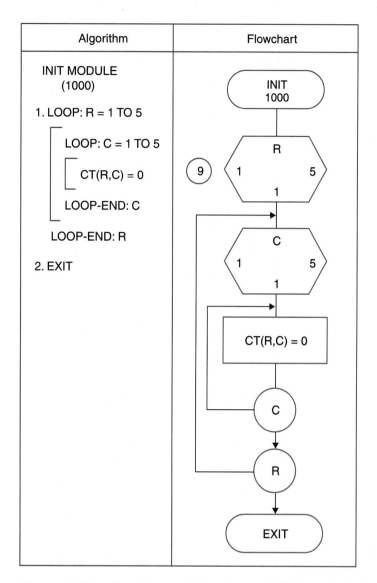

Algorithm	Flowchart
INIT MODULE (1000) 1. LOOP: R = 1 TO 5 ⎡ LOOP: C = 1 TO 5 ⎢ ⎡ CT(R,C) = 0 ⎢ LOOP-END: C ⎣ LOOP-END: R 2. EXIT	

Figure 9.19b Algorithm and Flowchart of INIT Module for Cross-Tabulation Solution

The CT array represents the output where row 1 represents females; row 2, males; column 1, yes for question 1; and column 2, no for question 1.

CT	1 Yes	2 No
1 Females	32	25
2 Males	51	42

The values of the four elements add up to the total number of questionnaires. The CT array shows that 32 females answered yes to question 1, 25 females answered no, 51 males answered yes, and 42 males answered no.

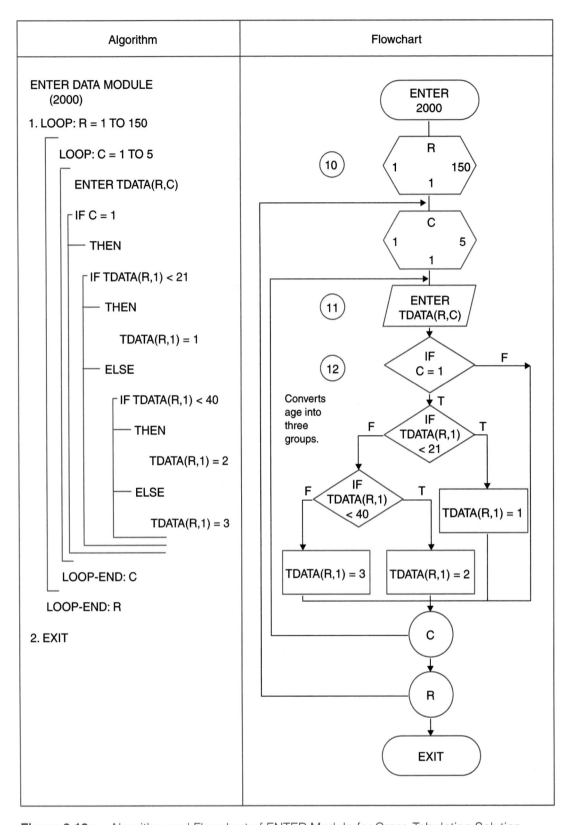

Figure 9.19c Algorithm and Flowchart of ENTER Module for Cross-Tabulation Solution

Algorithm	Flowchart

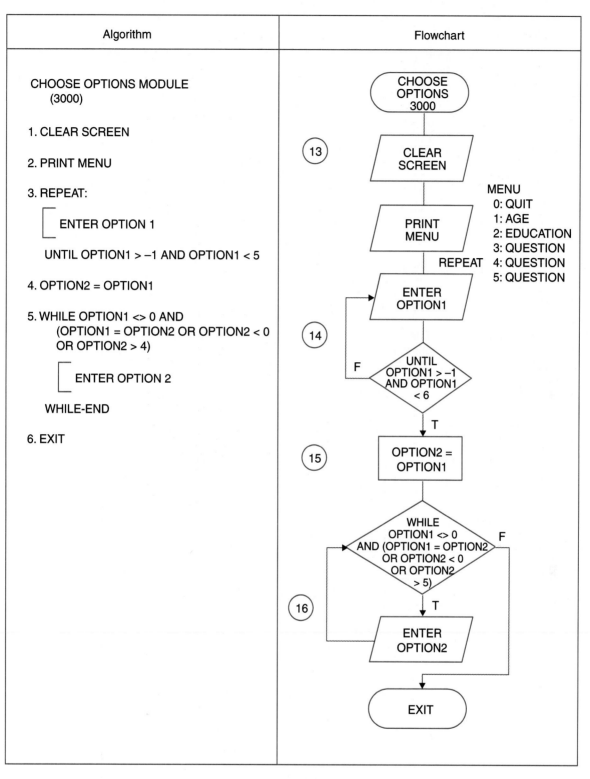

Algorithm (left column):

CHOOSE OPTIONS MODULE
 (3000)

1. CLEAR SCREEN

2. PRINT MENU

3. REPEAT:

 ENTER OPTION 1

 UNTIL OPTION1 > –1 AND OPTION1 < 5

4. OPTION2 = OPTION1

5. WHILE OPTION1 <> 0 AND
 (OPTION1 = OPTION2 OR OPTION2 < 0
 OR OPTION2 > 4)

 ENTER OPTION 2

 WHILE-END

6. EXIT

Flowchart (right column):

CHOOSE OPTIONS 3000

⑬ CLEAR SCREEN

PRINT MENU

MENU
0: QUIT
1: AGE
2: EDUCATION
3: QUESTION
4: QUESTION
5: QUESTION

REPEAT

⑭ ENTER OPTION1

UNTIL OPTION1 > –1 AND OPTION1 < 6 — F / T

⑮ OPTION2 = OPTION1

WHILE OPTION1 <> 0 AND (OPTION1 = OPTION2 OR OPTION2 < 0 OR OPTION2 > 5) — F / T

⑯ ENTER OPTION2

EXIT

Figure 9.19d Algorithms and Flowcharts of CHOOSE Options Module for Cross-Tabulation Solution

Algorithm	Flowchart

CALC MODULE
(4000)

1. RMAX = 0

2. CMAX = 0

3. LOOP: Q = 1 TO 150

 R = TDATA(Q,OPTION1)

 C = TDATA(Q,OPTION2)

 CT(R,C) = CT(R,C) + 1

 IF R > RMAX

 — THEN

 RMAX = R

 IF C > CMAX

 — THEN

 CMAX = C

 LOOP-END: Q

4. EXIT

RMAX = Maximum number of rows used in the cross-tabulation array.

CMAX = Maximum number of columns used in the cross-tabulation array.

Q is the questionnaire; the number goes to 150 becuase there are 150 questionnaires.

TDATA is the questionnaire data.

R is the pointer to the row of the cross-tabulation array.

C is the pointer to the column in the cross-tabulation array.

Calculates the size of the cross-tabulation array for printout purposes.

Figure 9.19e Algorithm and Flowchart of CALC Module for Cross-Tabulation Solution

Algorithm	Flowchart
PRINT MODULE (5000) 1. PRINT COLUMN HEADINGS 2. LOOP: R = 1 TO RMAX PRINT ROW HEADING LOOP: C = 1 TO CMAX PRINT CT(R,C) W/O {CR} LOOP-END: C PRINT {CR} LOOP-END: R 3. EXIT	

* {CR} = Cursor return for monitor or carriage return for printer.

Figure 9.19f Algorithm and Flowchart of PRINT Module for Cross-Tabulation Solution

Notes to Figure 9.19

The following notes explain the flowcharts in Figure 9.19a–f. The numbers of the notes correspond to the circled numbers in Figure 9.19a–f.

1. The INIT module initializes the variables.
2. The ENTER Data module enters the questionnaire data.
3. The CHOOSE OPTIONS module chooses the two items for the cross-tabulation.
4. This loop repeats the cross-tabulation until no more cross-tabulations are desired by the user.
5. The CALC module calculates the cross-tabulation.
6. The PRINT module prints the results of the cross-tabulation.
7. The INIT module, processed at this point, reinitializes the variables for another cross-tabulation on the same data.
8. This Choose Options module chooses the two items for the next cross-tabulation.
9. This nested automatic-counter loop sets the elements of the cross-tabulation array to zero.
10. These loops are set so that the rows (R) are the questionnaires (1 to 150) and the columns (C) are the items (1 to 5).
11. This instruction enters the questionnaire data, one element at a time.
12. This decision instruction checks to see if the column number is 1; if it is, then the instruction changes the ages to age groups: 1 = under 21, 2 = between 21 and 40, 3 = over 40.
13. These instructions clear the screen and print the menu of the questionnaire items.
14. This loop enters the first item number (OPTION1) for the cross-tabulation option. It repeats until the number entered is within the limits of the options of the menu (0 to 5).
15. This instruction initializes OPTION2 to prepare for the next loop.
16. This loop enters the second item number (OPTION2) and repeats until the answer is within the limitations of the number of items (1 to 5) and is not equal to OPTION1.
17. RMAX is the maximum number of rows used in the cross-tabulation. It is set to zero since the maximum number of rows has to be greater than zero.
18. CMAX is the maximum number of columns used in the cross-tabulation. It is set to zero since the maximum number of columns has to be greater than zero.
19. This loop is set up for the number of questionnaires (150).
20. R is the pointer to the row of the cross-tabulation array. The value comes from the value in the OPTION1 column for that questionnaire. OPTION1 is the first item number entered by the user.
21. C is the pointer to the column of the cross-tabulation array. The value comes from the value in the OPTION2 column for that questionnaire. OPTION2 is the second item number entered by the user.
22. This instruction increments the element in the cross-tabulation array specified by the pointers R and C.
23. These decisions calculate the maximum row number and the maximum column number for printing.

24. This instruction prints column headings.
25. This loop specifies the number of rows to print, determined by RMAX.
26. This instruction prints the row heading each time through the loop.
27. This loop specifies the number of columns to print, determined by CMAX.
28. This instruction prints an element without a carriage return, as specified by R and C.
29. This instruction returns the carriage after a row is printed.

Putting It All Together

This PIAT illustrates the use of arrays in the solution to a cross-tabulation problem.

Problem: A company has data on the number of days each employee was absent during the previous year. The data also include sex (1 = female, 2 = male), pay-type (1 = salary, 2 = hourly wage, 3 = commission), education (1 = no diploma or degree, 2 = high school diploma, 3 = associate degree, 4 = bachelor's degree, 5 = graduate degree), and age group (1 = below 30, 2 = 30–50, 3 = over 50). The personnel manager would like a cross-tabulation on these data giving the average number of days absent. The company has 55 employees (see Figure 9.20a–i).

In this example output, the CT array represents the average absences, where row 1 represents salary pay-type; row 2, hourly pay-type; row 3, commission pay-type; column 1, females; and column 2, males.

CT	1 Females	2 Males
1 Salary pay-type	2.4	5.6
2 Hourly pay-type	5.2	4.8
3 Commission pay-type	6.1	5.3

The CT array shows that average number of absences for females having salary pay-type is 2.4; males had 5.6. Females having hourly pay-types had an average number of absences of 5.2; males had 4.8. Females who were under commission pay-type had an average number of absences of 6.1; males had 5.3. In this case, to find the average

number of absences, the number of absences in each category is totaled in the CT array, and the number of people in each category are counted in the N array. The average is then calculated by dividing each element in the CT array by the corresponding element in the N array. The N array is tested for zero before dividing to avoid a division by zero error.

	Sex	Pay-Type	Education	Age Group	# of Days Absent
Employee 1					
2					
3					
. . .					
55					

Figure 9.20a Cross-Tabulation Problem

Given Data	Required Results
Per employee: Sex (1 = F, 2 = M) Pay-type (1 = S, 2 = H, 3 = C) Education (1 = N, 2 HS, 3 = C, 4 = B, 5 = G) Age group (1 - < 30; 2 - 30 to 50; 3 - > 50) # of days absent	Cross-tabulate for average number of days absent
Processing Required	
Calculation of average using cross-tabulation technique	
Solution Alternatives	
Must use cross-tabulation pointer technique	

Figure 9.20b Problem Analysis Chart

Interactivity (Structure) Chart

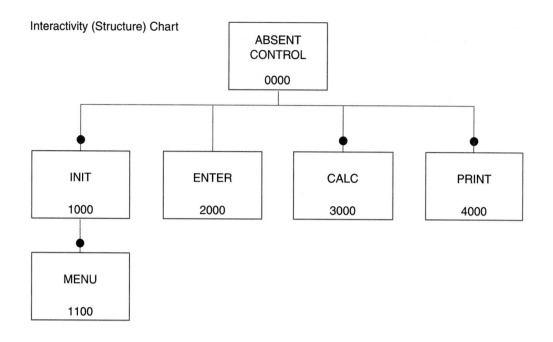

IPO Chart

Input	Processing	Module Reference Number	Output
For each employee	1. Set cross-tab array and counter array = 0	1000	Average # of days absent for any two: (Cross-tabulation)
1. Sex	2. Enter data	2000	1. Sex
2. Pay-type	3. Print menu	1100	2. Pay-type
3. Education	4. Choose options	1100	3. Education
4. Age group	5. Calculate cross-tabulation	3000	4. Age
5. # of days absent	6. Print cross-tab array	4000	
	7 Repeat 3–7 until option 1 = 0		

Figure 9.20c Interactivity (Structure) Chart and IPO Chart

Algorithm	Flowchart

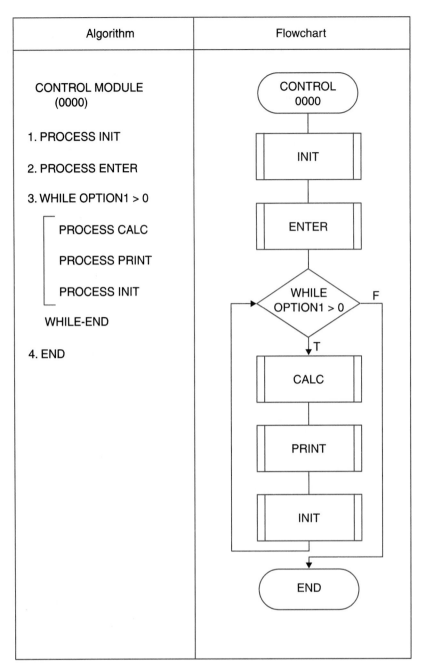

CONTROL MODULE
(0000)

1. PROCESS INIT

2. PROCESS ENTER

3. WHILE OPTION1 > 0

 PROCESS CALC

 PROCESS PRINT

 PROCESS INIT

 WHILE-END

4. END

Figure 9.20d Algorithm and Flowchart of CONTROL Module for PIAT Cross-Tabulation Problem

Algorithm	Flowchart

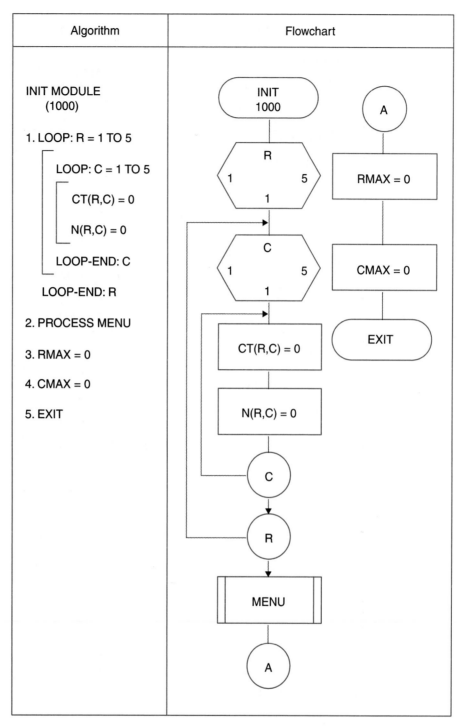

INIT MODULE
 (1000)

1. LOOP: R = 1 TO 5

 LOOP: C = 1 TO 5

 CT(R,C) = 0

 N(R,C) = 0

 LOOP-END: C

 LOOP-END: R

2. PROCESS MENU

3. RMAX = 0

4. CMAX = 0

5. EXIT

Figure 9.20e Algorithm and Flowchart of INIT Module for PIAT Cross-Tabulation Problem

Algorithm	Flowchart
MENU MODULE (1100) 1. CLEAR SCREEN 2. PRINT MENU 3. REPEAT ENTER OPTION1 UNTIL OPTION1 > –1 AND OPTION1 < 5 4. OPTION2 = OPTION1 5. WHILE OPTION1 <> 0 AND (OPTION1 = OPTION2 OR OPTION2 > 4 OR OPTION2 < 0) ENTER OPTION2 WHILE-END 5. EXIT	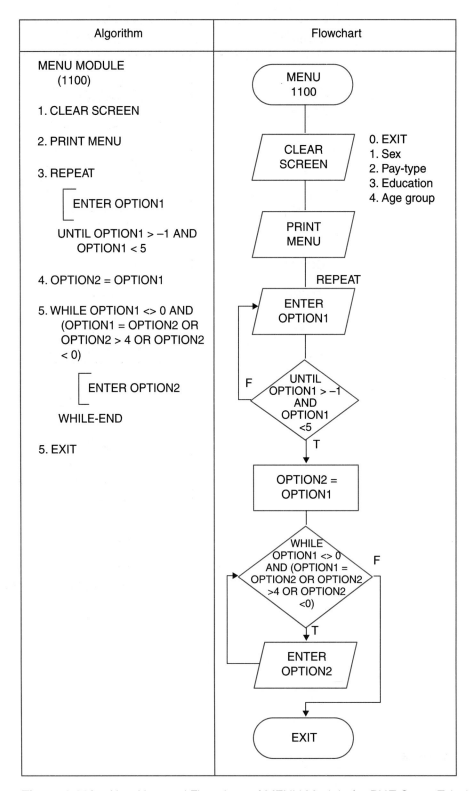

Figure 9.20f Algorithm and Flowchart of MENU Module for PIAT Cross-Tabulation Problem

Algorithm	Flowchart
ENTER MODULE (2000) 1. LOOP: R = 1 TO 55 LOOP: C = 1 TO 5 ENTER A(R,C) LOOP-END: C LOOP-END: R 2. EXIT	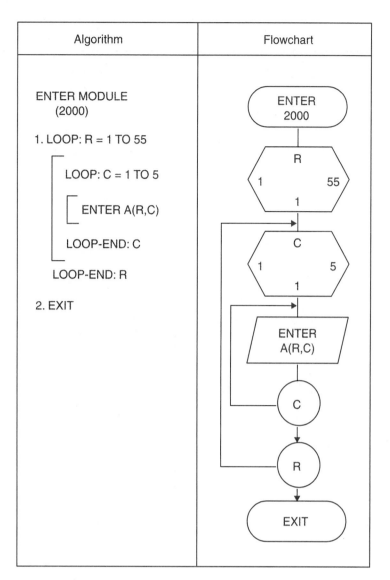

Figure 9.20g Algorithm and Flowchart of ENTER Module for PIAT Cross-Tabulation Problem

Algorithm	Flowchart

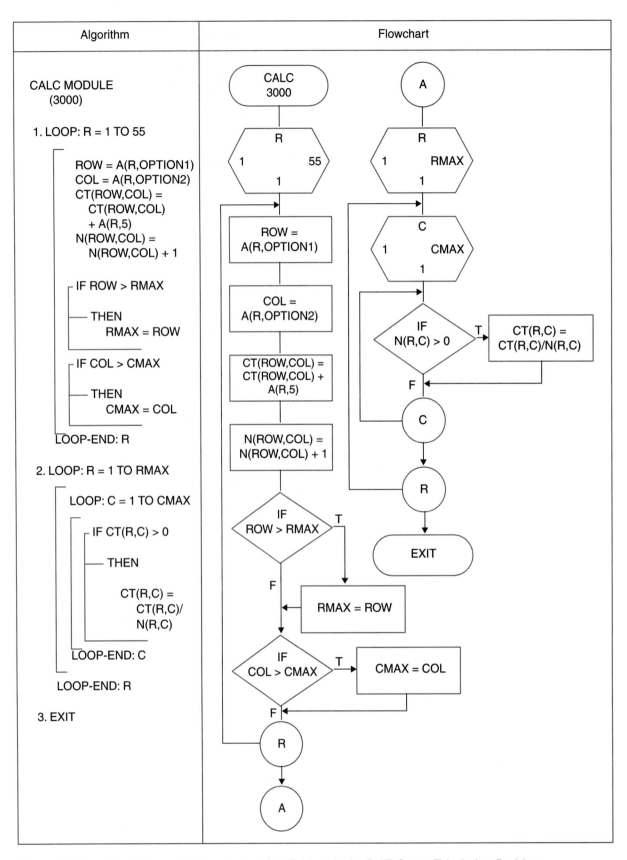

CALC MODULE
 (3000)

1. LOOP: R = 1 TO 55

 ROW = A(R,OPTION1)
 COL = A(R,OPTION2)
 CT(ROW,COL) =
 CT(ROW,COL)
 + A(R,5)
 N(ROW,COL) =
 N(ROW,COL) + 1

 IF ROW > RMAX

 THEN
 RMAX = ROW

 IF COL > CMAX

 THEN
 CMAX = COL

 LOOP-END: R

2. LOOP: R = 1 TO RMAX

 LOOP: C = 1 TO CMAX

 IF CT(R,C) > 0

 THEN

 CT(R,C) =
 CT(R,C)/
 N(R,C)

 LOOP-END: C

 LOOP-END: R

3. EXIT

Figure 9.20h Algorithm and Flowchart of CALC Module for PIAT Cross-Tabulation Problem

Algorithm	Flowchart
PRINT MODULE (4000) 1. WRITE COLUMN HEADINGS 2. LOOP: R = 1 TO RMAX WRITE ROW HEADINGS LOOP: C = 1 TO CMAX WRITE CT(R,C) W/O {CR} LOOP-END: C WRITE {CR} LOOP-END: R 3. EXIT	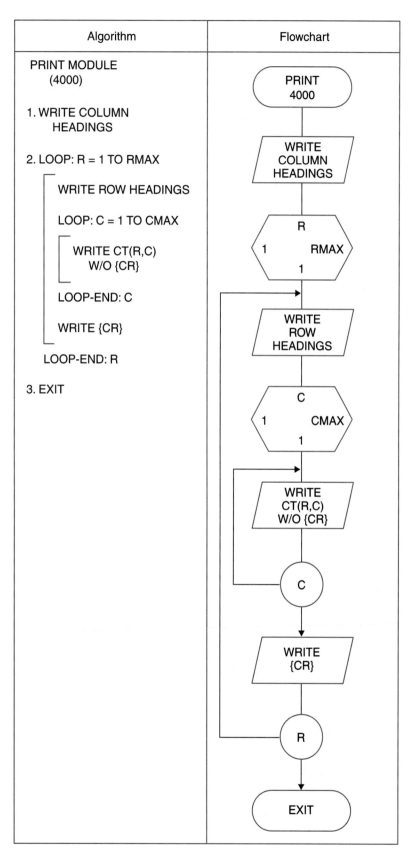

Figure 9.20i Algorithm and Flowchart of PRINT Module for PIAT Cros-Tabulation Problem

Summary

Arrays facilitate the processing of data that is needed more than once in the processing of a solution. Even though arrays are not difficult to learn, they are difficult to use well. Therefore, it is important that a student of programming learn how to use them in a solution. To use arrays, the student needs to learn to enter or load data into an array, to print an array, to sum the elements of an array, and to use an array to store data needed at a later point in the solution. There are three types of arrays: one-dimensional, two-dimensional, and multidimensional.

A technique that is important to the use of arrays is the pointer technique, wherein you use the value of an element in one array to point to the number of the element in another. It is used in frequency distributions and cross-tabulations.

New Terms

array	one-dimensional array
base one system	parallel arrays
base-zero system	pointer technique
binary search	sequential search
dynamic arrays	static array
multidimensional array	two-dimensional array

Questions

1. Draw and name a one-dimensional array that would hold 10 temperatures. Number the elements.

2. Draw and name four parallel arrays that would hold the following data:

Store	Sales	Customers	Region
A24	5000	500	A
A26	3000	200	A
C30	6000	550	C
B44	4000	560	B

3. Draw and name a two-dimensional array that would hold the data for 10 homework scores for 15 students. Number the rows and the columns.

4. Can all of the data in Question 2 be put into a two-dimensional array? Explain.

5. What is the pointer technique for finding data? Why is it used?

6. What are the problem requirements for using the pointer technique?

Problems

Complete the seven steps of solution development for the following problems.

1. Mr. Jones always gives True/False tests to his class. His tests always have 20 questions. The maximum class size is 35. He needs a program that will calculate the students' grades based on the best score.

 Grade
 A will range from the best score, to the best score minus 2.
 B will range from the best score minus 3, to the best score minus 4.
 C will range from the best score minus 5, to the best score minus 6.
 D will range from the best score minus 7, to the best score minus 8.
 F will be anything below the best score minus 8.

 Each student's ID and test answers will be entered. The output will be each student's ID, number correct, and grade, along with the single highest score for the class. Develop a solution for Mr. Jones' problem. Use four one-dimensional arrays—one for the correct scores and the other three for the needed output.

2. A restaurant manager wants to know how many employees are needed at the restaurant each hour of the day. The minimum number of employees needed at any hour is 3. After that, one additional employee is required for each 20 customers. The restaurant is open 24 hours a day. The manager has counted the number of customers each hour for 14 days. The manager will use the average number of customers for each hour over the 14 days to calculate the needed number of employees for each hour. Develop a solution to output the needed number of employees per hour. (There is no such thing as a partial employee.)

3. A company has 10 salespeople. The manager needs to know the average dollar amount of sales for each salesperson for a week, and the total dollar amount of sales for the store for each day and for the week. The store is open 7 days a week, and each salesperson gets 2 days off. The data are entered into a two-dimensional array with the days of the week as the columns and the salespeople as the rows. Develop a solution to output the needed information.

4. An instructor has 30 students in his class. Each student is identified by a number from 1 to 30. Grades are stored in a one-dimensional array. The instructor would like to enter a student number and have the student's test score printed on the monitor. Develop a solution to output the needed information.

5. An instructor has a class of 25 students. Each student is identified by a number from 1 to 25. All tests are stored in a two-dimensional array, with each column containing the grades for each test. The instructor would like to enter the student number and the test number and have the grade for that test printed on the monitor. Develop a solution to output the needed information.

6. The instructor in Problem 5 would like to enter the last name of the student instead of the student number. Modify your solution to allow the instructor to enter the student name instead of the student number.

7. A university has four undergraduate class levels and a graduate school. There are 7 majors, although not all students have chosen a major. There are 1,200 students

(Problem 7 continued on next page)

attending the university. The administration would like to know how many students are in each level, and how many students have each major. They also would like to know how many of each class level have declared each major. Develop a solution to output the needed information.

8. A questionnaire was sent to a random selection of the alumni of a college. In all, 95 questionnaires were returned. The questionnaire requested the following items:

age

sex

marital status

college major

salary

The administration would like to know the average salary, given any two sets of items. Develop a solution to cross-tabulate the items and output the needed information. (Remember, the computer cannot divide by zero.)

9. a. The student names and the grades for four tests for Mr. Smith's class have been placed in parallel arrays. Mr. Smith would like to have one student's name and test scores printed. Develop a solution that will enter the student's name, search for the name in the same array, and then print the name and test scores. Use the sequential-search method.

b. What would have to change in the solution in Problem 9a if we were to use a binary search instead of the sequential-search method?

c. Modify the solution in Problem 9a to use a binary search instead of a sequential search.

10. Modify Problem 4 in Chapter 8 (the Roman numeral problem) using arrays. Which method do you think is the most efficient for the programmer? Which method do you think is the most user friendly?

11. The human resources manager of XYZ Corporation would like to analyze the following characteristics of company employees:

a. the wages of women compared with those of men

b. the total number of employees in each of the 12 departments

c. number of women and men in each of the 12 departments

d. average age of the women and men in each department

Develop a solution for this problem.

12. John has a weather station in his house. He has been keeping track of the fastest wind speed for each day for two weeks. He would like to know the average wind speed over the two weeks, the days on which the highest wind speed and the lowest wind speed were recorded, and the difference between the highest wind speed recorded and each day's average wind speed.

Correct the following segments of algorithms:

1. S = 1

 WHILE EMPLOYEENO <> 0
 ENTER EMPOYEENO(S)
 S = S + 1
 WHILE-END

2. LOOP: R0W = 1 TO 5

 LOOP: COLUMN = 1 TO 3
 PRINT A(COLUMN, ROW)
 LOOP-END: ROW
 LOOP-END: COLUMN

3. LOOP: C = 1 TO 100

 ENTER NAME(EMPLOYEE_NUMBER)
 LOOP-END: C

Chapter 10

Data Structures

Objectives

When you have finished this chapter, you should be able to:

1. Explain the different types of data structures used in a computer.
2. Use several sorting techniques.
3. Use stacks in a solution.
4. Develop a solution to a problem using a linked list technique.
5. Develop a solution to a problem using linked lists to access data.
6. Explain how a binary tree is set up.
7. Develop a solution to a problem using a binary tree to search for data.

Data are stored in the computer in one of many logical ways in order to find needed values as fast as possible. The way data are stored is called the structure of data, hence the term *data structures*. You have already used several types of data structures in developing solutions for problems in this text. The simplest data structure is the single variable. This data structure was the first one you used. The single variable data structure includes single values of the data types presented in Chapter 2. Other data structures include strings, arrays, and records. The string is a type of array because it is a group of characters. An array consists of multiple values for one variable name. Data can be stored in one-dimensional, two-dimensional, or multidimensional arrays. The single variable and

the array are types of data structures that are used as methods of storing data in the internal memory during the processing of the data. Data often need to be saved from one processing time to the next. Also, with large amounts of data, it is impractical or impossible to store all of the data internally during processing. Therefore, data need to be stored in files on an external storage device. Files use records as the data structure. A record is an entity in a file that is designed to store data about that item. All records in a file have the same layout or structure. This chapter emphasizes methods of accessing data in a file, including linked lists and binary trees. Sorting techniques, and a brief description of other data structures, such as stacks, queues, and hashing, are also included in this chapter.

Beginning File Concepts

Chapter 3 presented the concept of how data are stored on an external storage device. To review, data are stored in files. Files contain information pertinent to one application. The file is divided into records. Each record contains information pertaining to one entity in the file. Each record is divided into fields. Each field contains one piece of data. All records in a file have the same number of fields. Each corresponding field in all records contains the same number of bytes, the same name, the same type of data, and the same format. That is, the first field in all records in a personnel file may contain the last name of an employee, the second field may contain the first name, the third field the employee's address, and so forth. The first field in all records is given attributes such as 25 bytes, the field name of LAST_NAME, string data type, and so on. Each field in all records has its own set of attributes. Only the field name must be a unique attribute. You might think of a file as a sort of table with the records listed vertically and the fields horizontally. Records are numbered from one to the maximum number of records in the file. Fields have names.

In order to find records, many techniques are used, including linked lists, binary tables, hashing, and others. Most of these techniques use a primary key to enable the program to find the needed record. The **primary key** is a field containing unique data for each record. Most techniques are ordered by the primary key. In that way a unique record may be found. The primary key may be an account number, a combination of last name and first name, the last name as long as there are no duplicates, or a computer-generated string value.

There are three new instructions that are used with files. They are the following:

1. OPEN logical file number, file name: The open instruction tells the operating system that an external file will be used. The operating system then sets up everything that is needed for the program to communicate with the external storage device. The logical file number is a reference number that the programmer gives to a particular file. The file name is the name of the file on the external storage device.
2. READ (logical file number), record (record number): When the logical file number is added to the READ instruction, the operating system will retrieve the data contained in one record from an external file referenced by the logical file number. The record is a data structure that designates the fields in the record and their attributes.
3. CLOSE logical file number: The CLOSE instruction tells the operating system that the program is finished using the file reference by the logical file number and to complete whatever is needed to stop communications between the program and the file on the external storage device.

There is more information about file concepts in Chapters 14–17.

Sorting Techniques

Most accessing techniques require data to be sorted. **Sorting is** the process of putting data in alphabetical or numeric order using one key field or a concatenation of two or more fields. Data are sorted according to the primary key. Once data are in order, they can be accessed by various means, including the sequential search and the binary search. These techniques were presented in Chapter 9.

There are many different techniques of sorting. Appendix B provides the algorithm and flowchart for three techniques, including the selection exchange sort, the bubble sort, and the shell sort. In this chapter we explain how these three techniques work, along with showing the algorithms and the flowcharts. Other popular sorting techniques include the quick sort and the heap sort. The best sorting method is determined by how large the array or file is and how close the file is to being in order. Ultimately the best sorting method is determined by the number of comparisons and switches that need to take place.

All sorting techniques require a comparison of two data items with a possibility of a switch. If the sort is in ascending order, the comparison would be to see if the primary key of data 1 is larger than the primary key of data 2. If it is, then data 1 is switched with data 2, and if not, they are left in the same position. The comparisons take place until data are in the required order.

The first technique is the *selection exchange sort*. Figure 10.1 shows the switches made during each pass. The theory is to find the smallest value of the remaining values and switch with the value found in the first position of the remaining elements. During the first pass (first time through the outer loop) the smallest value (15) is switched with the first element (65). During the second pass the smallest of the remaining values (25) remained as the second element (25). During the third pass the smallest of the remaining values (34) remains as the third element since no element was found that was smaller. This continues until the last position is in the right order. See Figure 10.1 for all passes. Figure 10.2a–b shows the algorithm and the flowchart for the selection exchange sort.

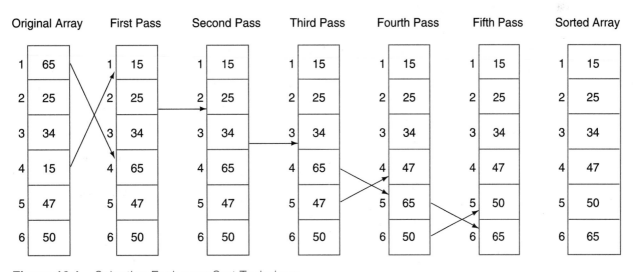

Figure 10.1 Selection Exchange Sort Technique

Algorithm	Flowchart

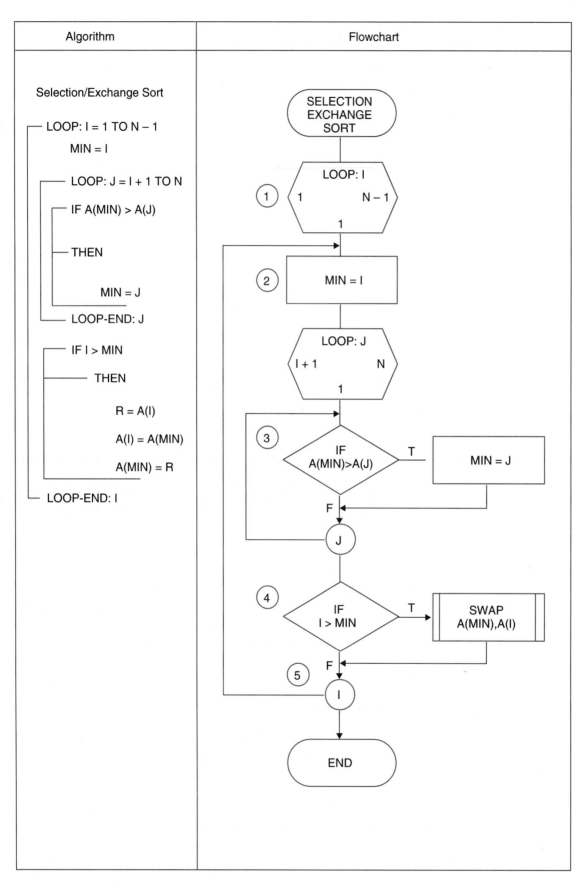

Selection/Exchange Sort

LOOP: I = 1 TO N – 1

 MIN = I

 LOOP: J = I + 1 TO N

 IF A(MIN) > A(J)

 THEN

 MIN = J

 LOOP-END: J

 IF I > MIN

 THEN

 R = A(I)

 A(I) = A(MIN)

 A(MIN) = R

LOOP-END: I

Figure 10.2a The Selection Exchange Sort

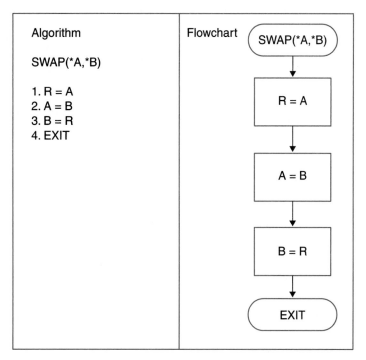

Algorithm	Flowchart

Algorithm

SWAP(*A,*B)

1. R = A
2. A = B
3. B = R
4. EXIT

Flowchart

SWAP(*A,*B)

R = A

A = B

B = R

EXIT

Figure 10.2b Flowchart and Algorithm for SWAP Module

Notes on Figure 10.2a

The following are notes explaining the circled numbers on the flowchart of Figure 10.2a.

1. This loop is set up to start with the first element and continue to the second-to-last element. The last element will be in the correct order when the second-to-last element is in order, because there is nowhere else for the last element to be placed.
2. The initial start value for the minimum value is the value of the lowest element number of the unordered elements.
3. This is the comparison test. It tests to see if this value is less than the previously saved minimum value. If it is, then the element number of the minimum value is set equal to the present element number. If not, then the flow continues to loop back either to make another comparison or to drop out of the loop.
4. This test is to see if the minimum value is in an element other than the smallest element of this segment. If it is, then the two values are swapped; if not, then the elements remain the same.
5. The end of this loop increments the element number.

The next sorting technique we discuss is the *bubble sort*. Figure 10.3a shows the values of the array elements at the end of each iteration of the J loop. In Figure 10.3a, the first element (65) is compared to the second element (25). The first element is greater than the second and therefore they are exchanged. When J = 2, 65 is now the second element. The second element (65) is compared to the third element (34). The second element is greater than the third and therefore they are exchanged. The same is true for each of the values of J: Figure 10.3b shows the switches made in each pass using the bubble sort. The theory of the bubble sort is to compare each element with the next element. If the element is larger than the next element, the elements are switched. If not, they remain the same. After the first pass, the largest element has "bubbled" its way to the bottom.

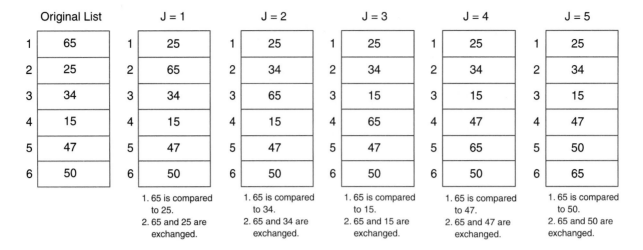

Figure 10.3a Example Pass, I = 1

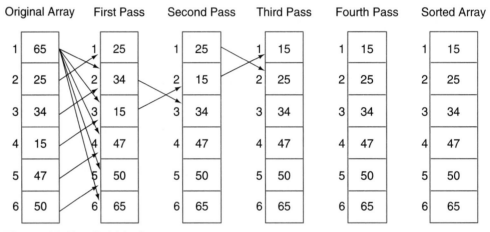

Figure 10.3b Bubble Sort

On the next pass elements 1 through $n-1$ are considered, because the nth element is the largest. The elements are again compared and the largest is bubbled down to the $n-1$ element. A flag is set during each pass. If a switch has been made, another pass is made; if a switch has not been made, then data are in order and no more passes are made. During the second pass, only 15 and 34 need to be switched. All others are either in order or nearly in order. During the third pass 15 and 25 are switched. The value 15 now has bubbled to the top. During the fourth pass there are no switches; the flag has not been set, which indicates that the data are in order and no more passes need to be made. Figure 10.4 shows the algorithm and the flowchart for the bubble sort.

Notes on Figure 10.4

The following are notes explaining the circled numbers on the flowchart of Figure 10.4.

1. The flag is set equal to 1. This primes the loop so the test will pass the first time through.
2. The number of elements being compared is set equal to the number of elements minus 1. Since the comparison is always with one element and the next element, the last element does not need to be tested.

Algorithm	Flowchart

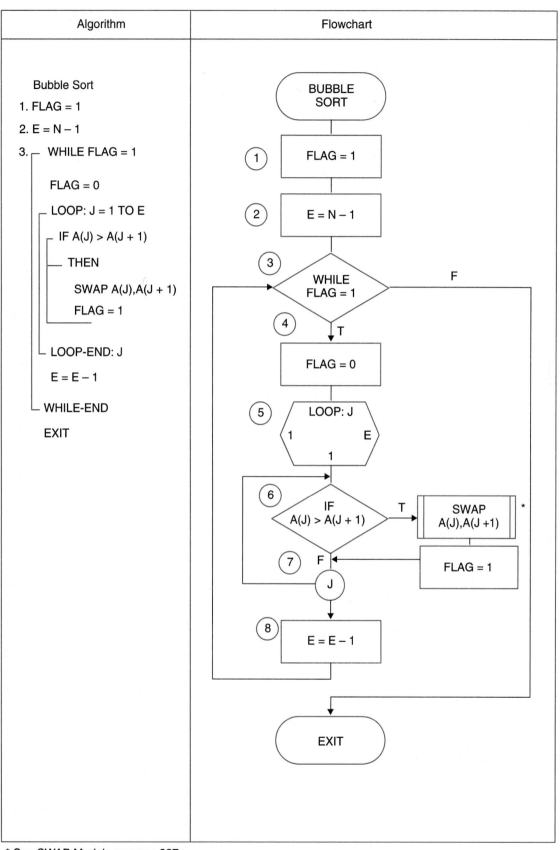

Bubble Sort
1. FLAG = 1
2. E = N − 1
3. ⌐ WHILE FLAG = 1

 FLAG = 0

 ⌐ LOOP: J = 1 TO E

 ⌐ IF A(J) > A(J + 1)

 └ THEN

 SWAP A(J),A(J + 1)

 FLAG = 1

 └ LOOP-END: J

 E = E − 1

└ WHILE-END

EXIT

* See SWAP Module on page 237

Figure 10.4 Flowchart and Algorithm for the Bubble Sort

3. The loop is set to continue as long as the flag is set equal to 1. This indicates that data are not in order.
4. The flag is set equal to 0. This says that data are in order until a switch is made and the flag is set equal to 1.
5. A loop is set up to compare the elements.
6. The comparison of the elements is made. When the present element is greater than the next element, the elements are switched and the flag is set equal to 1, indicating the data may not be in order as of yet.
7. The counter is incremented and the loop continues with the next element. If the elements have all been compared, the flow of the program drops out of the loop.
8. The number of elements to be compared is decreased by 1 since the last element of each loop is in the correct order.

The last sort technique to be presented is the *shell sort*. Figure 10.5 shows the switches that have been made in each pass using the shell sort technique. The theory of the shell sort is to compare each element with another element in the data set using a gap to separate the elements. The gap during the first pass is the number of elements divided by 2. The truncated value is used. In Figure 10.5 the gap is 3 for the first pass since there are 6 elements. In the succeeding passes, the new gap is the old gap divided by 2. Again, the truncated value is used. In the second pass, the gap is 1 since the previous gap was 3. When two elements need to be switched, and there are elements a gap above the present element, then elements are compared upward until elements are reached that do not need switching. During the first pass the following comparisons are made:

1. *element 1 to element 4*—a switch is made (there are no elements a gap above element 1 so the next comparison is in order)
2. *element 2 to element 5*—no switches
3. *element 3 to element 6*—no switches

During the second pass the following comparisons are made:

1. *element 1 to element 2*—no switches
2. *element 2 to element 3*—no switches
3. *element 3 to element 4*—no switches

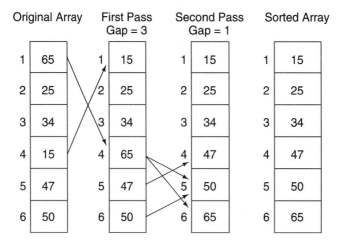

Figure 10.5 Shell Sort

4. *element 4 to element 5*—A switch is made and, because there are elements a gap above, the following elements are compared: the new value of element 4 to element 3—no switches; since no switch took place, the elements above the fourth element are assumed to be in order and the comparison continues in order.

5. *element 5 to element 6*—A switch is made, and other comparisons are made as above in 4.

At this time the data are in order. Figure 10.6 shows the algorithm and the flowchart for the shell sort.

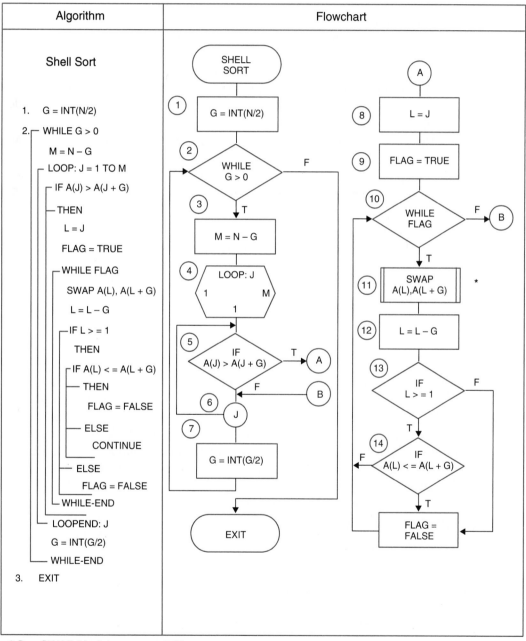

Algorithm	Flowchart

Shell Sort

1. G = INT(N/2)
2. WHILE G > 0
 M = N − G
 LOOP: J = 1 TO M
 IF A(J) > A(J + G)
 THEN
 L = J
 FLAG = TRUE
 WHILE FLAG
 SWAP A(L), A(L + G)
 L = L − G
 IF L > = 1
 THEN
 IF A(L) < = A(L + G)
 THEN
 FLAG = FALSE
 ELSE
 CONTINUE
 ELSE
 FLAG = FALSE
 WHILE-END
 LOOPEND: J
 G = INT(G/2)
 WHILE-END
3. EXIT

* See SWAP Module on page 237

Figure 10.6 Algorithm and Flowchart for the Shell Sort

Notes on Figure 10.6

The following are notes explaining the circled numbers on the flowchart of Figure 10.6.

1. The gap is set to the truncated value of the number of elements divided by 2.
2. A loop is set up to continue while the gap is greater than zero.
3. The number of elements to be compared with another element is set equal to the number of elements minus the gap.
4. A loop is set up to start at the first element and continue to the element number set in note 3.
5. The comparison is made. If the element is greater than the element a gap below, then a switch needs to take place; if not, then the flow continues to the end of the loop.
6. The element number is incremented and the flow returns to the loop set up in note 4.
7. The gap is set equal to the truncated value of the last gap divided by 2. The flow then returns to the loop set in note 2.
8. L is set equal to the element number. This allows the comparisons with elements above the present element.
9. The FLAG is set equal to TRUE to prime the loop so flow will continue through the loop at least once.
10. A loop is set up to continue as long as FLAG is TRUE. FLAG will be set equal to false when a switch is not necessary.
11. The switch is made.
12. L is set equal to the previous value of L minus the gap. This allows the checking to be made with elements above the present element.
13. This checks to see if there are any more elements to be compared.
14. If the elements need to be switched, the flow continues through the loop. If not, the FLAG is set to FALSE to stop the loop.

Stacks

A **stack** is a list of numbers, such as an array of numbers, to which all additions are at one end and all deletions are at the same end. This is a last-in, first-out procedure. A stack is like a stack of plates, the last one added on the top is the first one to be used. When a value (X) is added to the stack it is called **pushing the stack.** When a value (X) is used from the stack it is called **popping the stack.** Once a number is used, it cannot be used again. The stack has a number, called the *stack pointer*, which points to the last value pushed and the next value to be popped. When the stack is pushed, the pointer is incremented and the value (X) being pushed is placed in the stack at the element number in the pointer. When the stack is popped, the value in the stack at the element number of the pointer is placed in X, and the pointer is decremented. Figure 10.7 shows a stack to which the following procedures have been implemented:

PUSH 3

PUSH 4

PUSH 7

POP X

POP X

PUSH 6

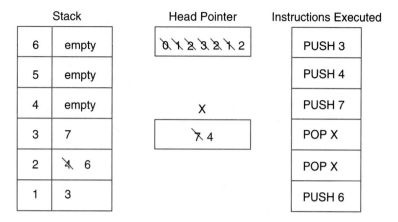

Figure 10.7 Example of a Stack That Has Been Pushed and Popped

Notes on Figure 10.8

Figure 10.8 shows the algorithm and flowchart to *push* the stack and to *pop* the stack. The following are notes explaining the circled numbers on the two flowcharts:

1. The ERROR flag is set to 0.
2. When the pointer is equal to the maximum number of elements in the stack, no more values can be pushed. An error message is printed, and the ERROR flag is set equal to 1.
3. When the pointer is not equal to the maximum number of elements, the pointer is incremented, and X is placed in the stack at the element number that is equal to the pointer.
4. The module is exited.
5. The ERROR flag is set to 0.
6. When the pointer is equal to 0, there are no values in the stack to be popped. An error message is printed, and the ERROR flag is set equal to 2.
7. When the pointer is greater than 0, X is set equal to the value of the element of the stack that has the element number of the pointer. The pointer is then decremented.
8. The module is exited.

Records As a Data Structure

The structure of a record is considered a data structure since a record is a way of storing data. The record is defined and given a name. It can then be referred to by the name or by its parts—the fields. It is defined by designating the fields and their attributes. Each field will have a name and can be used as an individual piece of data or the record may be manipulated as a group of fields given the record name and number. The record name corresponds to the file, the file name, and the logical file number. There is only one record name per file. The individual records in the file are referred to by the record number. There can only be one record in the internal memory at a time.

Algorithm	Flowchart

PUSH (X)
1. ERROR = 0
2. IF POINTER = MAX
 THEN
 PRINT "ERROR: STACK FULL"
 ERROR = 1
 ELSE
 POINTER = POINTER + 1
 STACK(POINTER) = X
3. EXIT

DEFINITIONS:
MAX = Maximum elements in stack
POINTER = Pointer to last element
 to be pushed or next element
 to be popped
STACK = Stack array
X = Value to be PUSHED
 or POPPED

POP (X)
1. ERROR = 0
2. IF POINTER = 0
 THEN
 PRINT "ERROR: EMPTY STACK"
 ERROR = 2
 ELSE
 X = STACK (POINTER)
 POINTER = POINTER –1
3. EXIT

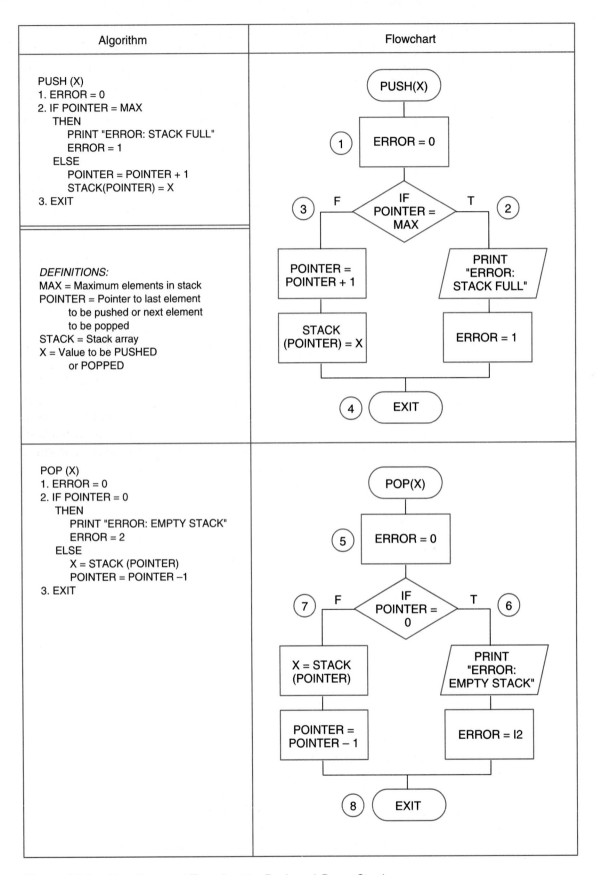

Figure 10.8 Algorithm and Flowchart to *Push* and *Pop* a Stack

A **linked list** is a file in which each record points to its successor, except for the last record, which has an end-of-list indicator. The method of pointing to the next record is a field, which contains the record number of the next record in order. This is called the *linking field*, since the value in this field links the records in proper order. Figure 10.9 shows a file that is a linked list. The file must contain a key field along with a corresponding link field. Notice that the link field of the first record points to the second record, which is the next to be processed. The link field of the second record points to the third record, which is the third record to be processed. The linking continues through the last record. The last record contains a 0, which indicates the last record in order.

The advantage of a linked list is in the addition and deletion process. A file does not have to be physically rewritten to add and delete records. When a record is deleted, it is put into an empty list. The additions are placed in the records that have been deleted or at the end if there are no empty records. An empty records linked list is kept in the link field to locate the records that are available in which to place the additions.

If there are no empty records, the addition is placed at the bottom of the file. Figure 10.10 shows a file that has had additions and deletions. Notice the two linked lists—the data linked list is processed in the following order: 2, 1, 3, 7, 8, 6. The empty list is as follows: 4, 5. The data list starts at 2, which is designated by the LISTPOINTER. The empty list starts at 4, which is designated by the empty pointer. The 0 indicates the end of the list. Notice there is a 0 in the link field for the end of each list. Figure 10.11a–k shows a file as records are added and deleted. Figure 10.11a shows the original file before any additions and deletions. Figure 10.11b shows the same file after all the following additions and deletions have been made. Each of the other Figures in 10.11 show one addition or deletion as indicated. The following additions and deletions are shown being completed:

1. Figure 10.11c: add 69
2. Figure 10.11d: delete 17
3. Figure 10.11e: delete 44
4. Figure 10.11f: add 37
5. Figure 10.11g: delete 29
6. Figure 10.11h: add 40
7. Figure 10.11i: add 90
8. Figure 10.11j: add 1
9. Figure 10.11k: add 16

Record No.	Key	Other Data	Link
1	5		2
2	10		3
3	17		4
4	25		5
5	30		6
6	35		0

LISTPOINTER

1

EMPTYPOINTER

0

0 indicates end of list.

Figure 10.9 Example of an Initially Created Linked List

Record No.	Key	Other Data	Link
1	15		3
2	3		1
3	17		7
4	14		5
5	30		0
6	35		0
7	25		8
8	28		6

LISTPOINTER

2

EMPTYPOINTER

4

0 indicates end of list.

Figure 10.10 Example of a Linked List After Multiple Adds and Deletions

Rec. #	Key	Other Data	Link
1	34		4
2	29		10
3	67		7
4	44		3
5	25		9
6	17		2
7	79		0
8	13		6
9	2		0
10	33		1

HEADPOINTER

8

EMPTYPOINTER

5

Comment: Notice that records 5 and 9 are records that have been deleted in the past and are included in the empty list, even though there are data in the record. The used list skips these records.

Figure 10.11a Original File

Rec. #	Key	Other Data	Link
1	34		4
2	40		3
3	67		5
4	37		2
5	69		7
6	90		0
7	79		6
8	13		11
9	1		8
10	33		1
11	16		10

HEADPOINTER

9

EMPTYPOINTER

0

Figure 10.11b File After the Additions and Deletions

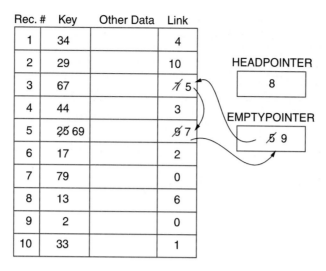

Rec. #	Key	Other Data	Link
1	34		4
2	29		10
3	67		~~1~~ 5
4	44		3
5	~~25~~ 69		~~0~~ 7
6	17		2
7	79		0
8	13		6
9	2		0
10	33		1

HEADPOINTER 8

EMPTYPOINTER ~~5~~ 9

Figure 10.11c Add 69

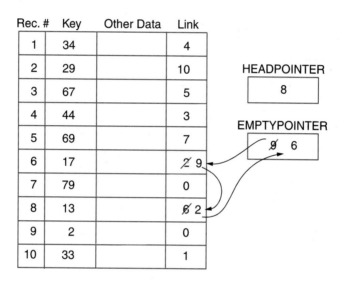

Rec. #	Key	Other Data	Link
1	34		4
2	29		10
3	67		5
4	44		3
5	69		7
6	17		~~2~~ 9
7	79		0
8	13		~~6~~ 2
9	2		0
10	33		1

HEADPOINTER 8

EMPTYPOINTER ~~9~~ 6

Figure 10.11d Delete 17

Rec. #	Key	Other Data	Link
1	34		~~4~~ 3
2	29		10
3	67		5
4	44		~~3~~ 6
5	69		7
6	17		9
7	79		0
8	13		2
9	2		0
10	33		1

HEADPOINTER 8

EMPTYPOINTER ~~6~~ 4

Figure 10.11e Delete 44

Rec. #	Key	Other Data	Link
1	34		3̶4̶
2	29		10
3	67		5
4	4̶4̶ 37		6̶ 3
5	69		7
6	17		9
7	79		0
8	13		2
9	2		0
10	33		1

HEADPOINTER
8

EMPTYPOINTER
4̶ 6

Figure 10.11f Add 37

Rec. #	Key	Other Data	Link
1	34		4
2	29		1̶0̶ 6
3	67		5
4	37		3
5	69		7
6	17		9
7	79		0
8	13		2̶ 10
9	2		0
10	33		1

HEADPOINTER
8

EMPTYPOINTER
6̶ 2

Figure 10.11g Delete 29

Rec. #	Key	Other Data	Link
1	34		4
2	2̶9̶ 40		6̶ 3
3	67		5
4	37		3̶ 2
5	69		7
6	17		9
7	79		0
8	13		10
9	2		0
10	33		1

HEADPOINTER
8

EMPTYPOINTER
2̶ 6

Figure 10.11h Add 40

Rec. #	Key	Other Data	Link
1	34		4
2	40		3
3	67		5
4	37		2
5	69		7
6	1̶7̶ 90		9̶ 0
7	79		0̶ 6
8	13		10
9	2		0
10	33		1

HEADPOINTER
8

EMPTYPOINTER
6̶ 9

Figure 10.11i Add 90

Rec. #	Key	Other Data	Link
1	34		4
2	40		3
3	67		5
4	37		2
5	69		7
6	90		0
7	79		6
8	13		10
9	2̶ 1		0̶ 8
10	33		1

HEADPOINTER
8̶ 9

EMPTYPOINTER
9̶ 0

Figure 10.11j Add 1

Rec. #	Key	Other Data	Link
1	34		4
2	40		3
3	67		5
4	37		2
5	69		7
6	90		0
7	79		6
8	13		1̶0̶ 11
9	1		8
10	33		1
(11)	16		10

HEADPOINTER
9

EMPTYPOINTER
0

Comment: In this case, since there were no more empty records, the new record was placed in an appended record.

Figure 10.11k Add 16

Figure 10.12 presents the algorithm and flowchart to create a linked list file and Figure 10.13 a–k presents the interactivity chart and algorithms and flowcharts used to modify the linked list file. Follow the flowchart and algorithms in Figure 10.13b–k to see how the additions and deletions are accomplished as a computer program. In general, the steps to add and delete are as follows:

To add

1. Search to find where the new record will be logically inserted (if New record > List record (R)). (You need to remember the previous record number.)
2. Add new record to the record number indicated by the EMPTYPOINTER value.
3. Put the value of the EMPTYPOINTER into the previous record link field of the previous record remembering the old value.
4. Put the value of the remembered old previous record link field (from step 3) into new record link field remembering the old value.
5. Put the value of the remembered new record link field (from step 4) into EMPTYPOINTER.

To delete

1. Search to find record to be deleted (if delete record = list record (R)). (You will need to remember the previous record number.)
2. Put the value of the EMPTYPOINTER into deleted record link field remembering the old value of the link field.
3. Put the remembered value of the old deleted record link field (from step 3) into previous record link field remembering the old value.
4. Put the remembered value of the old previous record link field (from step 3) into the EMPTYPOINTER.

Remember these are general instructions. Refer to Figure 10.13b–k for the algorithms and flowcharts.

Notes on Figure 10.12

Figure 10.12 shows the algorithm and flowchart to create a linked list. The following are notes explaining the circled numbers for this figure.

1. Process a READ module, which opens the data file, enters all records for the file, and counts the number of records (N). The file must have a key and an extra field for the link field.
2. The file is sorted in key order.
3. The HEADPOINTER is set to 1, since the first record would be the first record in order.
4. A loop is set up to process all records in order to add the correct link number in the link field. The loop starts with the first record and continues to the next to the last record, since the last record must have a 0 to indicate the end of the list.
5. A record is read. The LINK field is set equal to the record + 1, which would be the next record in order. The record is written to the file. The next record is then processed.
6. The last record is read. The LINK field is set equal to 0. The record is written to the file.

Algorithm	Flowchart

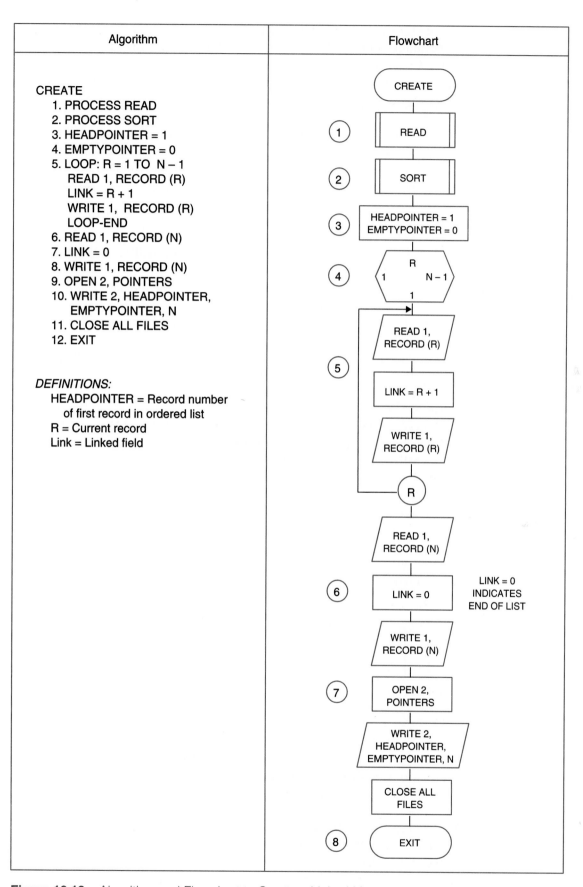

CREATE
1. PROCESS READ
2. PROCESS SORT
3. HEADPOINTER = 1
4. EMPTYPOINTER = 0
5. LOOP: R = 1 TO N − 1
 READ 1, RECORD (R)
 LINK = R + 1
 WRITE 1, RECORD (R)
 LOOP-END
6. READ 1, RECORD (N)
7. LINK = 0
8. WRITE 1, RECORD (N)
9. OPEN 2, POINTERS
10. WRITE 2, HEADPOINTER,
 EMPTYPOINTER, N
11. CLOSE ALL FILES
12. EXIT

DEFINITIONS:
 HEADPOINTER = Record number
 of first record in ordered list
 R = Current record
 Link = Linked field

Flowchart nodes:
- CREATE
- (1) READ
- (2) SORT
- (3) HEADPOINTER = 1 / EMPTYPOINTER = 0
- (4) R / 1 / N − 1 / 1
- (5) READ 1, RECORD (R) / LINK = R + 1 / WRITE 1, RECORD (R) / R
- READ 1, RECORD (N)
- (6) LINK = 0 — LINK = 0 INDICATES END OF LIST
- WRITE 1, RECORD (N)
- (7) OPEN 2, POINTERS
- WRITE 2, HEADPOINTER, EMPTYPOINTER, N
- CLOSE ALL FILES
- (8) EXIT

Figure 10.12 Algorithm and Flowchart to Create a Linked List

7. The POINTER file is opened. The HEADPOINTER, EMPTYPOINTER, and the number of records (N) is written to this file. ALL files are then closed.
8. The Create module is completed.

Notes on Figure 10.13a–k

Figure 10.13a–k shows the interactivity (structure) chart, the algorithms, and the flow-charts for a program that processes and updates a file using a linked list technique. Figure 10.13a shows the structure chart for this program. The following are the notes explaining the circled numbers in Figure 10.13b–k.

1. The Control module for the LINKED LIST program.
2. Process the INIT module.
3. This instruction primes the loop so the condition will be true the first time through the loop.
4. A loop is set up to continue to process or update the file until Q is entered for the choice in the MENU module.
5. The MENU module is processed, which prints the menu and allows the user to enter a value for CHOICE.
6. One of the modules is processed, according to the value of CHOICE.
7. The program is completed.
8. The INIT module initializes values for the program.
9. The files are opened.
10. The values of HEADPOINTER, EMPTYPOINTER, and N are read.
11. The POINTER file is closed, which ends the INIT module.
12. The MENU module prints the menu and allows the user to enter a value for CHOICE.
13. The screen is cleared and the menu is printed on the screen.
14. A loop is set up to make sure a correct value is entered for CHOICE. When a correct value is entered the MENU module is completed.
15. The ADD module allows the user to add a record to the linked list.
16. The new record number (NR) is set equal to the empty pointer. The data for a new record are entered through processing the READNEW module.

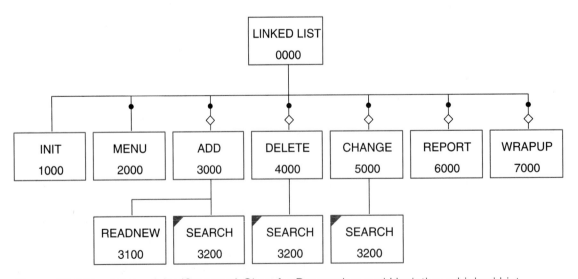

Figure 10.13a Interactivity (Structure) Chart for Processing and Updating a Linked List

Algorithm	Flowchart

LINKED LIST
1. PROCESS INIT
2. CHOICE = "A"
3. WHILE CHOICE < > "Q"
 PROCESS MENU
 CASE OF CHOICE
 "A": PROCESS ADD
 "D": PROCESS DELETE
 "C": PROCESS CHANGE
 "R": PROCESS REPORT
 "Q": PROCESS QUIT
 CASE-END
 WHILE-END
4. END

DEFINITIONS:
R = Present record
HEADPOINTER = First ordered record
EMPTYPOINTER = First empty record
CHOICE = Menu selection
S = Error flag
 = 1: Record found
 = 2: Record not found
LINK = Value of link field
KEY = Value of key field
SKEY = Search key
PR = Previous record

Figure 10.13b Algorithm and Flowchart of the LINKEDLIST (CONTROL) Module for Processing and Updating a Linked List

17. The search for the records that precede and succeed the new record is executed.
18. A test is made to see if there is a duplicate record. If there is, a message is printed and the ADD module is exited.
19. If there is no duplicate record, a test is made to see if there is an empty record in which to place the new record. If there is no empty record, a record is appended to the end of the file. The number of records is incremented by 1 and the new record number (NR) is set equal to the last record number, which is the total number of records.

Algorithm	Flowchart
INIT 1. OPEN 1, DATAFILE 2. OPEN 2, POINTERFILE 3. READ HEADPOINTER 4. READ EMPTYPOINTER 5. CLOSE 2 6. EXIT	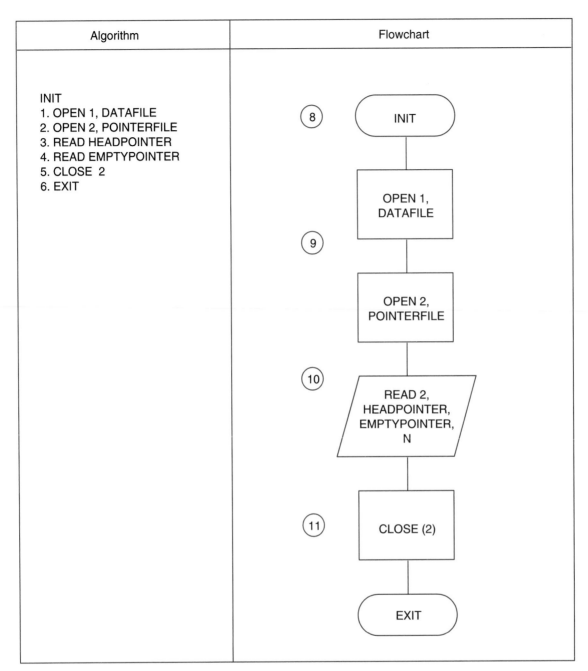

Figure 10.13c Algorithm and Flowchart of the INIT Module for Processing and Updating a Linked List

20. If there is an empty record, the empty record is read and the EMPTY-POINTER is set equal to the LINK field value.
21. A test is made to see if the new record should be placed before the first record in order. If so, the LINK field is set equal to the HEADPOINTER, the HEADPOINTER is set equal to the new record number (NR), and the record is written to the file.
22. If not, the link field of the previous record and the new record need to be updated. To do this, the new record is written to the file and the previous record is read. The LINK field value is saved and the LINK field is set equal to the new record number (NR). The previous record is then written to the file and

Algorithm	Flowchart

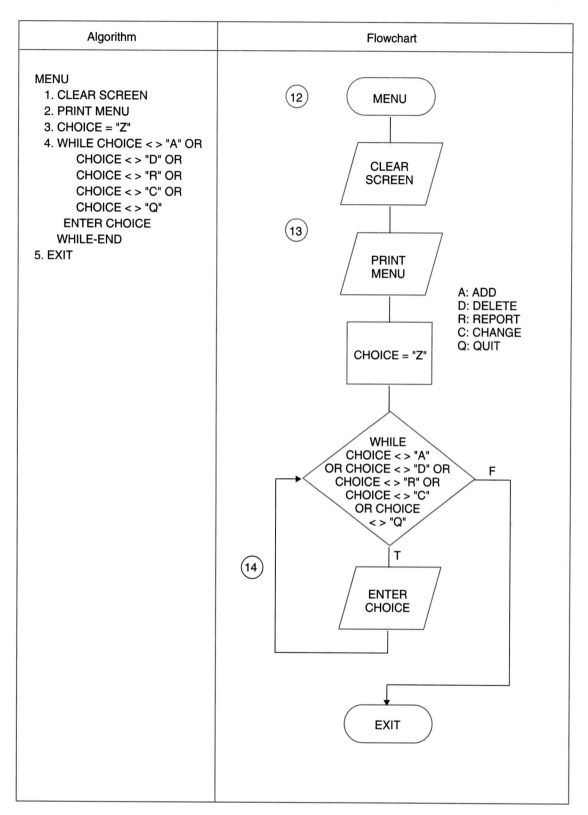

MENU
 1. CLEAR SCREEN
 2. PRINT MENU
 3. CHOICE = "Z"
 4. WHILE CHOICE < > "A" OR
 CHOICE < > "D" OR
 CHOICE < > "R" OR
 CHOICE < > "C" OR
 CHOICE < > "Q"
 ENTER CHOICE
 WHILE-END
 5. EXIT

Figure 10.13d Algorithm and Flowchart of the MENU Module for Processing and Updating a Linked List

Algorithm	Flowchart

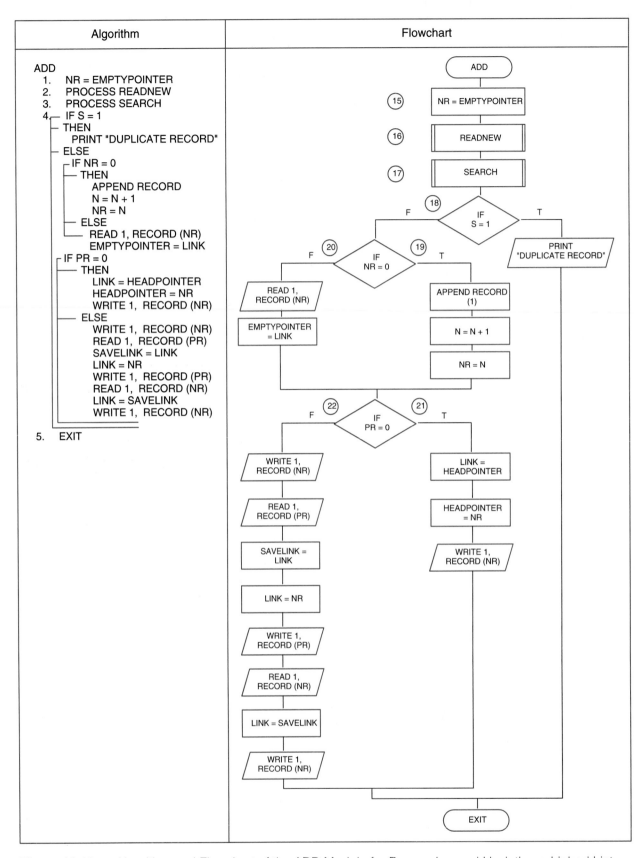

ADD
1. NR = EMPTYPOINTER
2. PROCESS READNEW
3. PROCESS SEARCH
4. ⌐ IF S = 1
 ├ THEN
 │ PRINT "DUPLICATE RECORD"
 ├ ELSE
 │ ⌐ IF NR = 0
 │ ├ THEN
 │ │ APPEND RECORD
 │ │ N = N + 1
 │ │ NR = N
 │ └ ELSE
 │ READ 1, RECORD (NR)
 │ EMPTYPOINTER = LINK
 ⌐ IF PR = 0
 ├ THEN
 │ LINK = HEADPOINTER
 │ HEADPOINTER = NR
 │ WRITE 1, RECORD (NR)
 └ ELSE
 WRITE 1, RECORD (NR)
 READ 1, RECORD (PR)
 SAVELINK = LINK
 LINK = NR
 WRITE 1, RECORD (PR)
 READ 1, RECORD (NR)
 LINK = SAVELINK
 WRITE 1, RECORD (NR)

5. EXIT

Figure 10.13e Algorithm and Flowchart of the ADD Module for Processing and Updating a Linked List

Algorithm	Flowchart
READNEW 1. READ NEW RECORD DATA 2. SKEY = NRKEY 3. EXIT	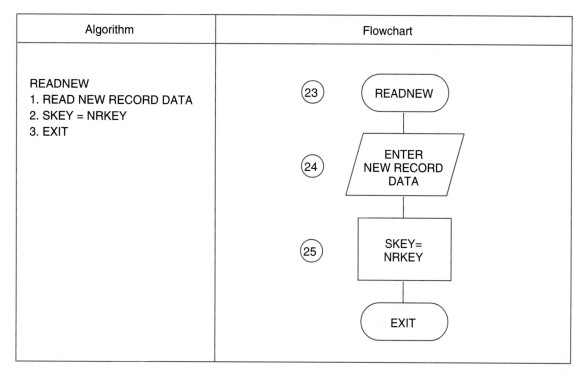

Figure 10.13f Algorithm and Flowchart of the READNEW Module for Processing and Updating a Linked List

the new record is read. The LINK field of the new record is set equal to the saved link value from the previous record and the new record is written to the file. This completes the ADD module.

23. The READNEW module enters data for the new record.
24. The new data are entered.
25. The search key (SKEY) is set equal to the new record key. The READNEW module is completed.
26. The SEARCH module searches for a record using a search key value (SKEY).
27. The record number is set equal to the HEADPOINTER. The error flag (S) is set equal to 0, and the previous record number (PR) is set equal to 0.
28. A loop is set up to process the file until a record is found (S = 1), or the key of the record is less than the search key (S = 2), or the end of the list occurs (S = 2).
29. The current record is read.
30. A test is made to see if the record has been found; if it has, S is set equal to 1 and the module is exited.
31. If the record has not been found, a test is made to see if the record key is less than the search key.
32. If it is not, S is set equal to 2 and the module is exited.
33. If it is, the record has not been found yet. The previous record (PR) is set equal to the present record number (R), and the end of the link is checked.
34. If the present record is at the end of the list, S is set equal to 2, R is set equal to 0, and the module is exited.
35. If present record is not at the end of the list, then the present record number (R) is set equal to the LINK field value and the loop continues.

Algorithm	Flowchart

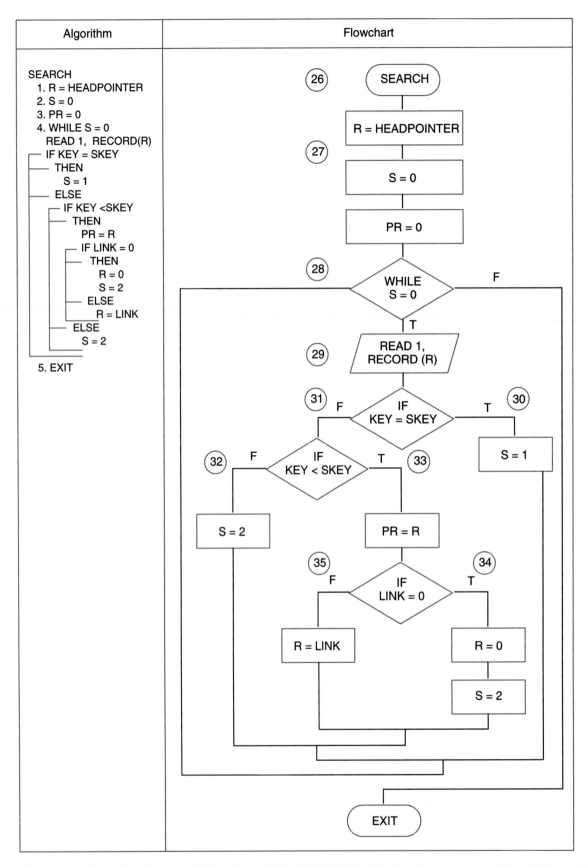

SEARCH
 1. R = HEADPOINTER
 2. S = 0
 3. PR = 0
 4. WHILE S = 0
 READ 1, RECORD(R)
 — IF KEY = SKEY
 — THEN
 S = 1
 — ELSE
 — IF KEY <SKEY
 — THEN
 PR = R
 — IF LINK = 0
 — THEN
 R = 0
 S = 2
 — ELSE
 R = LINK
 — ELSE
 S = 2
 5. EXIT

Figure 10.13g Algorithm and Flowchart of the SEARCH Module for Processing and Updating a Linked List

Algorithm	Flowchart

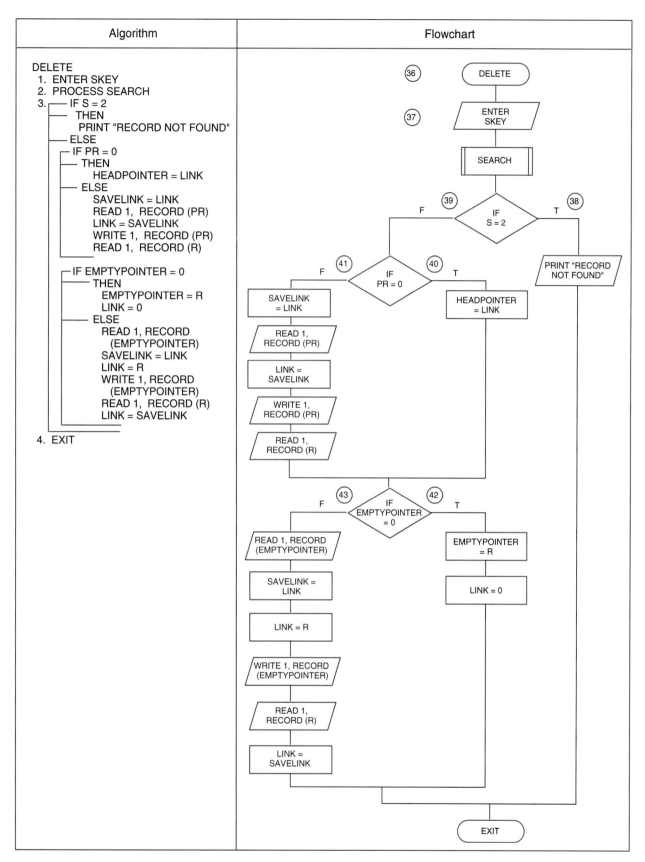

DELETE
1. ENTER SKEY
2. PROCESS SEARCH
3. ── IF S = 2
 ── THEN
 PRINT "RECORD NOT FOUND"
 ── ELSE
 ─ IF PR = 0
 ── THEN
 HEADPOINTER = LINK
 ── ELSE
 SAVELINK = LINK
 READ 1, RECORD (PR)
 LINK = SAVELINK
 WRITE 1, RECORD (PR)
 READ 1, RECORD (R)

 ─ IF EMPTYPOINTER = 0
 ── THEN
 EMPTYPOINTER = R
 LINK = 0
 ── ELSE
 READ 1, RECORD
 (EMPTYPOINTER)
 SAVELINK = LINK
 LINK = R
 WRITE 1, RECORD
 (EMPTYPOINTER)
 READ 1, RECORD (R)
 LINK = SAVELINK
4. EXIT

Figure 10.13h Algorithm and Flowchart of the DELETE Module for Processing and Updating a Linked List

Algorithm	Flowchart
CHANGE 1. ENTER SKEY 2. PROCESS SEARCH 3. IF S = 2 THEN PRINT "RECORD NOT FOUND" ELSE CHANGE RECORD (R) DATA WRITE 1, RECORD (R) 4. EXIT	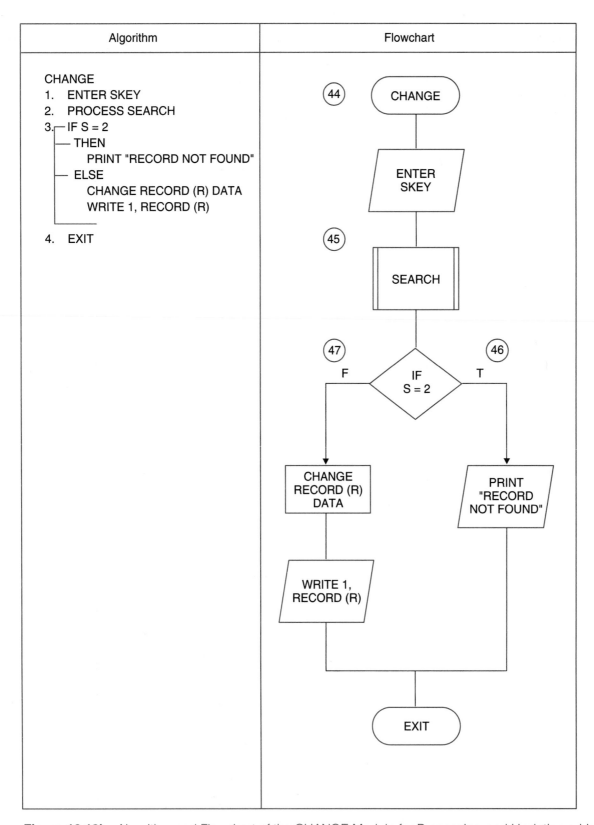

Figure 10.13i Algorithm and Flowchart of the CHANGE Module for Processing and Updating a Linked List

Algorithm	Flowchart
REPORT 1. R = HEADPOINTER 2. WHILE R < > 0 READ 1, RECORD(R) PRINT RECORD(R) R = LINK WHILE-END 3. EXIT	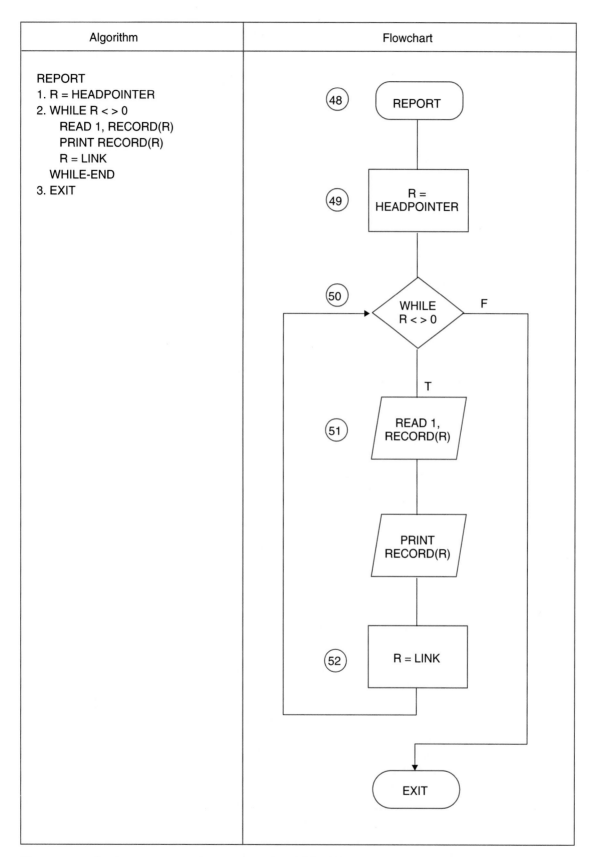

Figure 10.13j Algorithm and Flowchart of the REPORT Module for Processing and Updating a Linked List

Algorithm	Flowchart
WRAPUP 1. OPEN 2, POINTERS 2. WRITE 2, HEADPOINTER, EMPTYPOINTER, N 3. CLOSE ALL FILES 4. PRINT "RUN COMPLETE" 5. EXIT	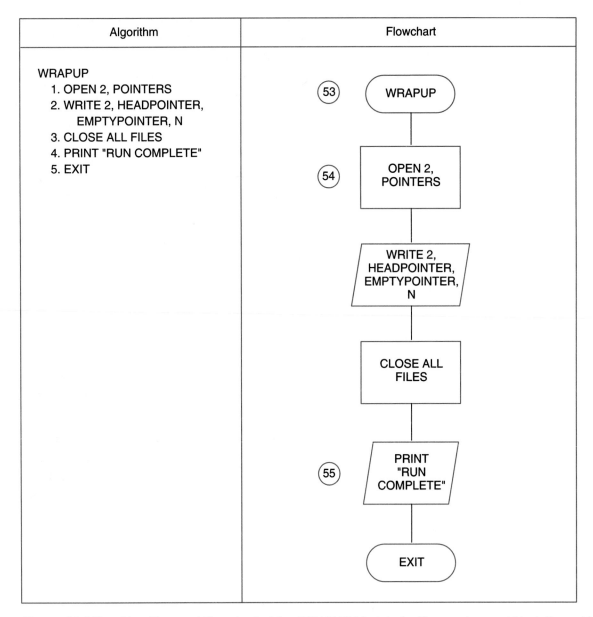

Figure 10.13k Algorithm and Flowchart of the WRAPUP Module for Processing and Updating a Linked List

36. The DELETE module allows the user to delete a record.
37. The search key of the record to be deleted is entered and the SEARCH module is processed to find the correct record.
38. A test is made to see if the record was found. If it was not, an error message is printed, and the module is exited.
39. If a record was found, the link fields need to be updated. A test is made to see if the deleted record is the first record in the ordered list.
40. If it is, the HEADPOINTER is set equal to the LINK field value.
41. If not, the LINK field value is saved. The previous record is read. The LINK field value is set equal to the saved link value. The previous record is written to the file, and the record to be deleted is read.
42. Next, the empty linked list needs to be updated. A test is made to see if the EMPTYPOINTER is equal to 0. If it is, the EMPTYPOINTER is set equal to the record number to be deleted and the LINK field is set equal to 0.

43. If the EMPTYPOINTER is not equal to 0, the EMPTYPOINTER record is read. The LINK field value is saved. The LINK field is set equal to R. The EMPTYPOINTER record is written to the file. The record to be deleted is read and the LINK field is set equal to the saved link. The record to be deleted is written to the file. The module is exited.

44. The CHANGE module allows the user to change data in a record.

45. A search key of the record to be changed is entered and the SEARCH module is processed to find the correct record.

46. A test is made to see if the record was found. If it was not, an error message is printed, and the module is exited.

47. If the record was found, the data are changed, and the record is written to the file. The module is then exited.

48. The REPORT module processes a linked list in order.

49. The current record number is set equal to the HEADPOINTER.

50. A loop is set up to continue until the next record to be processed is 0.

51. The record is read, processed, and printed.

52. The current record is set equal to the LINK field value and the loop continues until the value of the link field is 0.

53. The WRAPUP module completes the program.

54. The POINTER file is opened. The values of the HEADPOINTER, EMPTY-POINTER, and N are written to the POINTER file. All files are closed.

55. A message telling the user that the program run is complete is printed to the monitor. The module is exited.

The theory of linked lists can be expanded to include circularly linked lists, doubly linked lists, and multilinked lists. Circularly linked lists have no end. The last record is linked to the first record. This allows a search to continue from the last search, without starting at the first ordered record again. A doubly linked list has two linked fields. One is in ascending order, and the other is in descending order. Multilinked lists have multiple link fields, one for each way the file needs to be ordered.

Binary Trees

A tree structure uses a top down or hierarchical structure for the data. The tree structure looks like an upside down tree found in nature, with the root on top, the branches growing down and the leaves at the bottom. A genealogy chart is an example of a tree structure. Each record is called a *node*. The top position is called the *root*. In Figure 10.14, the node with the key value of "G" is the root node. A node at a higher level in the tree is called the *parent*. Referring again to Figure 10.14, the node with the key value of "D" is the parent of the nodes with the key values of "B" and "F". The nodes with the key values of "B" and "F" are called *children* since they are on a lower level in the tree. They are the children of "D". The bottom nodes are called *leaf nodes*. A branch consists of the chain of nodes from the leaf to the root. A subtree consists of the chain of nodes from a node to the leaves.

A **binary tree** is a tree in which each node has at the most two children. Each record of the file has all of the regular data, including a key, plus two other fields. One holds the pointer to the left child node (called the *left node*) and the other holds the pointer to the right child node (called the *right node*). Figure 10.15 shows the data file that accompanies the tree diagram in Figure 10.14. Note that the root is midpoint at

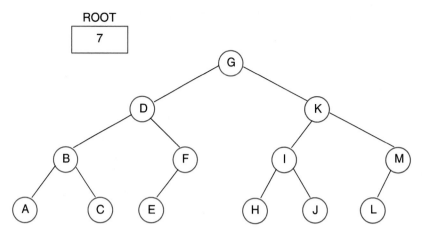

Figure 10.14 An Example of a Binary Tree

Record	Key	Data	Left Pointer	Right Pointer
1	A		*	*
2	B		1	3
3	C		*	*
4	D		2	6
5	E		*	*
6	F		5	*
7	G		4	11
8	H		*	*
9	I		8	10
10	J		*	*
11	K		9	13
12	L		*	*
13	M		12	*

Figure 10.15 File for Binary Tree in Figure 10.14

record 7. The left pointer for the root node is 4, and the right pointer is 11. These are shown as the children nodes of the root node in the tree diagram. The binary tree is in order when the left node is less than the current node and the right node is greater than the current node. Note on Figure 10.14 that this is always true.

Notes on Figure 10.16

Figure 10.16 shows the algorithm and the flowchart to create a binary tree. The following are the notes explaining the circled numbers on the flowchart for this procedure.

1. The CREATE module *creates* a data file with two pointer fields—the left pointer (pointer to the left node) and the right pointer (pointer to the right node).

Algorithm	Flowchart

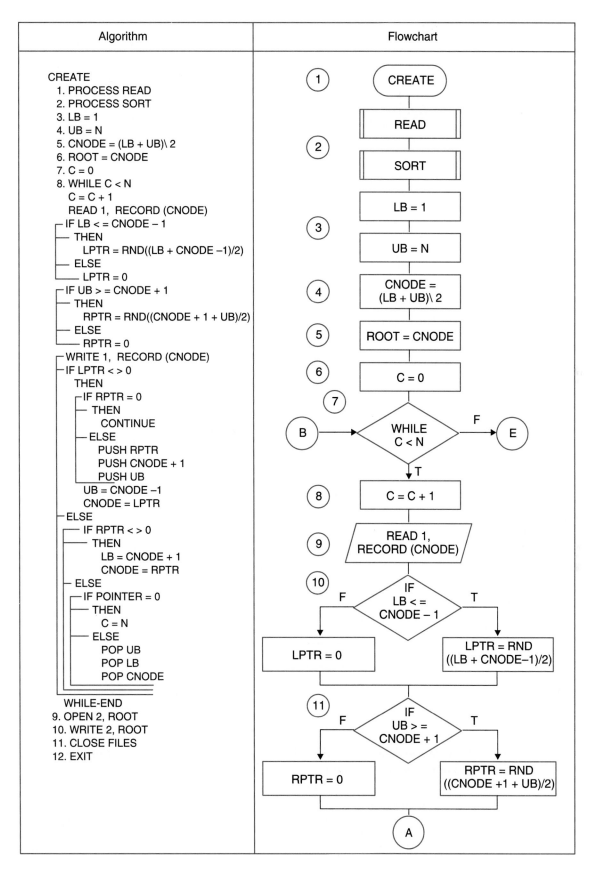

Algorithm

```
CREATE
  1. PROCESS READ
  2. PROCESS SORT
  3. LB = 1
  4. UB = N
  5. CNODE = (LB + UB)\ 2
  6. ROOT = CNODE
  7. C = 0
  8. WHILE C < N
     C = C + 1
     READ 1,  RECORD (CNODE)
   ┌ IF LB < = CNODE – 1
   ├─ THEN
   │      LPTR = RND((LB + CNODE –1)/2)
   ├─ ELSE
   └──── LPTR = 0
   ┌ IF UB > = CNODE + 1
   ├─ THEN
   │      RPTR = RND((CNODE + 1 + UB)/2)
   ├─ ELSE
   └──── RPTR = 0
   ┌ WRITE 1,  RECORD (CNODE)
   ┌ IF LPTR < > 0
   │    THEN
   │      ┌ IF RPTR = 0
   │      ├─ THEN
   │      │     CONTINUE
   │      ├─ ELSE
   │      │     PUSH RPTR
   │      │     PUSH CNODE + 1
   │      └     PUSH UB
   │       UB = CNODE –1
   │       CNODE = LPTR
   ├ ELSE
   │  ┌─ IF RPTR < > 0
   │  ├─ THEN
   │  │     LB = CNODE + 1
   │  │     CNODE = RPTR
   │  ├ ELSE
   │  │  ┌─ IF POINTER = 0
   │  │  ├─ THEN
   │  │  │    C = N
   │  │  ├─ ELSE
   │  │  │    POP UB
   │  │  │    POP LB
   └──┴──┴    POP CNODE
     WHILE-END
  9. OPEN 2, ROOT
 10. WRITE 2, ROOT
 11. CLOSE FILES
 12. EXIT
```

Figure 10.16 Algorithm and Flowchart to Create a Binary Tree *(continued on page 266)*

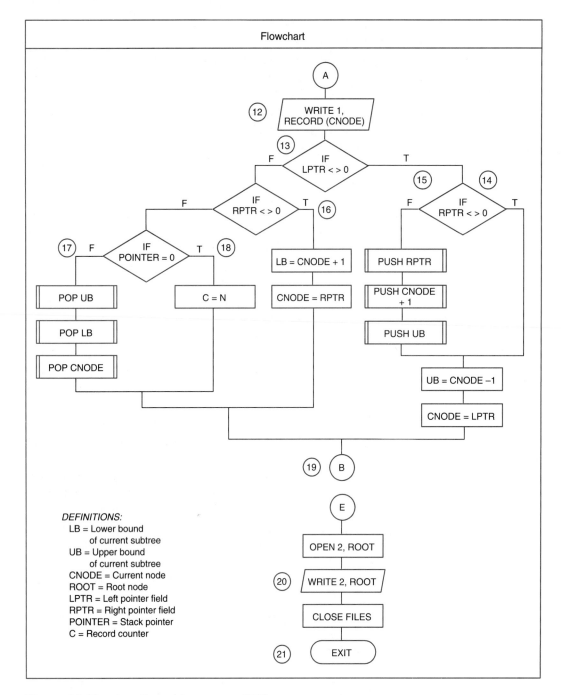

Figure 10.16 *(continued from page 265)*

2. Read the data for the file and count the number of records. Sort the file according to the key field.
3. Set the initial value of the lower bound (LB) to 1 and the upper bound (UB) to the number of records in the file (N).
4. Set the current node (CNODE) to the whole number average of the lower bound and the upper bound. This is an integer division and therefore always rounds down.

5. The root node (ROOT) is set to the current node.
6. The node counter (C) is set equal to 0.
7. A loop is set up to continue to process the nodes as long as the node counter is less than the total records in the file.
8. The node counter is incremented.
9. The current node is read.
10. A test is made to see if the current node has a left child. If it does, the left pointer is set equal to the integer average of the lower bound and the current node minus 1. If it does not, the left pointer is set equal to 0, which indicates there are no more left nodes.
11. A test is made to see if the current node has a right child. If it does, the right pointer is set equal to the integer average of the upper bound and the current node plus 1. If it does not, the right pointer is set equal to 0, which indicates there are no more right nodes.
12. The pointers have values, and the record can now be written to the file.
13. The next node needs to be found.
14. When the left pointer is not equal to 0 and the right pointer is not equal to 0, the new upper boundary is set equal to the current node minus one and the current node is set equal to the left pointer.
15. When the left pointer is not equal to 0 and the right pointer is equal to 0, the values of the right pointer, the current node plus one, and the upper bound need to be saved by pushing the stack. The new upper boundary is equal to the current node minus one and the current node is set equal to the left pointer.
16. When the left pointer is equal to 0 and the right pointer is not equal to 0, the lower bound needs to be set equal to the current node plus one and the current node is set equal to the right pointer.
17. When the left pointer is equal to 0 and the right pointer is equal to 0 and the stack pointer is equal to 0, all of the records in the file have been processed and the node counter is set equal to the total number of records in the file.
18. When the left pointer is equal to 0 and the right pointer is equal to 0 and the stack pointer is not equal to 0, the values of the upper bound, the lower bound, and the current node are popped from the stack.
19. The loop continues until the value of the node counter is greater than or equal to the number of records in the file.
20. The ROOT file is opened and the root record number is saved. All of the files are then closed.
21. The module is exited.

Figure 10.17 shows the tree diagram for the file found in Figure 10.18, which was created from the procedure in Figure 10.16.

One advantage of a tree structure is the *search* technique. Since all of the keys on the left of a node are less than the current node, and all of the keys on the right of a node are greater than the current node, the search can proceed down a branch, moving to the left or right according to the key. Using the file in Figure 10.18, the order of comparisons to find 30 would be 35, 15, 25, and 30. This would take four comparisons instead of six in a sequential search. To find the number 45, the order of comparisons would be 35, 55, and 45. This would take three comparisons instead of nine in a sequential search. As soon as the key value of the node equals the search key the procedure stops.

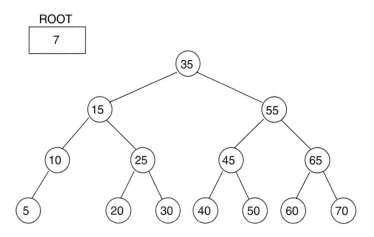

Figure 10.17 Binary Tree Created Using the Flowchart in Figure 10.16

Record	Key	Data	Left Pointer	Right Pointer
1	5		0	0
2	10		1	0
3	15		2	5
4	20		0	0
5	25		4	6
6	30		0	0
7	35		3	11
8	40		0	0
9	45		8	10
10	50		0	0
11	55		9	13
12	60		0	0
13	65		12	14
14	70		0	0

Figure 10.18 File for Binary Tree in Figure 10.16

Notes on Figure 10.19

Figure 10.19 shows the algorithm and the flowchart of the search module for a binary tree. The following are notes explaining the circled numbers in the flowchart of Figure 10.19.

1. The SEARCH module searches for a record given a search key.
2. The current node (CNODE) is set equal to the root node (ROOT).
3. The parent node (PN) is initialized to 0.
4. The search flag (S) is initialized to 0.
5. A loop is set up to process the file as long as the search flag is equal to 0.

Algorithm	Flowchart
SEARCH 1. CNODE = ROOT 2. PN = 0 3. S = 0 4.⌐ WHILE S = 0 READ 1, RECORD (CNODE) ⌐IF KEY = SKEY ── THEN S = 1 ── ELSE ⌐IF KEY < SKEY ─THEN CNODE = LPTR ─ ELSE CNODE = RPTR ⌐IF CNODE = 0 ─THEN S = 2 ─ ELSE CONTINUE WHILE–END 5. IF S = 1 THEN PRINT RECORD DATA ELSE PRINT "RECORD NOT FOUND" 6. EXIT *DEFINITIONS:* CNODE = CURRENT NODE PN = PARENT NODE S = SEARCH FLAG 1 = RECORD FOUND 2 = RECORD NOT FOUND KEY = KEY OF CURRENT RECORD SKEY = SEARCH KEY RPTR = POINTER TO RIGHT NODE	

Figure 10.19 Algorithm and Flowchart to Search a Binary Tree

6. The current node record is read.
7. If the key of the current node is equal to the search key, the record has been found. The search flag is set equal to 1, indicating that the record has been found.
8. If the key of the current node is not equal to the search key, the parent node is set equal to the current node and a new current node is located. (The value of the parent node is required if you are adding or deleting records.)
9. If the key of the current node is less than the search key, then the next node to be compared is found at the left pointer.
10. If the key of the current node is greater than the search key, then the next node to be compared is found at the right pointer.
11. If the new current node is equal to zero, the record could not be found. The search flag is set equal to 2, indicating that the record could not be found.
12. If the search flag is 1, the record data are printed; if not, a message is printed stating that the record could not be found.
13. The module is exited.

Records in a file need to be processed. If a report is needed, this processing will most likely need to be in order. However, a binary tree can be processed in any of several orders. The processing of each node once in an orderly manner is called tree **traversal.** Every *traversal* algorithm faces three choices at each node:

1. Process the node.
2. Follow the left branch (follow the left node).
3. Follow the right branch (follow the right node).

Whichever of these three choices is taken, the other two must be completed at a later time. It is the order of these choices that specifies the traversal method. Three traversal methods are as follows:

1. *Preorder:* The node is processed first, then the left branch, and then the right branch. This is abbreviated as NLR and represents the choices in 1, 2, 3 order.
2. *Inorder:* The left branch is processed first, then the node, and finally the right branch. This is abbreviated as LNR and represents the choices in 2, 1, 3 order.
3. *Postorder:* The left branch is processed first, then the right branch, and finally the node. This is abbreviated as LRN and represents the choices in 3, 2, 1 order.

The following list shows the order of the binary tree in Figure 10.14 when it is processed in each of the three methods:

1. *Preorder* (NLR): GDBACFEKIHJML
2. *Inorder* (LNR): ABCDEFGHIJKLM
3. *Postorder:* (LRN): ACBEFDHJILMKG

Notes on Figure 10.20

Figure 10.20 shows the algorithm and the flowchart to process an inorder traversal of a binary tree. The following are notes for Figure 10.20:

1. The REPORT module processes the binary tree using an inorder traversal method.
2. The stack is initialized to 0.
3. The current node (CNODE) is set equal to the root node.
4. The process flag is set equal to 0.

Algorithm	Flowchart

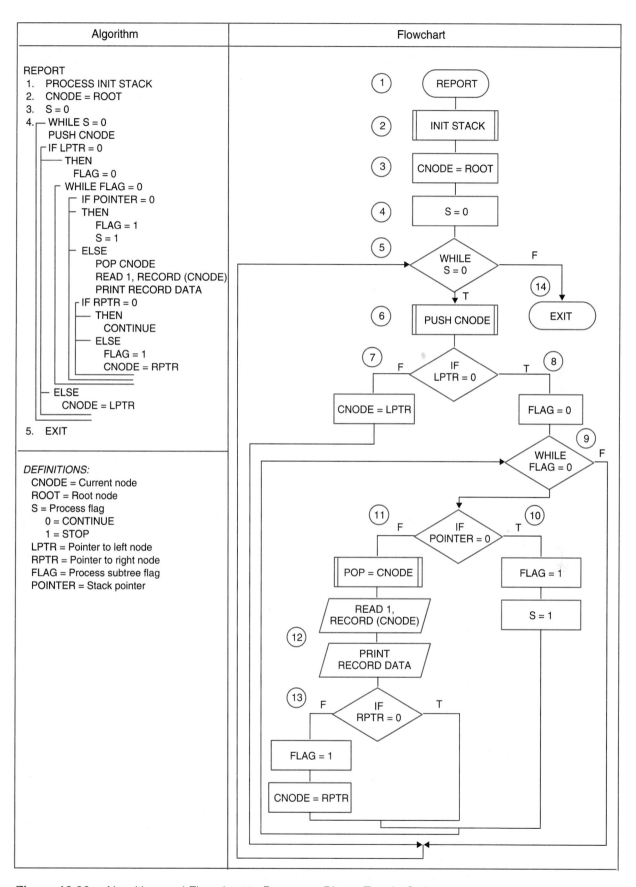

REPORT
1. PROCESS INIT STACK
2. CNODE = ROOT
3. S = 0
4. ┌─ WHILE S = 0
 │ PUSH CNODE
 │ ┌─ IF LPTR = 0
 │ ├── THEN
 │ │ FLAG = 0
 │ │ ┌─ WHILE FLAG = 0
 │ │ │ ┌─ IF POINTER = 0
 │ │ │ ├─ THEN
 │ │ │ │ FLAG = 1
 │ │ │ │ S = 1
 │ │ │ ├─ ELSE
 │ │ │ │ POP CNODE
 │ │ │ │ READ 1, RECORD (CNODE)
 │ │ │ │ PRINT RECORD DATA
 │ │ │ ┌─ IF RPTR = 0
 │ │ │ ├── THEN
 │ │ │ │ CONTINUE
 │ │ │ ├── ELSE
 │ │ │ │ FLAG = 1
 │ │ │ │ CNODE = RPTR
 │ ├── ELSE
 │ │ CNODE = LPTR
5. EXIT

DEFINITIONS:
 CNODE = Current node
 ROOT = Root node
 S = Process flag
 0 = CONTINUE
 1 = STOP
 LPTR = Pointer to left node
 RPTR = Pointer to right node
 FLAG = Process subtree flag
 POINTER = Stack pointer

Figure 10.20 Algorithm and Flowchart to Process a Binary Tree in Order

5. A loop is set up to process the file as long as the process flag is 0.
6. The current node is pushed into the stack. This is to keep track of where to return since the current node is the middle choice and needs to be processed.
7. If the left pointer is not zero, the current node is set equal to the left pointer.
8. If the left pointer is 0, either the end of the file is found, or a node is found that needs to be processed. A process branch flag is set equal to 0.
9. A loop is set up to process the nodes in the stack.
10. If the stack pointer is 0, all of the records have been processed. The process stack nodes flag (FLAG) is set equal to 1 as is the process stack flag (S).
11. If the stack pointer is not zero, then a node needs to be processed.
12. The stack is popped with the current node. The current node data are read and the data are printed.
13. If the right pointer of the current node is 0, the next node to be processed is in the stack. If not, the right node needs to be processed. The process stack nodes flag is set equal to 1, and the current node is set equal to the right pointer.
14. The REPORT module is exited.

The binary tree of Figure 10.20 is balanced since there is no more than one node difference between the number of nodes on the right from the number of nodes on the left. When nodes are added and deleted, the tree tends to become unbalanced. The more the imbalance, the less effective the tree becomes. Therefore, the tree either needs to be balanced as the nodes are added or deleted, or the tree needs to be balanced periodically. There are standard techniques for these processes. Because of the complexities of balancing a binary tree, this text does not present the algorithms and flowcharts to update a binary tree.

More complex trees use multilinked tree structure techniques. In a multilinked tree structure, one node can have many children. This fact adds much complexity to the techniques of processing and updating these files.

Other structures are also used with different types of lists. Included are **queues** and hashing techniques. A queue is a list that is added to at one end and used from the other end. That is, the first datum in is the first datum used. An example of a queue is waiting in line at the grocery store. The first person in the line is the first person out.

Another technique is **hashing.** An algorithm for a formula is devised to find the correct record to retrieve. The major problem with this method is that of creating an algorithm that produces a unique record number, since two records cannot have the same record number.

Summary

Many data structures are used in programming today. Those covered in this text include a single variable, arrays, and records in a file, which can be accessed sequentially or randomly. Sorting techniques are discussed in this chapter, since many of these data structures required ordering at one time or another. The techniques presented here include the exchange sort, the bubble sort, and the shell sort. Other data structures include stacks, linked lists, binary trees, etc. Stacks are lists of values that are entered at one end and used from the other end. A linked list is a file that has one or more fields that link the records in order. A binary tree file has two fields, one of which points to the left node and the other to the right node. The method of tree traversal is determined by how the nodes need to be processed. Other data structure techniques include multilinked trees, queues, and hashing.

New Terms

binary tree

hashing

linked list

queue

primary key

pushing the stack

popping the stack

sorting

stack

traversal

Questions

1. What is a stack?
2. How does a stack differ from a queue?
3. What is a linked list?
4. Does a linked list have to be in order? Explain your answer.
5. How does a circularly linked list differ from a singly linked list?
6. What is a binary tree?
7. Does a binary tree have to be in order? Explain your answer.
8. Under what circumstances would you use a stack?
9. Under what circumstances would you use a linked list?
10. Under what circumstances would you use a binary tree?

Problems

1. Identify the comparisons and the exchanges made to sort the following array using the selection exchange sort technique.

Number

23
45
13
5
56
74
10
30
65
40

2. Identify the comparisons and the exchanges made to sort the array found in Problem 1 using the bubble sort technique.

3. Identify the comparisons and the exchanges made to sort the array found in Problem 1 using the shell sort technique.

4. What would be the values found in a stack and the value of the stack pointer after the following executions:

PUSH 34

PUSH 54

PUSH 32

POP X

POP X

PUSH 10

PUSH 11

POP X

5. **a.** Using the following file and the Name field, create the linked list in the Link #1 field. Designate the EMPTYPOINTER and the HEADPOINTER for this link.

 b. Create a linked list for the debt in the Link #2 field. This will make the file into what is called a doubly linked file. Designate the HEADPOINTER for this link. Do you need another EMPTYPOINTER? Explain your answer.

Rec. #	Name	Debt	Other Data	Link #1	Link #2
1	DUNLAP	210			
2	JOHNSON	100			
3	WHITE	125			
4	GREEN	275			
5	SMITH	225			
6	ADAMS	25			
7	BLACK	400			
8	COOPER	50			
9	BENNET	350			
10	OLSON	75			
11	WILLIAMS	175			
12	NELSON	200			
13	HOLDING	335			
14	LITTLE	150			
15	GREY	250			
16	BROWN	325			
17	STONE	300			
18	SMART	40			

6. Show the following updates on the file from Problem 5:

Add:	Peterson	55
Delete:	Black	
Delete:	Adams	
Delete:	Stone	
Add:	Hudson	255
Add:	Neill	167

7. List the records that were compared when searching for Green. Use the file as updated in Problem 6.

8. Using the master file found in Problem 5, develop the two pointer fields for a binary tree. Draw the tree diagram for this file.

9. List the records that were compared when searching for Green. (Use the tree developed in Problem 8.)

Answer Problems 10–16 using the following data (add other data fields as desired):

1245	Hammer
1278	Screwdriver
1479	Saw
1098	Electric Drill
3001	Electric Saw
1174	Wrench
2023	Pliers
2488	Sander
5053	Sandpaper

10. Create a linked list using the inventory number for the order.

11. Add a linked list using the item description for the order.

12. Show the following additions to the file:
 a. 3423 Paint
 b. 3200 Paint Brush
 c. 6000 Drill Press
 d. 1000 Nails

13. List the records that would be compared when searching for
 a. 3200 Paint Brush
 b. 2488 Sander

14. Draw a binary tree using the both the data in the table and the additions in Problem 12.

15. Create a table for the binary tree in Problem 14.

16. Using the binary tree created in Problem 15, list the records that would be compared when searching for
 a. 3200 Paint Brush
 b. 2488 Sander

Chapter 11

Concepts of Relational Database Management Systems

Overview

RDBMS
Tables
Primary Keys
Normalizing Tables
Schema
Creating Tables
Queries
Interface Design
Reports
Planning a Solution Using a RDBMS

Objectives

When you have finished this chapter, you should be able to:

1. Understand the parts of a relational database created in a RDBMS.
2. Normalize relational database tables.
3. Create the schema for a relational database.
4. Design queries for a relational database.
5. Design forms for a relational database.
6. Design reports for a relational database.
7. Plan a relational database for a specific problem.

Relational Database Management System (RDBMS) software is a data organizer. The primary function of a RDBMS is to store large quantities of data, organize data, and print reports on all or selected parts of the data. This software is the key for many businesses. Business use of RDBMS includes inventory management, client and

sales tracking, and personnel management. Personal use of RDBMS includes lists of addresses, CD collections, recipe collections, home inventories, and book collections. When a problem arises that requires the organization of large amounts of data, the primary software solution is the use of RDBMS.

Programming is necessary to efficiently use a RDBMS. The language depends upon the database software in use. The modules that are necessary to develop depend upon the application and are tied to parts of the RDBMS. It is important for a programmer to understand database theory because databases are the key to many Internet and business problems and because it takes programming to efficiently implement a database as a solution to a problem.

RDBMS

A database is developed to store data. Just as the name implies, it stores data that needs to be retrieved at a later date. Items in the database that have one or more data values attached to it, such as an inventory item, an employee, a place, a thing, and so on, are called *entities*. Each entity has *attributes*. The attributes are data items that are required to be associated with the entity. For example, the attributes of a client entity might be last name, first name, street address, city, state, zip code, sales rep, and so forth. An entity set is a collection of related entities. An entity set may be inventory items, employees, clients, and so forth.

This terminology is standard for database theory. However, you will find other terms used informally by the developers of specific RDBMS. In this text we will refer to more informal names. An attribute is **a field.** An entity is **a record.** An entity set is a **table.** A database is a collection of related tables. For example, you might have a CD table. Each entity or record contains the attributes or fields about one CD. Each attribute or field contains one piece of data about the entity such as the name of the CD or its producer. Fields may contain entered numbers, text, images, hyperlinks, dates, or times, or may be results calculated using other fields.

Other components besides the tables are saved with the database, including queries, forms, reports, and macros (scripts). Queries are tables that contain selected data from the tables. Forms display data pertaining to one record on the monitor and may display interactive devices, such as buttons, to execute specific modules or requests pertaining to the database, such as next record, last record, print a report, and so forth. Reports are the printed output of the tables in the database. Macros or scripts are modules that are short sets of instructions to automate the database. These modules are activated through the use of button or other interactive device.

When starting a new database, you name the database before starting to develop it. Data are automatically saved as you add records to the tables. There is no need to save often or before you close. As you design your tables, forms, queries, and reports, you will automatically save each of them as part of the database. However, unless you want to save the database under another name, you do not need to go through a save procedure. All tables, queries, forms, reports, and macros are saved together under one database file name.

Tables

To reiterate the terminology of a database, data are organized into one or more tables or entity sets. **A table** is a group of related records or entities. Each record is an entity in the table. **A record** is a group of related fields or attributes. Each **field** or attribute is a

```
Database File Name: CD Collection

Table Name: CDs

Fields: → ID Number    Name         Producer         Date
```

Record 1	500	Winds	XYZ Corp	1998
Record 2	502	Whale Song	XYZ Corp	1990
Record 3	503	Songs of Ireland	ECL Publishers	2000
⋮				
Record N				

Figure 11.1 CDs Table

specific piece of data pertaining to the record. Each record in the table includes fields that are created in the design of the table. Each record is given a number. Each field is given a name. This name is the same for all records. Each table, query, report, and form is also given a name. You should name each according to what data they contain. The database is given a file name.

Let's examine an example of a database for a CD collection where only the CD names, producers, and dates of publication will be organized. See Figure 11.1. The database will be called CD Collection. There will be only one table called CDs. Each record, a horizontal row of data, will contain data in the four separate areas, called fields: the name of the CD, the artist, the producer, and the date of publication. These fields are named as follows:

Field Description	Field Name
Id Number	ID Number
Name of CD	Name
The Producer	Producer
Date of publication	Date

Records are numbered starting with one. Individual fields have the same number of characters in every record. For example, if 25 characters have been specified for the field called Name, then every record has a maximum of 25 characters to enter the name of the CD. You may have a derived field also called a computed field. The data in this type of field is derived, or computed from, other fields. For example, the retail cost may be computed from the wholesale cost and the percentage mark up.

Tables are holders of data to be organized and printed. In the planning stages of a database, it is important that all data required for the printed reports are part of one of the tables. Therefore, when you are developing a solution to a problem requiring a database, you should identify the information required for the reports. This sounds backwards; however, you need to know where you are going in order to know how to start. When you have your reports planned, then you may define your tables. The third step is

to plan how to get from the data to the report. Remember: plan your output, then your input, and finally your processing.

Primary, Secondary, and Foreign Keys

One field may be designated as a **primary key**. A primary key field contains unique data, that is, data that will not be repeated in the same field of another record. For example the CD id number (ID Number) would be a primary key in the CDs table. It is used in many projects to designate an identification number or an account number. You do not have to designate a primary key field. Many tables do not have a field that is unique. If you designate a field as a primary key, the RDBMS will not allow you to enter a value that is repeated from another record.

Secondary keys are not necessarily unique. They are primarily used to search for a specific record when you do not know the primary key. For example, in the CD example you may not know the ID number (ID Number), but you may know the name of the CD (Name). Therefore, you could search on the CD name.

A foreign key is a relational field to another table. The foreign key is a primary key in another table and determines the relationship between the tables. For example you may have information about the producers of the CD in second table. The producer's identification number would be the primary key in the Producers table. In order to find the producer information for a CD you need to have the producers identification number in the CDs table also. This would be a foreign key in the CDs table whereas it is a primary key in the Producers table. The field name is the same in the two tables. See Figure 11.2.

Normalizing Tables

For most problems, data should be separated into multiple but related tables. Let's revisit the CD COLLECTION database. This time there are more fields required, including the name, artist, and time length of each track on each CD. The address information for the publisher is also required. Figure 11.3 shows a single list of attributes, or fields, for a table if all data were to go into one table. This table has several problems. First, how many fields should be reserved for the music selections? Each CD has a different number of selections. Another problem arises when data is repeated in many records, such as the publishers' addresses. This repetition creates possible errors in entry as well as in table maintenance. Therefore, database tables should be normalized. **Normalizing** tables is separating data into multiple, related tables to increase efficiency as much as possible when entering, using, and updating data.

Other problems that may occur in database design are null data and data redundancy. Null data have no value, as all value is absent. Primary Keys cannot contain null data. A null data field is not a blank field because a blank field contains spaces, which are characters. Null values should be eliminated if at all possible.

Multiple values within a field occur when there is more than one value required for one field. Data repetition occurs when data is repeated in multiple records. Lost data occurs when pertinent data is deleted when a record is deleted. These three problems lead to three types of anomalies.

1. The insertion anomaly occurs when you cannot add a record because there is more than one value required for one field and the number of values for this field differ in each record, such as in the case of the music selections. If the

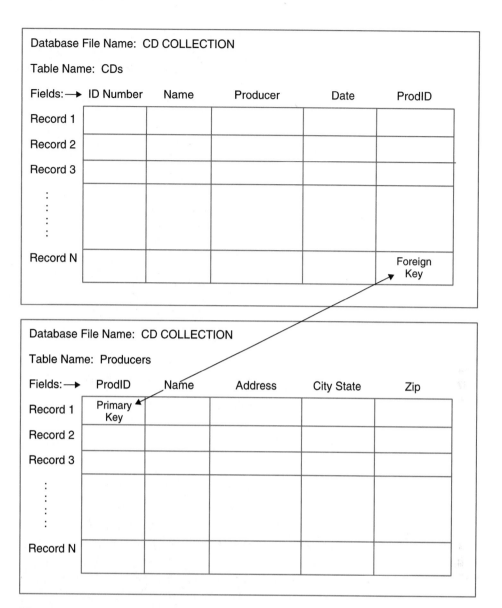

Figure 11.2 Primary and Foreign Keys for CDs Table and Producers Table

number of values for a field is the same for all records, the solution is to make multiple fields. However, as in the case of the music selections, there is a different number of values for each record. Therefore, you cannot designate different fields for each value because you do not know the number of values. This creates what is called an *unflat table*, a table that requires a different number of fields for each record. An unflat table is not allowed in a RDBMS. The solution is to create another table with a relational field tying the new table to the old table.

2. The update anomaly occurs when the same data must be changed in more than one record, as in the case of the publishers' addresses. Because there are considerably fewer publishers than CDs, there are many records to update when the address of a publisher changes. The solution is to create a table for the information on the publishers with a relational field tying the new table to the old table.

CDs Table

	CD Name	Selection	Artist	Producer	Date	Time	Address	City	State	Zipcode	Telephone
Record 1											
Record 2											
Record 3											
⋮											
Record N											

Attributes or Fields

CD Name
Selection
Artist
Producer
Date
Time
Address
City
State
Zipcode
Telephone

Figure 11.3 Attributes of the CDs Table

3. The deletion anomaly occurs when there is a loss of critical data when you delete a record. In the case of the CD collection, assume the publishers' information were to be kept in the original table. If only one CD had a specific publisher and that CD was deleted, you would lose the address information for that publisher. In contrast, if the address information is in a separate publisher information table, even if there were no CD's by that publisher in your collection, the address information would still remain in the publisher information table.

Normalization takes care of these problems. The first normalization takes care of unflat tables, or the insertion problem. The second normalization takes care of repeated data, or the update problem. The third normalization takes care of key data as non-key field of the file, or the deletion problem.

When you are normalizing a table, the first step is to create a list of fields or attributes, required for the output reports. See Figure 11.4a–c for the original field list, the list after the first normalization, and the list after the second normalization. There is no need for a third normalization. The original table shows all fields. Notice the fields that hold multiple data are simply named to represent the data, in this case Selections, Artist, and Time. The list is then divided into separate fields, such as the Artist into Artist first name and Artist last name, and grouped together like fields, in this case the CD data, the Selection data, and the Producers data.

The next step is to develop the first normalized form of your database. You will create a new table with one field that is identical to a field in the original table in order

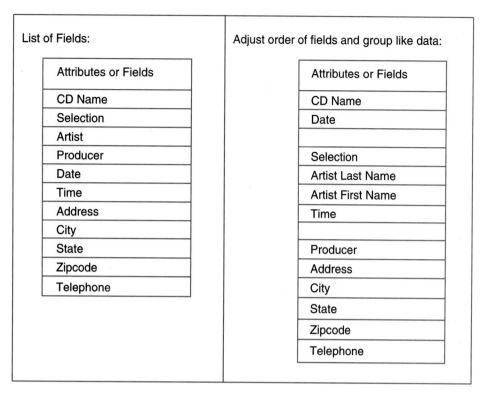

List of Fields:

Attributes or Fields
CD Name
Selection
Artist
Producer
Date
Time
Address
City
State
Zipcode
Telephone

Adjust order of fields and group like data:

Attributes or Fields
CD Name
Date
Selection
Artist Last Name
Artist First Name
Time
Producer
Address
City
State
Zipcode
Telephone

Figure 11.4a Normalization of CDs Table Original Data

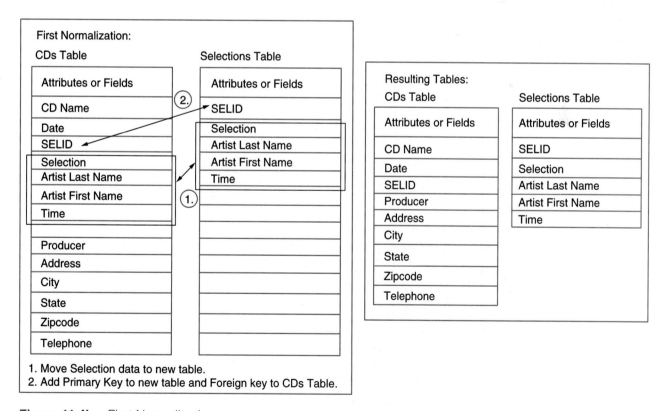

First Normalization:

CDs Table

Attributes or Fields
CD Name
Date
SELID
Selection
Artist Last Name
Artist First Name
Time
Producer
Address
City
State
Zipcode
Telephone

Selections Table

Attributes or Fields
SELID
Selection
Artist Last Name
Artist First Name
Time

Resulting Tables:

CDs Table

Attributes or Fields
CD Name
Date
SELID
Producer
Address
City
State
Zipcode
Telephone

Selections Table

Attributes or Fields
SELID
Selection
Artist Last Name
Artist First Name
Time

1. Move Selection data to new table.
2. Add Primary Key to new table and Foreign key to CDs Table.

Figure 11.4b First Normalization

1. Move Producer fields to Producers Table.
2. Add Primary Key of PRODID to Producers table and Foreign Key of PRODID to CDs Table.

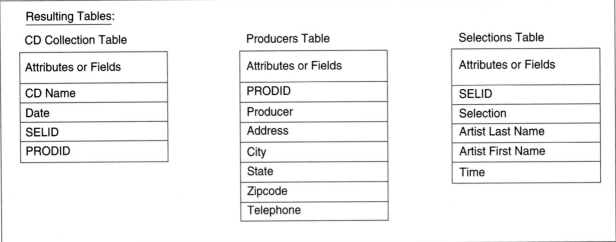

Figure 11.4c Second Normalization and Final Tables

to show the relationship between the two tables, in this case SELID. The field in the Selections table becomes the primary key. The field in the CDs table becomes a foreign key. The steps to complete the first normalization are as follows:

1. Move related fields that have more than one value to the new table. In this case move the fields associated with the selections to the new table, named Selections.
2. Add a primary key to the new table and a field with the same name as a foreign key to the original table. In this case add the primary key of SELID to the Selections table and the foreign key of SELID to CDs Table.

This will produce a new table and a modified table. See Figure 11.4b.

The next step is to develop the second normalized form of your database. You will be creating a third table with one field that is identical to a field in the original table to

show the relation between the two tables, in this case PRODID. See Figure 11.4c. The steps to complete the second normalization are as follows:

1. Move the related fields that are in more than one record to a new table. In this case move the fields associated with the Producers to Producers Table.
2. Add a field in the new table that will be a primary key and the same named field in the modified table as a foreign key. In this case add the primary key of PRODID to Producers table and foreign key of PRODID to CDs table.

There are now three tables.

The last step, if required, is to develop the third normalized form of your database. You would be creating a fourth table with one field that is identical to a field in the original table to show the relation between the two tables. Data to be moved from the first table are data that is not necessarily repeated in multiple records but is data that you want to save even if the record is deleted. To complete the third normalization, create a new relation field in both tables and move the required to the new table. In our CD COLLECTION database case, there is no data that requires a third normalization.

You now should have your tables in the most efficient form possible. If any of the normalizations require more than one new table, you would need to repeat the steps for that normalization. Complete all required new tables in each normalization before continuing on to the next normalization. You will probably rarely use all three normalizations for a solution to a problem; however, you should examine the original table for the possibility of all three.

Schema

The **schema** of your database shows the relationship between the tables. See Figure 11.5. Start with the normalized list of attributes for each table. Then draw a line between the tables connecting the relational fields. Add a one or an infinity sign on each end of the line to indicate the following:

1. One (1) to one (1) indicates that one record in one table relates to one record in another table. A one (1) appears on both ends of the line.
2. One (1) to infinity (∞) indicates that one record in the table on the one (1) side references to many records in the table on the infinity (∞) side. A one (1) is placed on the end of the line pointing to the table that references to one record and an infinity (∞) sign is placed on the other end.
3. Infinity (∞) to infinity (∞) indicates that there are many records in one table that are related to many records in the other table. An infinity sign (∞) appears on both ends of the line.

Figure 11.5 shows the schema for the CD COLLECTION database example. Notice that there are three tables, including the CDs table, the Selections table, and the Producers table.

Creating Tables

Data are entered into the tables through forms or a data view listing. The form shows one record on the monitor at a time. The data view shows multiple records but only as many fields as will fit on the monitor. You must scroll for additional fields. Tables must

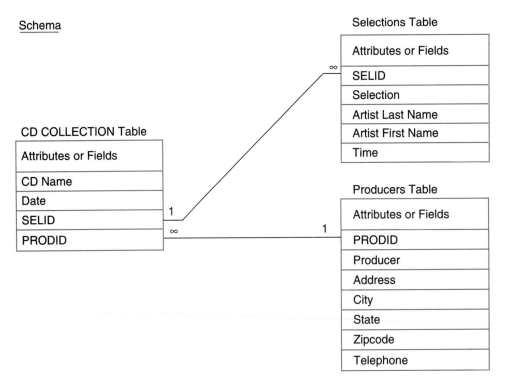

Figure 11.5 Schema of the CD COLLECTION Database

be created by designating the fields and their defining characteristics. Each field in a table has field characteristics, including the field name, the data type, the description of the field, the number of characters in the field (called the field size), the number of decimal places for numbers, format, input mask, and default value.

The name of the field should be long enough to readily identify the contents. However, it should be short enough so that the exact name can be remembered easily and the field header in the form or in the data view does not take up too much space. The field description describes the contents of the field. The field size is the maximum number of characters any entry will require. If the field contains a name, the field size may be 20 characters if all records that have names fewer than 21 characters. The computer will not allow you to enter more characters than the size of the field. If you have a name that is longer than 20 characters, you must increase the size of the field by changing the design of the table.

There are many data types for you to choose from, including text, number, currency, date, time, logical, memo, objects, and hyperlinks. A logical data type is a yes or a no in the field. You may use logical to indicate if your checks have been returned—yes if the check has been returned and no if not. A memo data type designates a field that can store more data than a text field. A memo field does not have a field size. An object, such as a picture, may be placed in a field. This would be an object data type. Hyperlinks to an Internet site may also be a data type. These data types allow for the organization of many different types of data. For example, a real estate office may want to include a picture of a house in their database of properties. Or you may be doing research and you want to save the hyperlink to a web site of the information your have found. There are limitations on some field sizes. The field size for the date, time, logical, memo, object, and the hyperlink data types have pre-determined field lengths.

The number of decimal places should be specified for a field designated as a number or currency data type. There are default values for the number of decimal places

Text	Standard Text Formats
Memo	Standard Text Formats
Number	General Number (3456.456)
	Currency ($3,456.45)
	Fixed (3456.45)
	Standard (3,456.45)
	Percent (34.56%)
	Scientific (1.34E+6)
Date	General (6/11/02 3:45:34 PM)
	Long (Sunday, August 20, 2000)
	Medium (20-Aug-00)
	Short (6/11/02)
Time	Long (3:45:34 PM)
	Medium (3:45 PM)
	Short (3:45:34)
Logical	Yes/No
	True/False
	On/Off

Figure 11.6 Database Formats for Fields

if you do not specify it. The format of the data may be specified. See Figure 11.6 for formats that may be available in your RDBMS.

An **input mask** allows the user to specify characters that are the same in a field in all records, such as the parentheses and dashes in a telephone number or the dashes in a social security number. An input mask decreases the number of keystrokes necessary for entering data. Default values may be specified, such as a state in an address list. For example, when most of the clients for a business are in California, then the default values for the State field would be set to CA. Again, this feature decreases the number of keystrokes in the data entry process.

Queries

Queries are the way to select fields and/or records to be processed. The result of a query is a query table that contains the selected data items required for an output report. When a query is made, you set up a query design that will select the required items from a specified table. Figure 11.7 shows an example of a query made on the

Field Name:	Name	Artist	Producer	Date
Table:	CDs	CDs	CDs	CDs
Sort:	X			
Show:	X	X	X	X
Criteria:				>1990
OR:				

Figure 11.7 Example of a Query

CDs table. Notice that the field names are in the top row, the table name is in the second row, the selected fields are in the third row, and the criteria "date > 1990" is in the fourth row. The fifth row is not used. This query will create a query table with records from the CDs table that have been published after 1990. You may then print this new query table through the use of reports.

The Query table is defined by using five rows of information. The first row contains the table name where the data for each field are found. The second row contains an X, which specifies the sort field for the new query table. The third row specifies the fields which should be included in the new table. The fourth and fifth rows specify the criteria for selecting records. All ANDs are in the same row, and the OR is in the fifth row. You will use relational operators to specify the criteria. Figure 11.8 shows examples of criteria. Example 1 shows one criterion: Cost is less than 25. Using this criteria, the new query table will include all records in which the cost is less than 25. Example 2 shows two criteria which both must be true in order to include a record in the function calculations. Notice that the criteria are on the same row. This is similar to the AND operator. The resulting table will include all records where the value of Cost is less than 25 and Quantity is less than 10. Example 3 also shows two criteria, but only one of these criteria must be true in order to include a record in the function calculations. Notice that the criteria are on separate rows. This is similar to the OR operator. The resulting table will include all records with Cost less than 25 or quantity is less that 10. Example 4 shows 3 criteria using both the AND and the OR logic. In this case, in order for the record to be included, the Cost must be >=10 and the Quantity must be less than 5, or the Cost must be < 10. The resulting table will include all records in which the Cost is less than 10 and all records in which the Cost is greater than or equal to 10 and the Quantity is less than 5. Notice the fourth row is the AND. The fifth row is the OR. When you want your selection to be based on two or more criteria and they all have to be met, then you place each criterion in the fourth row under the associated field name. When only one of the criteria must be met, then you place each criterion on different rows. Be very careful when you start mixing AND's and OR's to ensure you are asking for the correct records to be selected. Check your resulting table very carefully. If you identify more or fewer records than you had expected from the query, then you need to correct the design of the query.

Example 1:

Field Name	Item	Cost	Quantity
Table			
Sort	X		
Show	X	X	X
Criteria		<25	
OR			

Example 2:

Field Name	Item	Cost	Quantity
Table			
Sort	X		
Show	X	X	X
Criteria		<25	<10
OR			

Example 3:

Field Name	Item	Cost	Quantity
Table			
Sort	X		
Show	X	X	X
Criteria		<25	
OR			<10

Example 4:

Field Name	Item	Cost	Quantity
Table			
Sort	X		
Show	X	X	X
Criteria		>=10	<5
OR		<10	

Figure 11.8 Example of Criteria Ranges for Database Functions

The types of queries you will perform are determined by the project. You may want special sums, averages, or special updates. Just about any application that is used with a RDBMS requires some type of query.

Queries are developed through the use of an interface to the powerful Structural Query Language (SQL). SQL exists to facilitate the retrieval of selected data needed for processing. You may program your own methods to retrieving needed data that by learning to use SQL. Many RDBMS allow the users extensive use of SQL. This tends to decrease the user friendliness of the application for the developer while increasing its flexibility. The final results should create a database that is user-friendly for the end user. This is the job of the programmer. Other RDBMS use a programming language

such as Visual Basic to develop interfaces between the user and the database. The four logic structures are used to attach scripts or modules that contain short sets of instructions for interactive devices such as buttons or check boxes. An interactive screen may be developed to aid the user in the update and processing of the tables in a database. The developer of a database must know programming to be able to develop these scripts or modules.

Form Design

Forms allow the user to see all or some of the fields of a record on the monitor. The table is transversed through the use of buttons to go to the next, the previous, the last, or the first record. Derived fields may be added. Fields may be placed and moved to any location on the monitor. The data may be viewed or updated. New records may be added. Each field is labeled and has an input box that holds the data.

The form must be designed well to ensure the end user's productivity. Fields should be located on the form for easy reading and entering of data. If a source document is being used, the fields should be located on the monitor in the same order as on the source document. Key information should be located in one area. The form should be pleasant to look at and read, and pertinent information should be easy to locate. Add graphics only to enhance the locating and reading of data. Remember that the user of this form will be viewing it day after day. You want them to enjoy looking it.

Interface Design

Most solutions to business problems are designed to be accessed through an interface. A form with buttons is set up as an interface to automate the use of the database. This form is sometimes called a switchboard because it switches from one activity to another through the use of buttons. Chapters 12 and 13 present information on the design and implementation of interfaces. These interfaces are developed through the use of a programming language.

Reports

Reports are really the reason you have completed everything else. **Reports** are printed documents of the records in a specific database or query table. You may choose any or all of the fields to be included in your report. In the case of the CD COLLECTION database example, the report may include only the CD names and artists. You may sort them in any way you please. You may sort them by CD names, artists, date, or publisher depending upon the requirements of the problem. To design a report, you need to specify the report headings, the footers, the headers, the contents of each column, the width of the column, the format of the data, the column headings, the sorting criteria for the data, and the sub-totals which will be calculated. You also may specify images to be imported on the document and other line art as required by the specifications of the problem.

You may specify the column width as greater than, equal to, or less than the field length. If the data does not fit into the column it will word wrap within the column. To

avoid this, try to make the columns at least as wide as the field length. You do need to stay within the width of the paper on which you are creating the report. The number of characters across the page is dependent upon the font type, the number of characters, and the font size that you are using. Try to specify extra space for each column. You may have as many columns as you desire; however, the data must fit into the columns on the printed page.

Planning a Solution Using a RDBMS

This section presents forms to help you plan the development of databases. The first form allows you to list the required reports and the required data for each report. You should start with a list of reports and the corresponding data for the reports because this is how you will know what data is needed for the database. See Figure 11.9 for the report list form.

The next step is to list all of the data and create the field names. Figure 11.10 presents a form for entering the data items and field names. When you finish listing the data items and field names, you will need to normalize the original table to separate the data items into related tables. After you have normalized the database, then create the schema.

List of Required Reports:

Report Name	Data to be Included in Report

Figure 11.9 Lists of Required Reports Form

Original Table:		First Normalization New Table(s)		Second Normalization New Table(s)		Third Normalization New Table(s)	
Item	Field Name	Item	Field Name	Item	Field Name	Item	Field Name

Figure 11.10 RBDMS Table Development Form

Field Name	Field Name	Field Name	Field Name

Field Name	Field Name	Field Name

Figure 11.11 Schema Form

Figure 11.11 presents a form for developing the schema. Figure 11.12 presents a form for designing each table. In this step, the fields and their attributes are specified. You will need one of these forms for each table. After your tables are designed, you would design each of your database forms. You would designate the background, the fields that will be on the form, and their layout. Figure 11.13 presents a form to help design each form. Notice the place for the algorithms and flowcharts for your scripts that are attached to the form.

The next step is to design your queries. Figure 11.14 presents a form to help you. The last step is to design your reports. You would specify margins, headers, footers, report headings, column contents, column widths, and anything else that you want to specify for a specific report. See Figure 11.15 for a report layout form.

The last step is to develop your modules (macros or scripts). Use the algorithm and flowchart form to develop each module. Indicate on the algorithm and flowchart form the database form to which it is attached. An example of the form is found is Figure 11.16.

Table Name:_____

Field Name	Description	Data Type	Field Length	# of Decimals	Format	Input Mask	Default Value

Figure 11.12 Field Characteristics for One Table

Background:_____
Table:_____

Figure 11.13 Form Design

Table Name:_____

Field Name:							
Table:							
Sort:							
Show:							
Criteria:							
OR:							

Figure 11.14 Query Design Form

Report Layout:

Report Name: _____

 Table or Query Name:_____

 Margins:

 Top: _____ Top Printer: _____

 Bottom: _____ Bottom Printer: _____

 Left: _____ Left Printer: _____

 Right: _____ Right Printer: _____

 Header: _____ Top Printer: _____

 Footer: _____ Bottom Printer: _____

 Paper Size: _____Orientation: _____Color/B&W: _____

 Sort by: _____

 Subtotals on following fields: _____

 Images to be imported:

File	Description	Use	Location

Header Contents:

Footer Contents:

Report Heading:

Column Heading Field Name Column Width:					

Figure 11.15 Report Layout Form

```
┌─────────────────────────────────────────────────────────────┐
│ Macro Name: _____                      │
│ Attach to Form: _____                  │
├──────────────────────────────┬──────────────────────────────┤
│ Algorithm:                   │ Flowchart:                   │
│                              │                              │
│                              │                              │
│                              │                              │
│                              │                              │
│                              │                              │
│                              │                              │
│                              │                              │
│                              │                              │
│                              │                              │
│                              │                              │
└──────────────────────────────┴──────────────────────────────┘
```

Figure 11.16 Algorithm and Flowchart Form for Database Module

Example : Example RDBMS: Home Inventory

Problem: John has just purchased homeowner's insurance. His insurance agent has indicated that in case of a disaster he should have an inventory of everything he owns. He has just purchased a new computer with a RDBMS installed on it. John would like to use this for his inventory. He would like to keep the following information on his possessions:

1. Item
2. Original Cost
3. Room where it is located
4. Information about the store where the item was purchased
5. Appraised value of jewelry, art work, and antiques
6. Appraiser's name, address, and phone number

John would like the following reports:

1. An alphabetical listing of the items and all associated data
2. A list of the stores
3. A list of the appraisers
4. A form for the monitor to view each item
5. A list of items worth more than $500

Figure 11.17a–f presents the planning forms for this example.

This example involves an inventory for a household. This is a typical problem that arises for personal use of RDBMS. There are four reports and one screen form that are required. After normalization, there are three tables: Inventory, Store, and Appraiser. The schema is developed next. There are three tables with a relationship

Report Name	Data to be Included in Report
Inventory	item, original value, room, appraised value, store information, appraiser information
Stores	store information
Appraisers	appraiser information
Items Over 500	items, original value, appraised value
Form to show one item:	
Inventory	item, original value, room, appraised value, store, appraiser

Figure 11.17a List of Reports

Original Table:		First Normalization New Table(s)		Second Normalization New Table(s)		Third Normalization New Table(s)	
Item	Field Name	Item	Field Name	Item	Field Name	Item	Field Name
Item	ITEM	None		Table 2:		None	
Room	ROOM			Store Purchased	STORE		
Original Value	VALUE			Store Address	SADDRESS		
Appraised Value	APPVALUE			Store City	SCITY		
~~Store Purchased~~	~~STORE~~			Store State	SSTATE		
~~Store Address~~	~~SADDRESS~~			Store Zipcode	SZIPCODE		
~~Store City~~	~~SCITY~~			Store Phone	SPHONE		
~~Store State~~	~~SSTATE~~						
~~Store Zipcode~~	~~SZIPCODE~~						
~~Store Phone~~	~~SPHONE~~			Table 3:			
~~Appraiser~~	~~APPRAISER~~			Appraiser	APPRAISER		
~~App. Address~~	~~AADDRESS~~			App. Address	AADDRESS		
~~App. City~~	~~ACITY~~			App. City	ACITY		
~~App. State~~	~~ASTATE~~			App. State	ASTATE		
~~App. Zipcode~~	~~AZIPCODE~~			App. Zipcode	AZIPCODE		
~~App. Phone~~	~~APHONE~~			App. Phone	APHONE		
Rel. Field	SID						
Rel. Field	AID						

Figure 11.17b Table Development

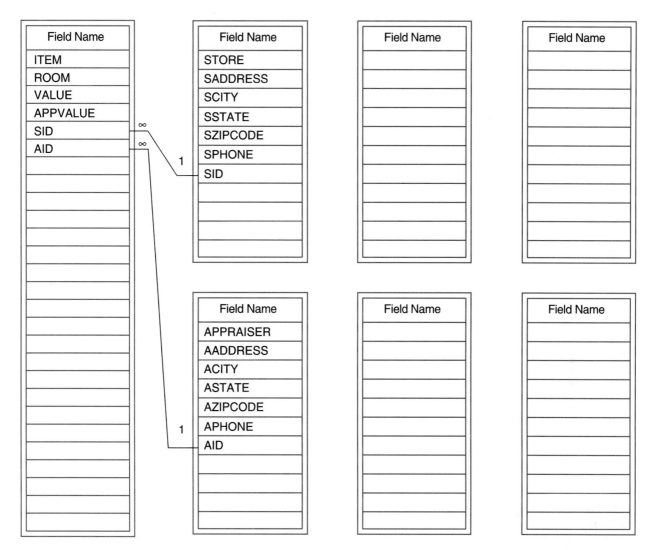

Field Name
ITEM
ROOM
VALUE
APPVALUE
SID
AID

Field Name
STORE
SADDRESS
SCITY
SSTATE
SZIPCODE
SPHONE
SID

Field Name

Field Name

Field Name
APPRAISER
AADDRESS
ACITY
ASTATE
AZIPCODE
APHONE
AID

Field Name

Field Name

Figure 11.17c Schema

between the Inventory table and the other two tables of many records to one record. There are many inventory items that were purchased at the same store and that had the same appraiser.

The fields and their characteristics for each table were developed next. Notice the input masks on the telephone numbers. The screen form for one inventory item was developed. The items were placed on the screen in the best viewing order. Report four required a query to select all items that cost or were appraised at more than $500. The reports then were planned. Report one lists the inventory and requires the Stores and Appraisers tables. Report two lists the store information from the Stores table. Report 3 lists the appraiser's information from the Appraisers table. Report 4 uses the query table to list the items that cost or were appraised at more than $500.

Databases must be maintained. This is a continuous and complex job which many times is handled by a person called Database Administrator or a Database Manager. The more complex the database, the bigger the need for a single person responsibile for maintaining it. Even at home, the database must be maintained. It is best to have one person in charge of that maintenance.

Field Name	Description	Data Type	Field Length	# of Decimals	Format	Input Mask	Default Value
Table: Inventory							
ITEM	Inventory Item	Text	10				
ROOM	Location	Text	10				
VALUE	Original Value	Number	10	2			
APPVALUE	Appraised Value	Number	10	2			
SID	Rel. Field	Text	5				
AID	Rel. Field	Text	5				
Table: Stores							
SID	Rel. Field	Text	5				
STORE	Store Name	Text	10				
SADDRESS	Store Address	Text	15				
SCITY	Store City	Text	10				
SSTATE	Store State	Text	3				
SZIPCODE	Store Zipcode	Text	10				
SPHONE	Store Phone	Text	15			(xxx) xxx-xxxx	
Table: Appraisers							
AID	Rel. Field	Text	5				
APPRAISER	Appraiser Name	Text	10				
AADDRESS	App. Address	Text	15				
ACITY	App. City	Text	10				
ASTATE	App. State	Text	3				
AZIPCODE	App. Zipcode	Text	10				
APHONE	App. Phone	Text	15			(xxx) xxx-xxxx	

Figure 11.17d Table Characteristics

Background:__Default____
Table:__Inventory/Store/Appraisers____

Item Name: | ITEM |

Room Where Located: | ROOM |

Original Value: | VALUE | Store Where Purchased: | STORE |

Appraised Value: | APPVALUE | Appraiser: | APPRAISER |

Figure 11.17e Form Design

Table Name:___Inventory_____

Field Name:	ITEM	ROOM	VALUE	APPVALUE	SID	AID	
Table:	Inventory	Inventory	Inventory	Inventory	Inventory	Inventory	
Sort:							
Show:	X	X	X	X			
Criteria:			>500				
OR:				>500			

Figure 11.17f Queries

Summary

Relational Database Management Systems (RDBMS) are data organizers. They allow the end-user to enter data, organize it, and print reports. Databases created through the use of a RDBMS have multiple parts, including tables that hold all the data, forms that display data, queries that select data, reports that output organized data, and macros that help to automate the database.

Tables are designed through analysis of the reports required. The data is then normalized into tables for the most efficient entry, use, and update of the database. A table is a group of related records. A record contains fields of individual pieces of data. The database schema shows the relationship between tables.

When designing a database, all of the parts are designed individually. The tables are designed by designating fields and their attributes. Forms are designed by creating a place on the monitor for each field. Queries are designed to select specific fields and records to be placed in a query table. Reports are designed in accordance with the problem requirements.

Problems that require the use of a RDBMS tend to be more complex than those that require other software. However, because the use of a RDBMS is key to any business, it is very important to learn how to design databases.

New Terms

Database

Field

Form

Input mask

Normalizing

Primary key

Query

Record

Relational Database Management Systems (RDBMS)

Report

Schema

Table

Questions

1. Name and explain the parts of a database.
2. What is a primary key?
3. When would you use a primary key?
4. Why would you not use a primary key? Give an example.
5. Why is a database table normalized?
6. Explain the process of normalizing your database table.
7. Explain a schema. Give an example of each of the types of table relationships.
8. Name and describe four attributes of a field. Include examples of each attribute in your answer.
9. Explain a query.
10. In Example 11.7, how would you change the query to reflect the following:
 a. Include in the report only the name, the artist, and the producer.
 b. Include only CDs by Tom Smith in the report.
 c. Include only Tom Smith CDs that were produced after 1985.
 d. Include all CDs that were produced after 1995 and all by Tom Smith.
 e. Include all CDs that were produced after 1995 and all Tom Smith CD's that were produced after 1985.

Problems

1. Bill Hammer owns a hardware store. He has a special catalog for customers that have indicated they would like to receive one. He has their names, addresses, and favorite department. He would like to have mailing labels printed for these customers. Bill also would like to have a list organized by favorite departments so he can add specials about the most popular departments. Plan a RDBMS that solves this problem.
2. Susan Students has an audio tape collection that she would like to catalog. She would like to have the following reports:
 a. An alphabetical listing of the names of the tapes and associated information (artist, and favorite listening place)
 b. An alphabetical listing of the artists and their tapes
 c. A listing of the tapes organized by where she most likes to listen to them (the car, bedroom, family room, while walking, or outside)
 d. A listing of all music selections and the tape on which each is found. Plan a RDBMS for this problem.
3. You are the secretary of the Computer Club. You would like to computerize your membership list. You have the name, address, telephone number, and e-mail address of each member. You also would like to add the committee to which each member is assigned and when the committee meets. You would like to have a report listing all members' information, a report that lists their names and e-mail addresses, a form that lists all information for each member, and mailing labels. Plan a RDBMS for this problem.

4. The New Plays Repertory Theater would like to computerize its ticket sales. The theater seats 500 people, offers seven plays each year, and sells season tickets and single-night tickets. The season tickets are all for performances at the same time during the week. Each play has 20 performances. The theater keeps a customer list of those people who buy tickets in order to send play information for advertising purposes.

The reports they would like to have are:
a. A listing of available seats
b. A listing of all plays and the actors in the plays
c. A listing of customers
d. Labels printed for all customers
e. A listing of season ticket holders

Plan a RDBMS for this problem.

5. You are planning a trip to France. You are compiling a lot of research on lodging, restaurants, and things to do in the cities you are interested in visiting. The data is beginning to overwhelm you. Therefore, you would like to organize it on the computer. The reports you would like include:
1. A listing of all lodgings and associated information
2. A listing of all restaurants and associated information
3. A listing of all activities and associated information
4. A listing of each city with the lodging, restaurants, and activities associated with it

Plan a RDBMS for this problem.

6. Sam Manager is the sales manager at the regional office of XYZ Gadgets, Inc. He needs to prepare an annual report that includes:
a. A listing of salespeople and their territory, sales, and number of clients
b. A listing of all clients
c. A listing of territories and the total number of sales, salespeople, and clients associated with each territory
d. A listing of the total sales per month

Plan a RDBMS for this problem.

Unit Three

Supplementary Exercises

Develop a complete solution, including a data dictionary, to the following problems:

1. Have the computer deal four hands of 13 cards each. The deck should be randomly set up. Check to make sure there are no duplicate cards in the deck. You will need to use the random number generator and arrays.

2. Given a two-dimensional table, look up a value given a number corresponding to the first row and a number corresponding to the first column. Accept only values that are equal to the values found in the table (see Chapter 9—Table Look-up Technique).

 Example: ROW_NUMBER = 25 COL_NUMBER = 100

 TABLE:

0	10	50	100	200	400
5	65	78	98	67	54
10	99	54	36	77	65
15	55	66	75	87	43
25	34	44	53	32	75
50	54	65	76	87	34

 THE ANSWER WOULD BE: 53

3. Modify the solution in Problem 2 to accept values of ROW_NUMBER and COL_NUMBER that are not exactly equal to one of the values found in the table.

4. The president of a college would like to know the age distribution of the students. Complete a solution using the frequency distribution technique to output the needed results.

5. Create a linked list of the following data. The data will be in ascending order of the temperature.

Day	Temperature
Nov. 28	33
Nov. 29	38

Nov. 30	30
Dec. 1	43
Dec. 2	48
Dec. 3	42
Dec. 4	40

Delete all November readings and add:

Dec. 5	35
Dec. 6	33
Dec. 7	40

6. Create a binary tree using all data found in Problem 5.

UNIT FOUR

OBJECT-ORIENTED PROGRAMMING

Chapter 12

Concepts of Object-Oriented Programming

Overview

Object-Oriented Programming
 Classes
 Objects
 Functions
GUI (Graphical User Interface)
Event-Driven Object-Oriented Programming
 Events
 Interactivity

Objectives

When you have finished this chapter, you should be able to:

1. Determine the difference between a procedural solution and an object-oriented solution.
2. Use object-oriented concepts in the design of a solution to a problem.

Computer hardware and software are becoming more and more complex as the industry moves into the new millennium. The graphic user interface (GUI) has added to this complexity. The faster computers process information and the more data they can manage, the more people expect of their computers and the more they demand user friendliness. As a result of GUIs and of increasing computer complexity, new strategies in programming design have been implemented. Object-oriented programming is one of these strategies.

Object-oriented programming (OOP) is a different way of approaching a solution to a problem. OOP methodology may be used to create procedural type programs, but more importantly, it lends itself to use with a GUI. Because the GUI is an interactive user interface that uses icons and menus, a nonlinear approach to programming is required. This is because the user has the option to choose any one of many things to do—save a

document, print a document, insert a picture, change attributes of a document, and so forth. Object-oriented programming allows a programmer to develop a nonlinear approach to a solution. When working with OOP instead of thinking about the various processes or the procedures of how to solve the problem, one looks at parts of the problem in terms of data: the properties of data, the attributes of data, how data work together, and how to develop functions that describe how data behaves. You will still work with modules; however, the modules will be related to the data parts and not the procedural parts of the solution. The result is a nonprocedural and a nonlinear solution to a problem.

This chapter will introduce new concepts and terms used in object-oriented programming. The chapter is divided into two parts. The first part will present concepts and terms used in object-oriented languages such as C++. The second part will present concepts and terms used in event-driven object-oriented languages such as Visual Basic and Lingo. Both are driven by data objects, messages, and events sent to the modules to complete the required processing, but the basic problem solving is different.

Object-Oriented Programming

Program design using object-oriented programming languages, for example C++, allows the programmer to look at the solution in terms of data parts instead of procedures. Consider a library book checkout system. In a procedural program, you would design the program with the procedures in mind:

- Input of the data—book name, book number, date, client, and so on
- Checking out the book
- Checking in the book
- Printout of all outstanding books
- Printout of all books
- Add, delete, and update book information

In an object-oriented program, you would design the program with the data parts in mind:

- Books
- Clients

Each part then would be assigned responsibilities:

- Books:
 Know its ID code
 Know the author, title, and other information about a book
 Check self out
 Check self in
 Know location of book
 List all books
 Add, delete, and change data on a book

- Clients:
 Know client ID code
 List books checked out for each client
 Add, delete, and update client data

There may be other data parts and other responsibilities, but you should be able to see the difference in the strategy of the solution. In order to complete the responsibilities, each part needs to interact with other parts. Now let's put some common names to the ideas presented.

Classes

A **class** is a collection of like items or objects. The class defines the members (objects) in the class. When you create a class you are essentially creating a new data type. Accounts in a bank would be a class; cars in a car dealership would be a class. In our library problem, the classes are BOOKS and CLIENTS. Associated with each class is a set of properties, structures, functions and/or behaviors. The BOOKS class contains the property definitions that define objects of that class: an ID code, location in the library, and so on. In addition, the class defines operations (functions) that can manipulate class objects such as the ability to check in and out, and so forth. The CLIENTS class would have the properties of client name, client address, client ID code, and others. An object is an instance of its class. For example, *Tom Sawyer* by Mark Twain is a book object, which belongs to the BOOKS class. Figure 12.1 shows a diagram of the BOOKS class. Superclasses may be created in which all subclasses have inherited characteristics. If our library example were to be expanded to include magazines, tapes, CDs, and other media types, we would create a superclass of library items. Each subclass adds characteristics and therefore becomes more specific. Another example might be the superclass of mammals that have specific characteristics that define a mammal. Humans are a subclass of the superclass of mammals and have inherited all the characteristics of the superclass of mammals, but have more characteristics that distinguish a human from another mammal. Figure 12.2 shows a diagram of the MAMMALS class, its members

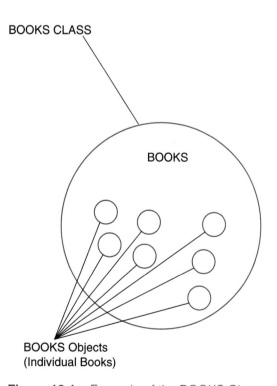

BOOKS CLASS

BOOKS

BOOKS Objects
(Individual Books)

Figure 12.1 Example of the BOOKS Class and Its Objects (instance of a class)

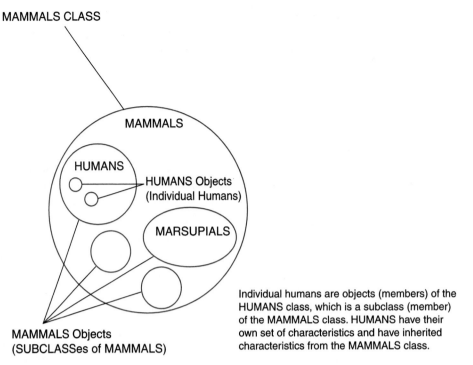

MAMMALS CLASS

MAMMALS

HUMANS

HUMANS Objects
(Individual Humans)

MARSUPIALS

MAMMALS Objects
(SUBCLASSes of MAMMALS)

Individual humans are objects (members) of the
HUMANS class, which is a subclass (member)
of the MAMMALS class. HUMANS have their
own set of characteristics and have inherited
characteristics from the MAMMALS class.

Figure 12.2 Example of the MAMMALS Class, Its Objects, Which Are Subclasses, and the Objects of the Subclasses

(objects) (HUMANS, MARSUPIALS, etc.), that are subclasses of the MAMMALS class, and the members or objects of the HUMANS class. Notice that there are classes (HUMANS, MARSUPIALS, etc.) that are members of another class (MAMMALS). This makes the HUMANS class, the MARSUPIALS class, and others, subclasses of MAMMALS.

The class also defines operations to manipulate class objects. These will be defined through the use of functions. The functions will be executed whenever they are called within a program. The function call includes the object and any data (parameters) required by the function. A function may be executed often in a single execution of the program and never used in another. For example, you may print several times when using a word processor on one occasion and never at another time. It is left to the user to decide what functions are needed during the running of the application and it is left up to the programmer to make sure all needed functions are part of the program. In our library checkout system, functions defined in the BOOKS class will define how to mark that the book is checked out, how the book is marked as checked in, how to update book data, how to add and delete books, and how to list the book's attributes. Functions defined in the CLIENTS class will define how to list all the books checked out for each client, how to print client data, how to update client data, how to add and delete clients, and so forth. These are all of the things that can be done with a specific book (a BOOKS object) or a specific client (a CLIENTS object).

Objects

An **object** is an **instance** (member) **of a class.** In our library checkout system example, each book would be an object of the BOOKS class and each client would be an object of the CLIENTS class. An object consists of one or more data values that define the object. An object may be a graphic, a person, a device, a magazine, a book, a tape, a video clip,

a sound clip, or some other data value, depending upon the class of which it is a member. It is what you, as a programmer, work with, in the the same manner as you work with a specific record in a file.

Functions

Developing the solutions for each function follows exactly the same steps you have been using for procedural programming, since each individual function is a procedural module. Each function is cohesive, and data are coupled in the same manner as was done in prodedural programming. While an object-oriented solution may be set up as procedural in nature, the power of object-oriented programming is in the way the program is controlled. In object-oriented programming, the user has almost complete control of the execution and the execution is controlled mainly through the use of icons, menus, and a pointing device. When an icon or a menu item is selected, the function for that item is executed. You may print a document at any time, not just when the document is completed. You may calculate at any time, even if there are no data. You may change your data at any time, before or after the calculation, not just at the beginning of the program. All this is possible when using object-oriented programming methods and a graphical user interface.

As a programmer, part of your job is to identify the classes and objects, and their attributes and functions. It also is your job to design the various interface screens and menus. Many functions are activated through the use of a menu. Because the solution is no longer procedural, that is, executed from the start of the solution to the last and controlled by the computer, it is your job to make the solution as user friendly as possible, allowing the user to execute functions whenever necessary. It also is your job to make sure the user does not execute a function at the wrong time which would produce destroyed data or garbage. You may want to develop safety windows that allow the user to back out of a function if he or she entered it at the wrong time. This non-procedural approach takes a different type of analysis and thought process. Once you are at the function level and designing the instructions within the function, the logic structures are the same as in a procedural solution.

GUI (Graphical User Interface)

A **graphical user interface (GUI)** is a computer screen designed for easy interaction between the user and the computer. Each item on the screen is an object with related responsibilities. A GUI may be designed using a language such as C++ or an event-driven language such as Visual Basic or Lingo. Chapter 13 presents information on interface design. Most applications today are driven by a graphical user interface because of an increase in the use of graphics and the speed of the computer. Computers are becoming more graphic oriented and people are demanding that the computer be increasingly user friendly.

Event-Driven Object-Oriented Programming

Event-driven object-oriented languages have many built-in capabilities to assist in the development of a GUI. Part of the language includes a relatively easy way to add and design interactivity tools, such as buttons, menus, input boxes, and so on, along with a

relatively easy method of adding scripts to these tools. Examples of these types of languages are Visual Basic and Lingo. These languages are easier to learn than an object-oriented language such as C++; however, they are not as flexible in the development of a solution to a problem.

Event-driven object-oriented programs use actions or events to trigger the execution of a set of instructions called a script. A script is another name for a module. The GUI is designed so that each item on the screen is an object with assigned properties. There are no programmer-defined classes. The object may be in the form of a button, a graphic, an user input box, an output box, a pull-down menu, a text box, a video clip, a sound clip, or anything else that has been placed on the screen. The properties of these objects may be such things as location on the screen, the name of the object, color, text formats, color palette used, designation of editable text, size, speed of video or sound, and so on. A script may be attached to each object. There is an overall program script that designates any actions that refer to the entire application, such as initializations, beeps of the mouse, or designating global variables (called public variables). Once the application starts, the user has complete control over the order of processing through the use of events, the GUI, a pointing device, and a keyboard.

Events

An **event** is something that happens in the computer. Some typical events are the following:

on mousedown
on mouseup
on keydown
on keyup

Each language has its own set of events. The only time the script for an event is executed is when the event happens. When a button is clicked, a set of instructions attached to the clicked object is executed when the mouse button goes down. A different set of instructions may be executed when the mouse button comes up. When a specific key is depressed or released, another set of instructions may be executed. In this case, the computer saves the value of the key. The programmer may use this information in the script. The program is executed though the use of events, many of which are directed by the user.

Interactivity

When using event-driven languages, **interactivity** exists between screens and/or between the user and the computer. Each screen becomes an entity to be designed and programmed. Interaction makes something happen. Interaction may cause something to happen on the screen, take the user to a new or a modified screen, stop the program from running, or create a printed document. This interaction produces a nonlinear application. For example, let's look at a problem to calculate the interest of an amount of money over several years. The user has to input the amount, the interest, and number of years. In a procedural solution, the program specifies the order in which the user enters the data and changes may or may not be allowed without starting over. In an event-driven object-oriented solution, there would be three user input boxes on the screen. The user may input the data in any order and change any of the data as necessary. The calculation does not take place until the user requests the answer through the use of a button—an event.

Interactivity requires that **navigational tools** be built into the program. These tools include hot spots, hot objects, time, different types of buttons such as a back button, a forward button, a jump button, a menu button, a quit button, and so on. It is very important

to plan navigation carefully. The user needs to navigate through the program with ease and with as little frustration as possible and must be able to quit at any time.

Summary

Object-oriented programming requires a different approach to the development of a solution to a problem. The programmer must look at the data parts and the events of the problem and not the processes of the problem. New terminology is used with object-oriented programming. A class is a group of items with designated attributes and functions. An object is an instance of the class. Subclasses and objects inherit the properties of the class of which it is a member. Classes must interact with each other to obtain needed data for functions to complete the responsibilities of the class. Event-driven object-oriented programming uses screen objects that are assigned properties and scripts. Interaction between screens is accomplished by events that happen when the program is executed.

New Terms

class

event

event-driven object-oriented
 programming

graphical user interface (GUI)

inheritance

instance of a class

interactivity

navigational tools

object

object-oriented programming

Questions

1. What is the relationship between a class and an object in object-oriented programming?
2. What is the difference between a procedural solution and an object-oriented solution to a problem?
3. What is a GUI?
4. Why is a GUI important to the development of solutions to problems today?
5. What is an example of an event?

Problems

1. Explain what classes you would use in the solutions to the following problems:
 a. A restaurant wants a scheduling system for their employees. The software will allow the restaurant owner to schedule employees for up to 40 hours in a seven day week. The restaurant is open from 6:00 AM to 10:00 PM. The system should include a check-in and a checkout system for the employees, a printout of employees' work hours, and the ability to change hours if necessary.

(Problem 1 continued on next page)

b. A college computer lab manager wants a system to check in and check out students using the lab. The system should include the ability to check student in and out of the lab, to check lab materials in and out, to keep track of students' hours, and to track all materials that are checked out. The system should also have the ability to check to see if a computer is open before allowing the student to check in.

Chapter 13

Object-Oriented Program Design

Overview

Designing an Object-Oriented Application
Interface Design
Designing an Event-Driven Object-Oriented Application

Objectives

When you have finished this chapter, you should be able to:

1. Determine the classes, properties, and functions for an object-oriented solution.
2. Design an object-oriented programming solution.
3. Design a functional, user-friendly user interface.
4. Design the storyboards for an event-driven object-oriented solution.
5. Determine the objects and the scripts needed for an event-driven object-oriented solution.

Designing object-oriented applications requires somewhat different tools and problem-solving strategies. Modules and interactivity take on a different meaning within an application. New tools need to be explored. These tools have not been standardized as much as the procedural tools that were introduced at the beginning of this text. The tools presented in this chapter include the output and input requirements chart; the user interface screen design chart; the class, properties, and functions chart; the attributes of class properties chart; interactivity diagram; storyboards; the object dictionary; and the interactivity chart for an event-driven object-oriented program.

Many people use UML (Unified Modeling Language) to design a solution using an object-oriented language. An overview of UML is presented in Appendix F. You may want to use that approach rather than the approach presented in this chapter. It consists of the development of several diagrams describing the solution. Both methods present ways of helping you design the most efficient solution to complex problems. Remember that these methods help you design the overall view of the solution and the way the final

program is executed. Each individual function uses the four logic structures and the various data structures presented in Chapters 4–11.

Designing an Object-Oriented Application

When working with an object-oriented language such as C++, there are certain steps you need to take in the basic design of the application. These steps are as follows:

1. Determine the requirements of the program in terms of output and needed data requirements.
2. Determine the classes of data, including the properties and their attributes, and the functions (operations) of each class.
3. Determine the interactivity between the classes and between the functions within classes.
4. Determine the instructions to accomplish each function.

A more detailed explanation of these steps follows, including a description of the tools designed to help develop each step.

Step 1: Determine the requirements of the program in terms of output and needed data requirements. From the given information, you will need to determine first what the output requirements are and design the user interface. Until you know what output is required and what the interface screens will look like, it is impossible to continue with an object-oriented solution. The output requirements will determine needed data to execute the program. The first part of this step is to use the *output and input requirements chart* to list all output requirements and all input values along with how they will be formatted when entered. Figure 13.1a shows a blank output and input requirements chart. Figure 13.1b shows this chart as it might be filled in for the library checkout system introduced in Chapter 12. The output and input requirements chart is similar to the problem analysis chart introduced earlier; however, with OOP you are not concerned with the processing as much as you are with the data and overall functions associated with the data. This chart is divided into three parts. In the first column, you should list the output requirements. From the output, determine the input required and list these items in the second column. In the third column, list how you will enter the input

Output Requirements	Input Requirements	Input Format

Figure 13.1a The Output and Input Requirements Chart

Output Requirements	Input Requirements	Input Format
List of Books	Book Data:	Screen for adding/deleting
Individual Book Data	ISBN	String, 15 characters, file
Change data—check in and check out books	Library Number	String, 15 characters, file
	Name	String, 25 characters, file
Add and delete books	Author(s)	String, 30 characters, file
List of Clients	Publisher	String, 15 characters, file
	Year of Publication	Date, 10 characters, file
	Area Location in Library	String, 10 characters, file
	Checked in or out	Logical, 1 character, screen, file
	Client ID checked out to	String, 5 characters, screen, file
	Client Data:	Screen for adding/deleting
	Client ID	String, 5 characters, file
	Name	String, 25 characters, file
	Address	String, 25 characters, file
	City	String, 15 characters, file
	State	String, 2 characters, file
	ZIP Code	String, 10 characters, file
	Telephone	String, 15 characters, file
	Books checked out: Library number and ISBN	Multiple entries (maximum 10) String, 15 characters, screen, file String, 15 characters, screen, file

Figure 13.1b The Output and Input Requirements Chart for the Library Checkout System

values—in an input box, as a result of a query, or some other method, along with any other information about the data.

The next part of this step is to design the user interface screen(s). A **screen** is the image on the monitor. You will need to place all labels, input fields, output fields, and

any graphics, text, and other formatting that will enhance the aesthetics of the screen(s). You may need to design a control panel or use buttons or hot spots or your output may look very similar to that of a procedural solution. Figure 13.2a shows a form you can use to draw a thumbnail sketch of the design. Figure 13.2b shows one idea of how a library checkout system screen might look. You may have designed the screen with a very different look. Design concept of user interfaces will be discussed later in this chapter. Without knowing anything about these design concepts, how would you change the thumbnail sketch to improve readability, ease of use, and aesthetics?

Step 2: Determine the classes of data. The second step that takes place in the design of the object-oriented programming solution of a problem is to break the data into classes. Figure 13.3a shows the *class, properties, and functions chart* used to define the classes along with the properties and functions for each class. The boxes in the first column are used to list each class as defined by the developer. Each box should contain one class name. The boxes in the second column are used to list the properties of class in the first column. There will be one or more properties for each class. The third column is used to list the functions for each class. Figure 13.3b illustrates a class, properties, and functions chart for the library checkout system. Notice that the two classes are BOOKS and CLIENTS. The properties are the data that are required for each BOOKS object or each CLIENTS object. The functions are a list of things that each object is required to accomplish. One of the functions of the BOOKS class is to be able to add or delete a BOOKS object. Another is to be able to register as part of the properties of the BOOKS class whether an object is checked out or not. Can you think of other properties or functions that are needed for this system? The properties and functions define the class.

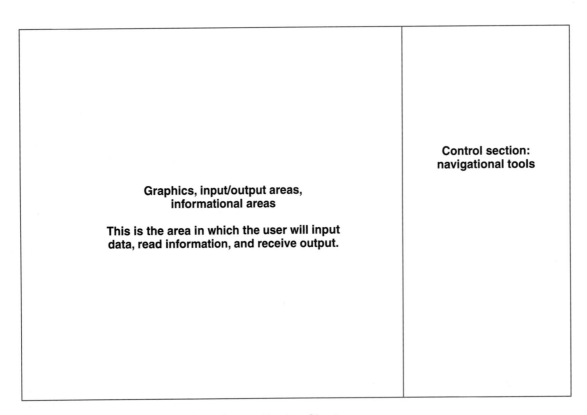

Figure 13.2a The User Interface Screen Design Chart

Figure 13.2b The User Interface Screen Design Chart for the Library Checkout System

Each property has attributes, such as the data type, the string length, and so on. Figure 13.4a shows the *attributes of class properties chart.* An attributes of class properties chart is created for each class. On the chart the class is listed at the top and the properties are listed on the left. The attributes for each property are listed in the area to the right of the properties. Figure 13.4b shows the library checkout BOOKS class attributes of class properties chart.

Step 3: Determine the interactivity between the classes and between the functions within classes. Interactivity will occur within the functions of the class and also between the classes. When a client is checking out a book, the CLIENTS class function of book checkout is executed. During this process, the BOOKS class function of updating data must be executed. The classes are interacting with each other. When the BOOKS class function of listing the BOOKS' object's properties and the book is checked out, the BOOKS class must interact with the CLIENTS class to obtain information about who has checked out the book. Also, the functions within the class may interact with each other. In the process of the function of book checkout in the CLIENTS class, the CLIENTS object's properties must be updated. Figure 13.5a shows a generic *interactivity chart for an object-oriented solution* to a problem. Notice that each of the classes is separate. The functions that define each class may interact within that class or they may interact directly with a function from another class. If a function from class 1 needs to interact with a function from class 2, the function from class 1 may directly access the class 2 function. Figure 13.5b shows the interactivity chart for the library checkout system. Notice the functions for each class and how the functions interact within the class and between classes.

Class	Properties	Functions

Figure 13.3a The Class, Properties, and Function Chart

Class	Properties	Functions
BOOKS	Library ID Number Name Author ISBN Number Publisher Year of Publication Checked out Client Checked out to Area in Library	Check out book Check in book List each book data Add and delete books Update book data
CLIENTS	Client ID Name Address City State ZIP Code Telephone Books Checked out	Add and delete books checked out Add and delete client Update client List client data

Figure 13.3b Example of a Class, Properties, and Function Chart for the Library Checkout System

Class: _____

Property	Attributes

Figure 13.4a The Attributes of Class Properties Chart

Class: _____BOOKS Class_____

Property	Attributes
Library ID Number	String, 15 characters, file, (Screen to add or delete or change)
Name	String, 15 characters, file, (Screen to add or delete or change)
Author	String, 25 characters, file, (Screen to add or delete or change)
ISBN Number	String, 30 characters, file, (Screen to add or delete or change)
Publisher	String, 15 characters, file, (Screen to add or delete or change)
Year of Publication	Date, 10 characters, file, (Screen to add or delete or change)
Checked out	String, 10 characters, file, Changed from Client Input
Client Checked out to	Logical, 1 character, screen, file, Changed from Client Input
Area in Library	String, 5 characters, screen, file, (Screen to add or delete or change)

Figure 13.4b Example of an Attributes of Class Properties for the BOOKS Class

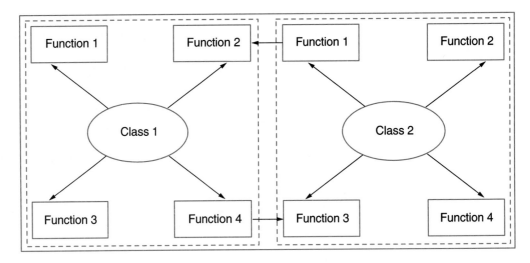

Figure 13.5a Interactivity Diagram for an Object-Oriented Solution to a Problem

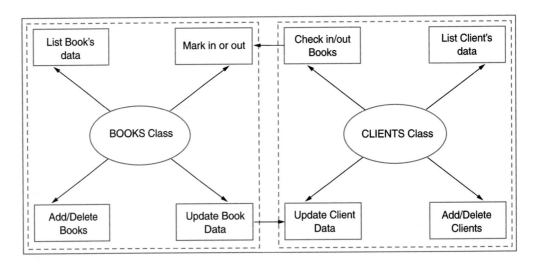

Figure 13.5b Interactivity Diagram for the Library Checkout System

Step 4: Determine the instructions to accomplish each function. Determining instructions to accomplish each function is the last step in the design of an object-oriented application. These instructions are similar to the instructions presented in Units One, Two, and Three of this text. Each language may have some specialized instructions. However, the four logic structures will be used in the same manner as in a procedural solution. The instructions for each function are small sets of procedural instructions. Each function is a cohesive module. The same rules hold true when using coupling methods such as global variables, parameters, and return values just as with the procedural approach to solution design.

Interface Design

A graphical user interface (GUI) is usually developed when using a language such as C++, and it must be developed when using an event-driven object-oriented language

such as Visual Basic or Lingo. The success of an application many times depends on the GUI design. There are five very basic rules you must follow when developing an interface design.

Rule 1: Never lose sight of the purpose or the message. The problem should have a purpose. Never lose sight of that purpose. Interface screens should be designed with this purpose in mind. The use of appropriate artwork, video clips and sound clips assures the enhancement of the message. If an appropriate object can not be found, leave it out. If you are designing an interface that introduces Alaska and you can only find palm trees, leave the trees out. Consider finding something else that is appropriate, such as a bear or a seal.

Rule 2: Keep it simple. With so many choices of clip art, photoclips, video clips, and other artwork available, it is easy to overdo the amount of material on the screen. The message is always the key to a successful application. If the message is lost in the glitz of other material on the screen, the application is a failure. Text should be kept at a minimum. People will not read a lot of text on a screen, even in a scroll bar text box. To increase readability, spread the text over several screens instead of cramming everything on one screen. In our complex world, many times simplicity is a welcome format.

Rule 3: Be consistent. Consistent does not mean constant. Screens should have the same look and feel throughout the application, but should not be boring. Subtle changes can be very attractive and eye catching. Different screens should use the same or look-alike fonts. A *font* is the typeface of the text. Fonts vary greatly in appearance. There are formal fonts, informal fonts, text fonts, title fonts, and so on. You want to choose your font carefully in conjunction with the purpose of the application. You would not want to use a different set of fonts on each screen. The user will get frustrated. Also, be consistent with colors. Develop a color scheme and use it throughout the application. Change the colors only when the change enhances the message.

Rule 4: Design the navigation with the user in mind. A program must be easy to navigate through. Be sure to include, on every screen, buttons to exit the program, to go back to the main screen menu, and to go back to the previous screen. It is important for the user to be able to go back to "home" when they get lost.

Rule 5: Understand when you can break the rules. Rules can be broken, but only for good reasons. If you break one of the first three rules, make sure the message has been enhanced and interest in the topic has been increased.

These rules allow the designer to create a functional, user-friendly interface. This is important, as a user will be more satisfied with a well-designed interface.

The application should have a visual style that is consistent with the purpose of the project. If the project is formal, the style should be formal. The fonts should be traditional fonts, the artwork should be serious and in conjunction with the theme of the project. If the project is informal, then more fun fonts can be used and the artwork can include some cartoons. However, even in the most formal project, a moment of levity many times is appreciated.

Each screen needs to be designed using the following elements: backgrounds, windows, navigational tools such as buttons and hot spots, input and output boxes, graphics, text, video, sound, and animation. The background of each screen should be similar to all other screens. Again, subtle changes add interest. Navigational tools should be placed in the same area on the screen. Navigational tools do not need to be buttons. Leaves on a tree, fish in a pond, rocks on the beach, houses on a street, all can be navigational tools. Buttons tend to be boring. Try to create new types of navigational tools. User input areas need to be large enough for the user to see and in a convenient and logical place on the screen to enter the data. The same holds true for the output. Text

should be limited to a few lines per screen or placed in a scrolled text box with a contrasting background for easy reading. Graphics, video, and sound should be added with caution and only to enhance the message or purpose of the screen. These items should add interest or should add information to data already presented.

A well-designed interface is pleasant to look at, functional, and provides easy access to all of the functions of the program. A well-designed interface allows the user to navigate through the program in a nonlinear mode, and should include the ability to quit the program at any time.

Designing an Event-Driven Object-Oriented Application

The steps to create an event-driven object-oriented project are as follows:

Step 1. Define the audience and the environment.
Step 2. Create the storyboards.
Step 3. Define all objects.
Step 4. Define the interactivity between screens.
Step 5. Define the scripts.

Step 1. Define the audience and the environment. In addition to the problem definition as part of the problem-solving steps as described in Chapter 1, the audience and the environment in which the project will be viewed must also be defined. The audience needs to be defined in terms of size, age, and knowledge. In general, the larger the audience, the fewer objects on the screen. More text and more objects can be placed on the screen if the viewer is sitting directly in front of the monitor as opposed to a situation in which the viewer is part of a 500-person audience. Terminology and graphics differ according to age and knowledge of the subject. Buzz words may be used if the audience is knowledgeable, but should not be used for an audience with minimal subject knowledge. The environment in which the project will be viewed is also important to analyze. Questions that need to be answered include what hardware and software will be used, what types of equipment are required, and what is the viewing area? Audience and environment analysis can make the difference between a successful project and one that is unsuccessful. The audience using the library checkout system would be the librarian on duty. She would know about the library and about the system; therefore, the interface can be simple, and buzz words may be used. Only one person at a time will be using the system; therefore, more information may be placed on the screen. The environment will be the library; therefore, the equipment needs to be quiet—no sound and a very quiet printer if needed. The system needs to be networked to the main computer files and needs to be dedicated to the library checkout system during all hours that the library is open.

Step 2. Create the storyboards. **Storyboards** define what is placed on the screens and the order of development of the screens. Figure 13.6a presents an example storyboard. Figure 13.6b shows a storyboard for the main screen in the library checkout system. In the box, the developer sketches the objects to be placed on the screen. Fine detail is not necessary, but all objects should be placed in enough detail so the programmer can use this document as a plan to start the computerization of the application. The lines allow for notes on the screen such as interactivity guidelines, object sources, animation effects, and so on. Each storyboard should have a reference number. Storyboards allow the developer to organize data, objects, and ideas.

Step 3. Define all objects. Screens are made up of objects. These objects may be developed in the software package or they may be imported from another source. Just as we needed a data dictionary when using a procedural language, we need an object dictionary for event-driven object-oriented languages. Figure 13.7a (also found in Appendix D) shows an object dictionary form. Figure 13.7b shows an example object dictionary for the library checkout system. Data that pertain to the object need to be placed in the object dictionary. This includes the name of the object, the source and/or file name, the file type and extension, the screens in which it appears, whether the object is a hyper-object or input object or output object, and whether or not there is a script attached to it. The name of the object identifies it as an object and can later be referred to by the name. This is similar to naming constants. The source and/or file name column identifies where the object was created or obtained. The object may be created within the program, or it may be obtained from a CD-ROM, or created in another package such as a paint package, or an animation package, or from a digital camera, or an audio tape or from many other sources. The name of the file is necessary so the file may be located. The file type and extension column defines the type file, such as a bitmapped file or a data file. The file extension defines the format file. The next column lists the storyboards or screens in which the object is found. The last two columns are check columns. The first indicates if this object is a hyper-object, an input object, or an output object, indicated by an H, I, or O. The last column indicates if there is a script attached to the object. A yes or a check mark indicates a script. The programmer can refer to this list when creating the screens on the computer. This will save time in trying to locate objects.

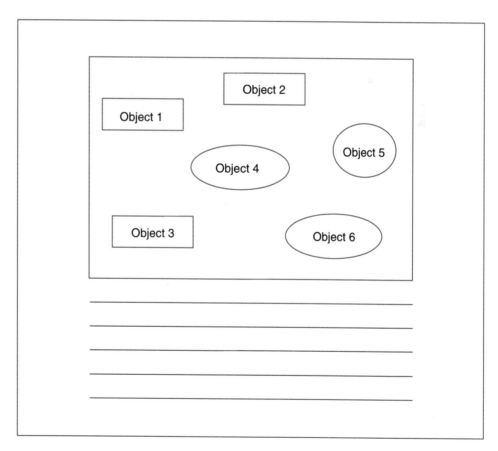

Figure 13.6a The Storyboard

Figure 13.6b A Storyboard for the Library Checkout System

Step 4. Define the interactivity between screens. The next step is to design the interactivity between the screens. Figure 13.8a shows an interactivity chart for an event-driven object-oriented solution to a problem. Figure 13.8b shows an example interactivity chart for the main screen in the library checkout system. An interactivity chart starts with the storyboard. There will be an interactivity chart for each screen. The storyboard is the center of the interactivity chart. Surrounding the storyboard are rectangles indicating the scripts and to which object each is attached. If a script calls another script,

Object Name	Source/File Name	File Type/Extension	Screen Number(s)	H/I/O*	Script

* H = Hyper, I = Input, O = Output

Figure 13.7a Object Dictionary Form

Object Name	Source/File Name	File Type/Extension	Screen Number(s)	H/I/O*	Script
Client ID			Main	I	Y
Book ID			Main	I	Y
Client Data			Main	O	N
Book Data			Main	O	N
Check In Button			Main	H	Y
Check Out Button			Main	H	Y
Add/Delete Client Button			All	H	Y
Add/Delete Client Button			All	H	Y
List Books Button			All	H	Y
List Clients Button			All	H	Y
Next Book Button			All	H	Y
Next Client Button			All	H	Y
Exit Button			All	H	Y
Man/Newspaper Graphic	ManReading	BMP	All		N
Background	Library Picture	TIF	All		N

* H = Hyper, I = Input, O= Output

Figure 13.7b Sample Object Dictionary for the Library Checkout System

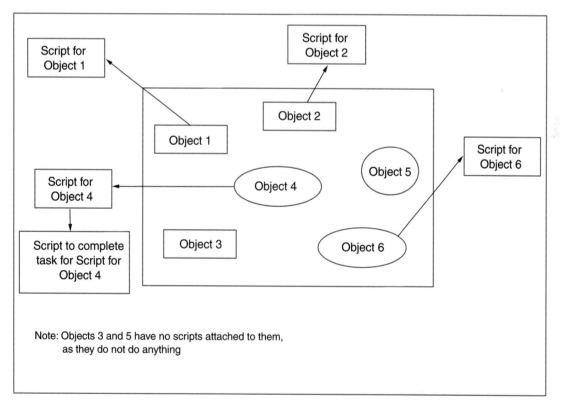

Note: Objects 3 and 5 have no scripts attached to them,
 as they do not do anything

Figure 13.8a Interactivity Chart for an Event-Driven Object-Oriented Solution

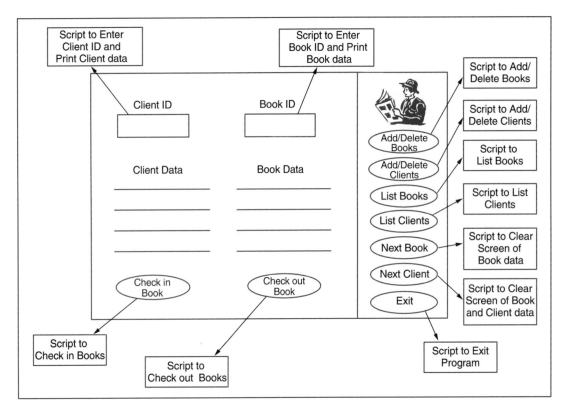

Figure 13.8b Example Interactivity Chart for the Library Checkout System

the interactivity between these scripts is shown (see Figure 13.8a). The interactivity diagram will not show the scripts that are separate from the interactivity, such as a script that clicks the mouse when a mouse button is pushed.

Step 5. Define the scripts. Scripts are short sets of instructions attached to objects or to the entire solution. These individual instruction sets are procedural in nature. However, the only time the instruction sets are executed is when an event happens. The logic structures presented in the earlier chapters in this text hold true for these scripts. Variables are made public (global) or they are kept as local variables. If the data is needed in another script, the variables must be public. There is no other way of coupling scripts.

Object-oriented programming allows easier handling of solutions for complex projects. It is a different way of approaching a solution to a problem. Object-oriented programming has grown in popularity and will continue to grow. However, procedural programming design is still prevalent in many industries. Students should learn to use both techniques and evaluate which is the best technique for the problem at hand.

Summary

There are several steps a programmer should follow when designing an object-oriented program. They include the following:

1. Define the classes.
2. Define the responsibilities for each class.

3. Determine the interactivity between the classes.
4. Determine the instructions to accomplish each responsibility.

Most programs today use a graphic user interface so the user can easily guide the program. A user interface needs to be functional, nice to look at, and easy to use. When designing an event-driven object-oriented program, follow these steps:

Step 1. Define the audience and the environment.
Step 2. Create the storyboards.
Step 3. Define all objects.
Step 4. Define the interactivity between screens.
Step 5. Define the scripts.

Tools introduced in this chapter include the output and input requirements chart; user interface screen design chart; class, properties, and functions chart; attribute chart; interactivity diagram; storyboard; object dictionary; and interactivity chart for an event-driven object-oriented program. A well-designed program allows for better efficiency in programming and maintenance.

New Terms

screen	storyboard

Questions

1. What is the purpose of the following charts:
 a. output and input requirements chart
 b. user interface screen design chart
 c. class, properties, and functions chart
 d. attributes for class properties chart
 e. interactivity chart for an object-oriented programming solution
 f. storyboard
 g. object dictionary
 h. interactivity chart for an event-driven object-oriented programming solution.
2. What are the steps to design an object-oriented solution to a problem?
3. What are the steps to design an event-driven object-oriented solution to a problem?
4. What are the rules to design a user interface?

Problems

1. Design a solution to the following problems using the four steps to design an object-oriented solution.
 a. A restaurant wants a scheduling system for their employees. The software will allow the restaurant owner to schedule employees for up to 40 hours in a seven day week. The restaurant is open from 6:00 AM to 10:00 PM. The system should include a check-in and a checkout system for the employees, a printout of employees' work hours, and the ability to change hours if necessary.

(Problem 1 continued on next page)

 b. A college computer lab manager wants a system to check in and check out students using the lab. The system should include the ability to check a student in and out of the lab, to check lab materials in and out, to keep track of students hours, and to track all materials that are checked out. The system should also have the ability to check to see if a computer is open before allowing the student to check in.

2. Design a solution to each of the problems in Problem 1 using the five steps to design an event-driven object-oriented solution.

Unit Four

Supplementary Exercises

1. Design a solution to the following problems using the four steps to design an object-oriented solution.
 a. A retail store wants an inventory system to track inventory in and out of the warehouse. The system should allow inventory to come into and out of the warehouse from outside the store and from within the store. It should allow a numbering system and have the ability to reorder items as needed.
 b. A catalog company wants to have a customer ordering system. This system should allow clients to phone in an order, have the computer process the order, update the inventory, and print out an order form that is also electronically sent to the warehouse.

2. Design a solution to each of the problems in Problem 1 using the five steps to design an event-driven object-oriented solution.

UNIT FIVE

FILE PROCESSING

Chapter 14

File Concepts

Overview

Sequential-Access Files
Random-Access Files
Buffers
Primary and Secondary Keys
Algorithm Instructions and Flowchart Symbols
Systems Flowcharts
Designing Records
Processing Sequential-Access Files
The Primer Read
Processing Random-Access Files

Objectives

When you have finished this chapter, you should be able to:

1. Explain the concepts of records, fields, and files.
2. Explain the concept of a primer read.
3. Explain the concept of sequential files.
4. Explain the concept of random-access files.
5. Develop a systems flowchart.
6. Design an input record.

Most large business solutions require record processing of some kind. One type of record processing is sequential access, designed so the computer can access records one after another in a sequence. Payroll, inventory, sales analysis, and other kinds of problems require sequential access to data. Another type of record processing is random-access or indexed sequential-access. Random-access processing is used in

query situations, which require that one record out of many be accessed without going through all of the records before it. This type of access is used to find information about a single record, such as the balance in an account, information on one sales receipt, or information on one inventory item.

This chapter presents file concepts for both sequential-access and random-access files. Chapter 15 presents techniques of using sequential files. Chapter 16 presents techniques of updating these files. Chapter 17 presents techniques of processing and updating random-access files.

Sequential-Access Files

To process **sequential-access files,** the programmer arranges the files in the order in which they are to be processed. Putting data in order is called *sorting*. Data can be sorted in a variety of ways. The most common sort modes are alphabetical order and ascending numerical order, sorting on one field. Another way to sort data includes ordering on two or more fields—one a major or primary field, the others, minor or secondary fields. For example, a mailing list usually needs to be alphabetical within the ZIP code. In this case the ZIP code would be the major field, and the last name (of persons on the list) would be the minor field. There are many processing methods of sorting. Appendix B presents several standard techniques for sorting, both for internally stored data (within the computer's internal memory) and also for externally stored data (on an auxiliary memory device). This chapter does not deal with sorting. We assume that the file is in the order to be processed. The file is usually sorted during the update process. The processing order of the data should be included in the internal documentation.

Random-Access Files

Random-access files are used for interactive and online processing. Random-access files enable the user to interact with the computer in order to retrieve the most current information in answer to specific inquiries as they arise in the workplace.

Random-access files can be updated interactively, or sequentially using transaction files, as with sequential files. If it is important for the data to be current at all times, then the update program must be interactive. Applications such as airline reservations and sales at ticket agencies must be interactive; the user must continually update the files with information about how many and which seats have been sold. Other applications, such as bank accounts, need to have the data available to the users at all times, but the files can be updated periodically. In these cases, the files need to be accessed interactively but can be updated either sequentially (through the use of a transaction file) or interactively (through random-access update techniques, presented in this chapter). Random files on microcomputers are normally updated interactively.

Buffers

All input and output operations work through buffers. Records are read from an auxiliary storage device into a buffer. Input can be used from a buffer or read from a buffer into the internal memory of the computer according to the designated language. Output is moved first to a buffer and then to the screen or printer. To save time, many languages

are able to direct the computer to retrieve values straight from the buffer. Changes to records are made in a buffer. This often involves extra or special *language-specific* instructions, which you will learn when you learn the syntax of particular languages. For problem solving, it is only necessary to realize how input and output instructions work.

Primary and Secondary Keys

A **key** is a value that is used in the solution either to arrange the file in order or to search through the file for a specific record. Each file has one or more key fields. A *key field* is the field in the record that is designated for the key. For example, the key field for a client file might be the field designated for the client number.

The *primary key* is the main key. This key is required to find a record in a random-access file. It should be unique so the computer can find an individual record. If there were two or more records with the same value for the key field, the computer would not know which record in that file the user wanted. Many times a field that is part of the data for the record will be unique, such as the client number for a client file. In this case, the programmer can designate this field as the key field. In other cases, such as a mailing list file, there are no fields that are unique. If no fields are unique, a primary key can be made up of a combination of all or parts of two or more fields. In a mailing list file, for example, none of the fields are necessarily unique. However, a combination of parts of the ZIP code, the last name, and possibly the first name would make the key unique, such as 95521SPRM. The programmer would use concatenation to combine these fields.

Sometimes a record is accessed by a key other than the primary key. These other keys are called **secondary keys.** They may or may not be unique. In the case of the mailing list, a user may need a list of people living in a particular ZIP code area. In this case, the ZIP code would be a secondary key.

In designing a record for a random-access file, it is important for the user to designate the primary key and secondary keys. The keys, both primary and secondary, are usually designed by the application and the needed output requirements of the problem. The keys are designated within the solution through the use of the fields; they do not have any special designation in the file. For example, the output requirements of a client-tracking problem might be, first, to search for one individual client; second, to output an alphabetical list of clients; and, third, to print out address labels in ZIP code order. The primary key would be the client number, since this is unique and would be used to search for an individual client. The secondary keys would be the name and the ZIP code, since they give the order in which the two output lists are to be arranged. In this case, the secondary keys probably would not be unique since more than one client may have the same name, and certainly more than one client would have the same ZIP code.

Algorithm Instructions and Flowchart Symbols

Several algorithm instructions and flowchart symbols have been covered so far in the text. In this chapter three more instructions will be added.

1. The OPEN LFN file name instruction tells the computer that the data in the file called *filename* is to be used, starting at the first record. LFN stands for **logical file number,** a reference number to be used in the solution. It is used in the READ instruction and the CLOSE instruction and refers to the file name in the OPEN instruction, which has the same logical file number. The

LFN is used since there can be more than one file open at one time and the computer has to know which file to read from and which file to close. It is easier to use the LFN than to write the name of the file each time you want to reference it.

2. READ (LFN) RECORD copies the next record into the internal memory or into a small holding memory called a *buffer memory*. The use of the buffer memory speeds up the input processing.

3. CLOSE LFN tells the computer that the file will not be used anymore. If a file is not closed, the data in the files may be lost. Once closed, the file may not be used again until it is reopened.

Table 14.1 includes all of the instructions and flowchart symbols covered so far.

Table 14.1 Algorithm Instructions and Flowchart Symbols *(continued on page 339)*

Algorithm Instruction	Flowchart Symbol	Definition (What the instruction tells the computer)
START module name (module number)		Start processing a module.
ENTER list of variables names		Enter data from a data block or from the keyboard into the following variable memory locations.
variable = expression		Assign the value of the following expression to a variable location. Called the *assignment instruction*. A module may be processed as part of an expression, if there is a return value.
WRITE list of variable names		Write to the printer the values represented by the following variable names.
PRINT list of variable names		Print on the screen the values represented by the following variable names.
PROCESS module name (list of parameters)		Execute the instructions in the named module and then return to execute the next instruction.
END/EXIT		Stop the processing of the module. Processing ceases with the End; with the Exit, processing continues with the next instruction after the PROCESS.

Table 14.1 *(continued on page 340)*

Algorithm Instruction	Flowchart Symbol	Definition (What the instruction tells the computer)
RETURN (variable)		Stop the processing of the module, place the value of the variable in the module name, and continue with the instruction continuing the module call.
IF/THEN/ELSE		When the condition is TRUE, use one set of instructions; when the condition is FALSE, use another set of instructions.
CONTINUE	or no box	A placeholder. Does *no* processing and continues to next instruction.
COUNTER = COUNTER + 1 (C = C + 1)		Increments the counter C.
SUM = SUM + VARIABLE (S = S + V)		Accumulates the values of the variable V.
WHILE/WHILE-END		Repeat a set of instructions as long as a condition is TRUE.
REPEAT/UNTIL		Repeat a set of instructions as long as a condition is FALSE.
LOOP: C = BEGIN TO END STEP S LOOP-END: C		Repeat a set of instructions using a counter C that starts at A(BEGIN) and increments by S(STEP) and C is greater than B(END).

Table 14.1 *(continued from page 339)*

Algorithm Instruction	Flowchart Symbol	Definition (What the instruction tells the computer)
CASE OF variable END-OF-CASE	◇	Select a set of instructions to execute according to the value of the variable.
OPEN LFN, FN	▭	Tells the computer the data in the file named FN is needed starting with the first record. (FN is given a reference number called the LFN).
READ (LFN) RECORD	▱	Copy the next record into the internal memory from the opened file LFN.
CLOSE LFN	▭	Tells the computer that the file corresponding to the LFN will not be used again.

Systems Flowcharts

A **systems flowchart** indicates the flow of data throughout the entire computer system. This includes where the data are coming from (keyboard, disk drive, etc.), the name of the program to be processed, and where the information is to go (disk drive, tape drive, printer, etc.). Figure 14.1 includes some of the more common systems flowchart symbols. These include the keyboard, a disk drive, a tape drive, a printer, a program, and a monitor. Most of these symbols can be found on a flowchart template. Figure 14.2 shows an example of a systems flowchart. This chart shows that the data are read from the disk drive and processed by the report program. The results are printed on a printer and also written to the disk drive.

Designing Records

Recall that a file is composed of records. A record contains all of the data on a given subject, such as all of the employment and benefit information for one employee. Records are composed of individual items, called *fields*—for example, one employee's withholding status for income tax.

A programmer beginning to design a solution to a problem must first identify the data to be put in the files. It is important to design the record carefully to make sure all of the necessary data are included and the lengths of the fields are correct. The length of

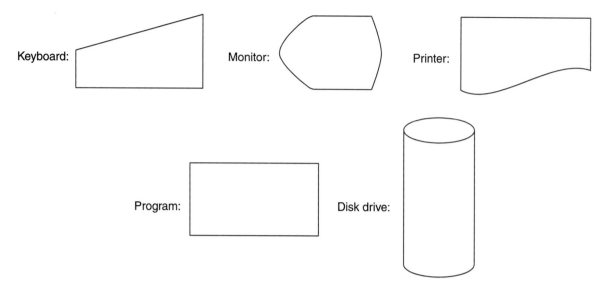

Figure 14.1 Systems Flowchart Symbols

Keyboard:

Monitor:

Printer:

Program:

Disk drive:

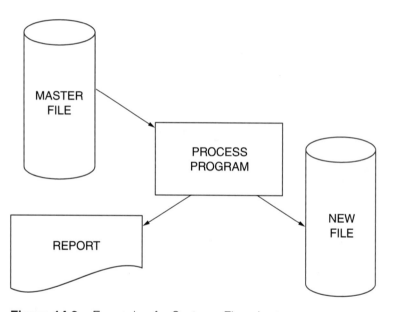

MASTER FILE

PROCESS PROGRAM

REPORT

NEW FILE

Figure 14.2 Example of a Systems Flowchart

a field is measured by the number of character spaces an item in the field occupies when it is entered in the computer. For example, the length of the field for the field name, ZIP-CODE_IN (*in* for *input*), would be 10 characters long, allowing space in the field for a nine-digit ZIP code plus the underscore. The order of the fields is also important. When you fill out the Given Data section of the problem analysis chart, you will identify all of the data needed to process the file. Use the Record Data Dictionary and Record Layout (Appendix D, Figure D.6) to identify the name, length, data type, and so forth of each data field, and to design the record. These fields will become the record. Figure 14.3 shows the record data dictionary and the design of a record, the record layout, for a mailing list file. All records in a file will have the number of fields identified in the record data dictionary, so you design only one record.

Record Data Dictionary

File Name: Mailing List

Field Number	Item Description	Field Name	Field Length	Data Type	# of Decimals	Comments
1	Last Name	LNAME_IN	15	C		
2	First Name	FNAME_IN	10	C		
3	Street Address	STADDRESS_IN	20	C		
4	City	CITY_IN	20	C		
5	State	STATE_IN	2	C		
6	Zip Code	ZIPCODE_IN	10	C		
7	Telephone Number	TELENO_IN	12	C		

Record Layout

1 15	16 25	26 45	46 65	66 67
LNAME_IN	FNAME_IN	STADDRESS_IN	CITY_IN	STATE_IN

68 77	78 89			
ZIPCODE_IN	TELENO_IN			

Figure 14.3 Record Data Dictionary and Record Layout for a Mailing List File

When you are designing a record, there are many items to consider: the number of fields and their order, names, data types and lengths, and the total record length.

1. The number of fields is the number of items you listed under the heading *Item Description* in the record data dictionary.
2. The order of the fields is determined by the order of importance of the data. Normally, the identifying fields will be first—those fields that will determine the order of the file, or the field that will be used to search for a specific record. Fields are also generally in the order of data entry, although this is not as important since the data-entry screen can be reordered.

 The problem will dictate the field used to order the file. If it is a mailing list, the field for ordering the file will be the ZIP code or the last name. The ZIP code field would be arranged numerically; the last name field, alphabetically. Also, the problem will dictate which fields will be used for a search. If the name of an individual is used to find the record, then the last name would

be the search field. The fields used to order the file and the search fields are usually at the beginning of the record. The order of data entry would then be considered for the remaining fields.

3. The names of the fields should correspond to the item descriptions of the data. Part of the name should also indicate that it is input. The total hours worked might be written as HOURS_IN, or the rate of pay as PAYRATE_IN. The field name is up to the programmer. Any name is acceptable as long as it is an understandable name and can be read and interpreted easily by another programmer. The field name is used as a variable name in the solution.

4. The data type of each field is determined by the requirements of the solution. If the data type is numerical, then specify the number of decimal places. Field data types include numerical (numbers), character string (alphanumeric), and logical (TRUE and FALSE).

5. The length of the field is largely determined by the number of characters in the field. When the field is string character data, the length is determined by the longest piece of data that would be entered. For example, the length of the field *Salesperson name* would be at least as long as the longest name that would be entered into that field—such as Gullickson or Stanislovsky, at 10 and 12 characters long, respectively. String data are normally left justified in the field with the remainder of the field on the right filled in with blanks. If the programmer did not design enough space in the field and the data are longer than the field, then the computer will truncate the data on the right.

 When the data type is numerical, the length of the field must be large enough to accept the number of digits in the largest number that might be placed in the field. The length must include spaces for the decimal point and a negative sign. If the field is not large enough for the number, the computer will either return an error or truncate the number on the left. If the number is to be in scientific notation, four spaces must be included for the magnitude of the number, which is E + nn. The letter E stands for *times ten to the power of.* It is followed by a plus or minus sign and a two-digit number indicating the power of ten. Numbers are normally right justified in the field.

 When the data type is logical, normally the field length is one character—for T or F. If the date data type is used, the field length is predetermined, normally in the form of MM/DD/YYYY.

6. Records are normally stored in external memory in blocks or sections of a specific length. Normally the length of a block is 256 characters, but systems may differ. It is more efficient to transfer sections of data or blocks of data at one time than to transfer a character at a time from external memory to internal memory. The length of a block is determined by the operating system of the computer system you are using, and therefore, the programmer has little or no control over this number. The total record length should be equal to or less than the number of characters in a block or else multiples of this number. That is, the total length of a record or block of records should be less than or equal to 256 or some multiple of 256. Planning record length this way saves space on the disk drive. If the length of a record is 260, for example, then each record will take two blocks of memory for storage. If the fields can be modified so the total length of the record is 256, which is only 4 characters less, the file will take half the space on the auxiliary storage device.

The following problem is used in many of the figures in this chapter and in Chapter 15 to illustrate record-processing techniques.

Problem: The vice president in charge of sales for a company needs to know the company's total sales during the last period. The sales are made by individual salespeople. The output needed is a list of all individual sales and the total sales for the company. Figure 14.4 illustrates the record data dictionary and the record layout for this problem.

Record Data Dictionary

File Name: Sales

Field Number	Item Description	Field Name	Field Length	Data Type	# of Decimals	Comments
1	Sales Date	DATE	8	STRING		MM/DD/YY
2	Sales person #	SN	3	STRING		
3	District #	DN	3	STRING		
4	Region #	RN	3	STRING		
5	Salesperson Name	SNAME	20	STRING		Last name first
6	Customer name	CNAME	20	STRING		Last name first
7	Customer Street	CSTREET	10	STRING		
8	Customer City	CCITY	8	STRING		
9	Customer State	CSTATE	2	STRING		
10	Customer Zip Code	CZIPCODE	10	STRING		
11	Customer Phone	CPHONE	14	STRING		(xxx)-xxx-xxxx
12	Sales Amount	SALES	10	REAL	2	

Record Layout

1 ... 8	9 ... 11	12 ... 14	15 ... 17	18 ... 37	38 ... 57
1 DATE	2 SN	3 DN	4 RN	5 SNAME	6 CNAME

58 ... 67	68 ... 75	76 ... 77	78 ... 87	88 ... 101	102 ... 111
7 CSTREET	8 CCITY	9 CSTATE	10 CZIPCODE	11 CPHONE	12 SALES

Figure 14.4 Record Data Dictionary and Record Layout for Company Sales Problem

Processing Sequential-Access Files

To process the records in a file by sequential access, either the WHILE/WHILE-END or the REPEAT/UNTIL loop is used. The processing is very similar with both of these loop structures. The WHILE/WHILE-END loop logic structure will be used throughout this chapter and Chapter 15.

When the user enters file data, a whole record is copied either into the computer's working memory or into a buffer memory. In sequential-access record processing, after a record is read into memory, it is processed and then the next record is read. This sequence continues until the computer encounters the **END-OF-FILE (EOF)** marker, at which time the processing stops. The EOF marker is a signal to the computer indicating that there are no more records in the file. When the END-OF-FILE marker is encountered, a status flag, or indicator, is set internally in the computer. This indicator is then tested as a logical data type function. The function name is EOF(LFN). The value of the EOF(LFN) function is FALSE until the END-OF-FILE marker is encountered. At this time the value of the EOF(LFN) becomes TRUE. When the EOF(LFN) function has a value of TRUE, all processing of records must stop. The END-OF-FILE is encountered during the READ instruction. The user should not do any processing after the EOF because there are no more records to be processed. If the EOF marker is processed as a record, the computer will either return an error, or the last record will be processed twice since no new data replaced the old. Therefore, within the loop, the Read cannot be the first module to be processed. It must be processed just before the processing of the condition of the EOF(LFN); therefore, it has to be the last module processed in the loop. This is the case when the READ statement actually sets the EOF status flag at the time of the execution of the READ. However, some languages look-ahead for the EOF. In this case, during the execution of the READ, the last record to be processed would be read and the status flag set at the same time. When the language uses the look-ahead method of finding the EOF, the primer READ is not used, and the READ statement is the first statement in the loop. This book will present file processing using the primer Read.

The Primer Read

Since the Read must be the last module processed, the computer needs data to process the first time through the loop. Therefore, there has to be a Read before the loop begins. This Read is called the *primer READ*. It primes the loop with data just as a pump is primed with water. Without this primer Read there would be no data to process the first time through the loop. If the READ were at the beginning of the loop, the EOF(LFN) would be processed as a record.

Figure 14.5 shows the Control module algorithm and flowchart for record processing. Notice the primer Read just before the loop. Also, notice the use of the logical EOF function. The condition to be processed is NOT EOF(LFN). The resultant of this condition is TRUE when the value of the function EOF(LFN) is FALSE, or when the end-of-file flag has not been encountered. This is because NOT FALSE is TRUE. When the END-OF-FILE marker is encountered, the value of the function is TRUE. The resultant of the condition, therefore, is FALSE since NOT TRUE is FALSE. The loop condition is read as WHILE NOT END-OF-FILE. The three modules processed in the loop are the Calc, the Print, and the Read modules. An alternative method is to incorporate these three modules into one module called the Proc module. The Proc module would

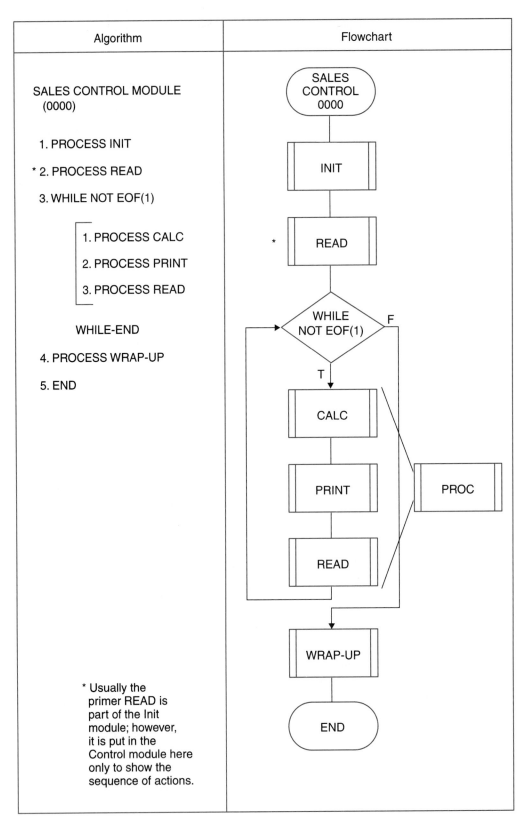

Algorithm	Flowchart
SALES CONTROL MODULE (0000) 1. PROCESS INIT * 2. PROCESS READ 3. WHILE NOT EOF(1) 1. PROCESS CALC 2. PROCESS PRINT 3. PROCESS READ WHILE-END 4. PROCESS WRAP-UP 5. END * Usually the primer READ is part of the Init module; however, it is put in the Control module here only to show the sequence of actions.	

Figure 14.5 The Control Module Showing the Primer READ

then process these three modules. In some languages this method is preferred because of the syntax of the loop structure.

Processing Random-Access Files

Random-access files are usually updated and processed (producing output reports through calculation and printing of data from each record) in the same program. Entire books have been written about how to store data in order to find it again in the shortest possible time. The method presented in this book is that of using index files. Other methods include:

1. *Linked lists.* The records are linked through an extra field in the record. This field points to the next record to be processed. A header number specifies the first record to be processed. The last record to be processed either has an end marker or points back to the first record again.
2. *Directories.* The file is subdivided into parts. Another file, which consists of two fields, holds the values of the record number of the first record in each subdivision and the key value of the last record in the division. Within the subdivisions the records are processed sequentially.
3. *Hashing.* Hashing methods require an algorithm using one or more of the fields of the record to calculate the record number.

These concepts were discussed in Chapter 10.

Summary

Record processing is a very important aspect of business applications on the computer. Record processing of sequential files embodies basic problem-solving concepts for the business programmer. The concepts of record processing are used in both the programming and the applications areas. When designing a record for a sequential-access file, there are many items to consider: the number of fields, the order of the fields, the names of the fields, the data type of each field, the length of each field, and the total record length.

Random-access files are becoming more and more important as business demands more up-to-date information. Random-access files are used any time data might be needed at a moment's notice or when data must be kept as current as possible. The program must be interactive because the user needs to "converse" regularly with the computer through the keyboard and monitor.

For the business user of computers, concepts of sequential-access file processing and random-access file processing are vital if the computer is to be used as a productivity tool.

New Terms

END-OF-FILE (EOF)

key

logical file number (LFN)

random-access file

secondary key

sequential-access file

systems flowchart

Questions

1. What is primer Read?
2. What is sorting?
3. What is the difference between a random-access file and a sequential-access file?
4. What is the LFN? Why is it used?
5. What is an EOF?
6. Can you efficiently process a sequential-access file randomly? Explain.
7. Can you process a random-access file sequentially? Explain.
8. Define a buffer and explain why it is used.
9. Define a key and explain why it is used.

Problems

The following problems will be presented again in Chapters 15 and 17. You are to design the input records and system flowcharts to be used in Chapters 15 and 17.

1. The Eureka Lumber Company would like to calculate the gross pay for its employees. The input records should include the following: name, address, Social Security number, pay type (H = hourly, S = salary), pay rate, and hours worked.

2. A sales manager wants a program that lists all of the company's customers according to the salesperson assigned to each. The input file contains the following data: the customer's name, address, and telephone number, and the salesperson's name.

3. The Up-to-Date Department Store needs to convert its inventory from a manual system to a computerized system. The input record should include the following: the inventory number, the department number, the store number, the item description, the quantity, the wholesale price, the markup percentage, and the vendor.

4. A sales manager would like to be able to track the clients of each of his salespeople. He would like to know when each customer was last contacted, the outcome of the contact (not interested, sale made, call back later, no contact made), the sale amount if a sale was made, and the name and number of the salesperson.

5. Design the input record for a video/sound inventory for a TV or radio station. (LP records, cassettes, CDs, and VCRs are all used by the station.) How would this file be ordered?

Chapter 15

Sequential-Access File Applications

Overview

Designing Output Reports
Headings and Line Counters
Control-Breaks
Multiple Control-Breaks
Using Indicators for Program Control
Error Handling
Null Files

Objectives

When you have finished this chapter, you should be able to:

1. Design an output record.
2. Develop a solution to process a sequential-access file.
3. Explain the concept of page control, including headings and page breaks.
4. Explain the concept of control breaks.
5. Explain the concept of indicators.
6. Explain the concept of error handling.

Most business computer applications require some type of file processing. In many such cases reports are generated for managers. These reports greatly aid the company's managers in making decisions for the present and future well-being of the company. Therefore the design of these reports should be well thought out. Some rules for good report design are presented in this chapter. Chapter 14 introduced the concept of sequential files. This chapter presents techniques of processing sequential files for business applications. Such techniques include page breaks, page headings, control-breaks, multiple control-breaks, indicators, error handling, and null files.

The problem to be used throughout this chapter was introduced in Chapter 14. The problem is as follows:

> The vice president in charge of sales for a company needs to know the company's total sales during the last period. The sales are made by individual salespeople. The output needed is a list of all individual sales and the total sales for the company. Each new record-processing technique covered in this chapter will be added to the previously developed solution.

Designing Output Reports

Output reports should be designed with the user in mind. When you are working on a problem solution, you record the information needed on the report in the Required Results section of the problem analysis chart. Where to place specific information within a report depends on the user's needs. The report should be designed so that the user can glean all the information needed in a matter of minutes. The basic rule for developing good reports is to plan to print only the information relevant to the person using the report. Any very important information should be highlighted in some way, so it is easily spotted. Time is valuable, and the computer is a productivity tool. The report should increase the productivity of the person for whom it was designed.

Within the report there are different types of lines. Heading lines are found at the top of the report or at the top of each page. The information found in these lines includes such items as company name, date and time of the execution of the program, programmer name, column headings, page number, manager's name, and the like. Detail lines print the specific information for each record in the file. Summary lines print information about groups of like records; summary lines are printed during a control-break and at the end of the report. Exception lines indicate an error in the data of the record. Exception lines can be either printed as part of the report or as a separate report called an *exceptions report*.

Report headings can also include relevant information about the company, such as its address, the division, the department and so on; the name of the report; the date; and any other pertinent information, such as the name of the person who will be using the report, the programmer, and so on, which is listed only once on the report. Print this report heading either on a page by itself or on the top of the first page. On the first page, as on all succeeding pages, print the column headings. Number all pages at the top, except for the first page with report headings. Using the Screen Design and Printer Design forms (Figures D.7 and D.8 in Appendix D), you can design output reports. The design of reports should be accomplished with the help and approval of the user.

The computer screen has 80 columns and from 24 to 26 rows. All of it can be used for input and output and for communication with the user. The screen design form can be used to design whatever has to be placed on the screen. Generally, the input and all inquiries and communication with the user are printed on the screen. All reports are generally written to the printer. Printers use paper that is normally 66 lines long and anywhere from 80 to 144 characters wide, although other paper sizes can be used. The printer form can be used to design the output reports.

Headings and Line Counters

To implement the report design, the programmer designs a few new modules. One is the Report Heading module, which writes the report heading. Another module writes the column headings. A third processes what needs to be done when one page is full and another page is to begin. This module is called a **page-break,** since there is a break in the processing in order to change the page.

The first thing to do is to incorporate each of these modules into the structure chart. The Report Heading and the primer READ are processed by the Init module. The Page-Break module is executed by the Print module since it is part of the printing process (see Figure 15.1).

In the Init module, the page number needs to be initialized to one and the report headings processed. Figure 15.2b shows the algorithm and the flowchart of the Init module. The sales file is opened first. Then the sales total is set equal to zero, the maximum line count (the number of lines to the bottom margin) is set equal to 60, and the page number is set equal to 1. Next, the primer READ is processed, and finally, the report headings are written to the printer.

In the Print module, the line count has to be tested to see if the end of the page is encountered. If it is, then the Page-Break module is executed. Figure 15.2e shows the algorithm and the flowchart of the Print module. Every time a line is written to the printer, it is counted in the line counter (LINECT). In the Print module, before anything can be written to the printer, the line count must be compared to the maximum line count. When the line count is greater than or equal to the maximum line count, then the page-break must be processed before the record is written to the printer. If it is not, the record is written to the printer and one is added to the line count.

Each of the additional modules—the Report Heading, the Page Heading, the Column Heading, and the Page-Break—must be developed. The Report Heading module gets the data from the system, prints the lines of the report heading, including the column headings, and initializes the line counter to the number of lines already used by the Report Heading. See Figure 15.2c to see the order of processing. Notice that the report headings are written first, and then the system date is retrieved and printed, the line count is initialized to 6, and the column headings are written to the printer.

The page-break adds one to the page counter, prints the lines of the page headings, and reinitializes the line counter to the number of lines used by the page headings. The Page-Break module also ejects the page, writes the Page Heading, and processes the Column Heading module. The Page-Break module is shown in Figure 15.2f. Notice that the page is ejected first. Next, the page number is incremented, and then the page headings, which include the page number, are written. The line count is set to the number of lines used by the page headings. Finally, the column headings are written.

The Column Headings module prints the column headings and increments the line counter by the number of lines the column headings use (see Figure 15.2d). Be sure to include the spacer lines in the total line count. Spacer lines are lines in which nothing is printed. These lines allow extra white space for readability.

Keep in mind the function of each module as you analyze what is processed in it. Figure 15.2a–g shows all the algorithms and flowcharts for the sales problem, including the headings and the page-breaks.

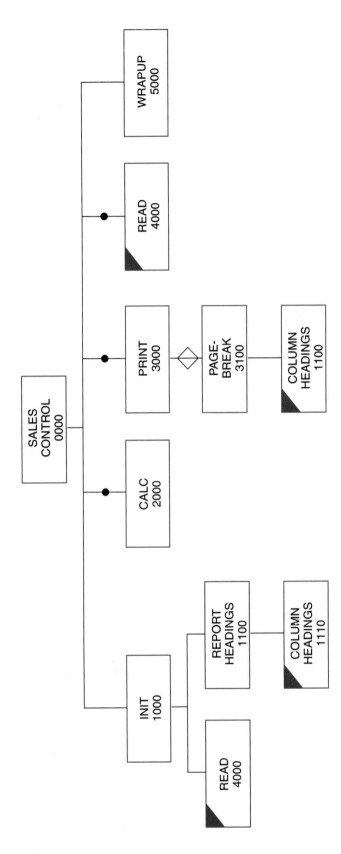

Figure 15.1 Interactivity (Structure) Chart for Sales Problem

Algorithm	Flowchart

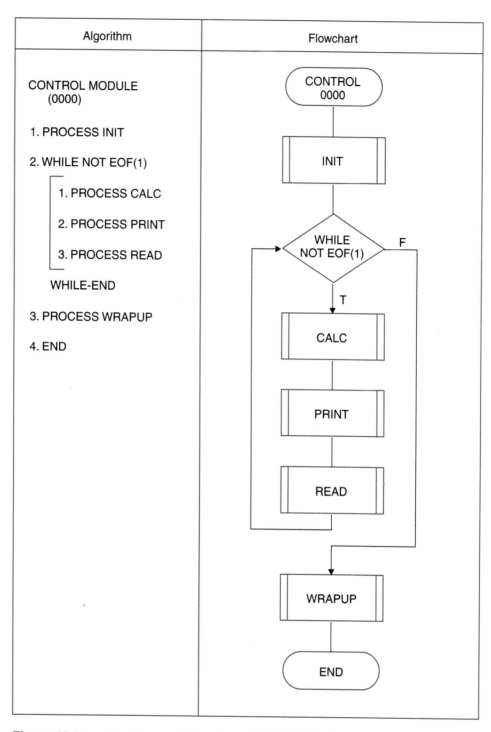

CONTROL MODULE
(0000)

1. PROCESS INIT

2. WHILE NOT EOF(1)

 1. PROCESS CALC

 2. PROCESS PRINT

 3. PROCESS READ

 WHILE-END

3. PROCESS WRAPUP

4. END

Figure 15.2a Algorithm and Flowchart of CONTROL Module for Sales Problem

Algorithm	Flowchart
INIT MODULE (1000) 1. OPEN 1, SALESFILE 2. SALESTOTAL = 0 3. MAXLINECT = 60 4. PAGENO = 1 5. PROCESS READ 6. PROCESS REPORT HEADINGS 7. EXIT	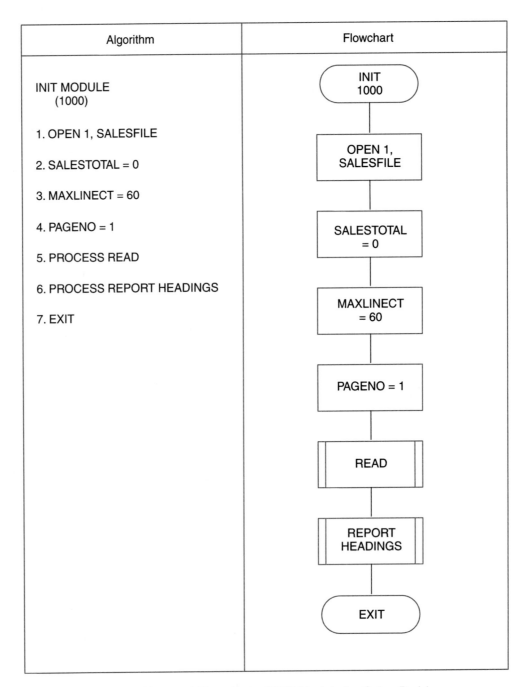

Figure 15.2b Algorithm and Flowchart of INIT Module for Sales Problem

Algorithm	Flowchart
READ MODULE (4000) 1. READ RECORD 2. EXIT	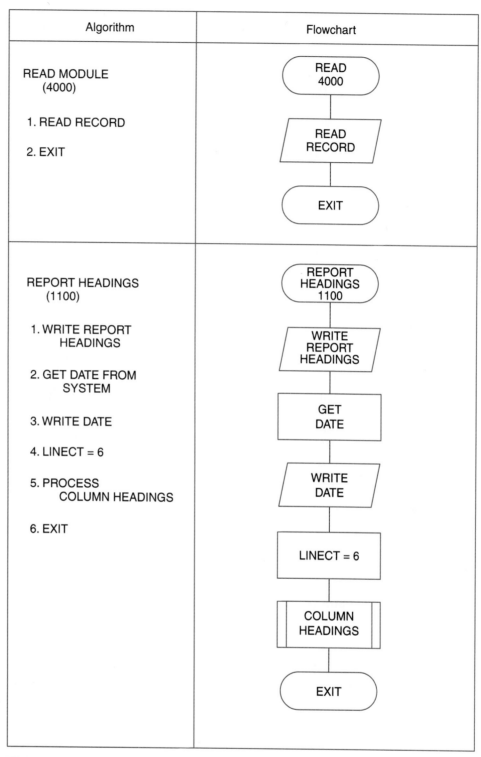
REPORT HEADINGS (1100) 1. WRITE REPORT 　　HEADINGS 2. GET DATE FROM 　　SYSTEM 3. WRITE DATE 4. LINECT = 6 5. PROCESS 　　COLUMN HEADINGS 6. EXIT	

Figure 15.2c Algorithms and Flowcharts of READ and REPORT HEADINGS Modules for Sales Problem

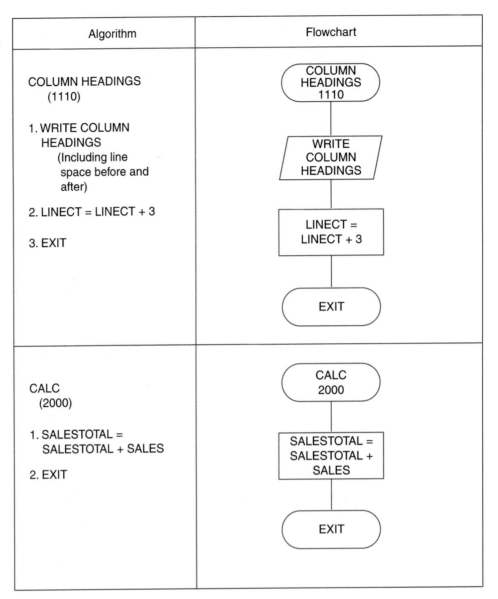

Algorithm	Flowchart
COLUMN HEADINGS (1110) 1. WRITE COLUMN HEADINGS (Including line space before and after) 2. LINECT = LINECT + 3 3. EXIT	
CALC (2000) 1. SALESTOTAL = SALESTOTAL + SALES 2. EXIT	

Figure 15.2d Algorithms and Flowcharts of COLUMN HEADINGS and CALC Modules for Sales Problem

Control-Breaks

Control-breaks are used to give subtotals for a group of similar records. For example, the requirements for a report might be to list all sales, with totals for each salesperson and a grand total for the company. The processing must temporarily break and write each salesperson's total to the printer before proceeding to the next salesperson.

In order to design a program that processes control-breaks, designate one field, such as salesperson's number, as the control field. The file must be previously sorted using this field. The file must be in order of the salesperson's number so that each salesperson's records will be together in the file. If the file has not been sorted by salesperson's number, then the control-break cannot work; that is, the program will not process totals for each salesperson, much less compile the necessary information for each.

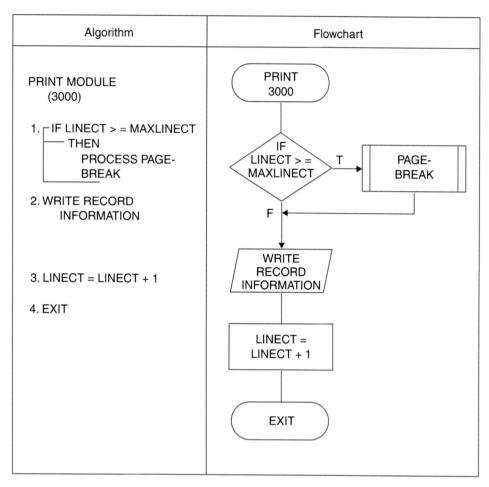

Algorithm	Flowchart
PRINT MODULE (3000) 1. IF LINECT > = MAXLINECT THEN PROCESS PAGE-BREAK 2. WRITE RECORD INFORMATION 3. LINECT = LINECT + 1 4. EXIT	

Figure 15.2e Algorithm and Flowchart of PRINT Module for Sales Problem

Keep the value of the control field of the first record, or the first record after the control-break occurs, in a holding variable memory location. This holding variable contains the first value of the control field of each group of records. In the example in Figure 15.3a–d, the holding variable would be where the initial value of each salesperson's number is kept. The variable name used in the example is SNHOLD, meaning *salesperson number hold value*. The value of the field is the same for all records for a salesperson; therefore, only the first value has to be kept. The salesperson's number does not change until a new salesperson is encountered. At that time the control-break is processed and the hold value is assigned a new value. The value of the control field of each record is compared to the hold value, and when they are no longer equal, the control-break occurs.

The Control-Break module processes all instructions that the computer has to follow each time the value of the control field changes. The instructions for the sales example stated previously in this chapter would include the following: Print the subtotals, add the subtotals to the total for the company, set each subtotal to zero, and set the hold value to the value of the new control field. In Figure 15.3, the control-break is illustrated. The salesperson's total is written to the printer. Next, the line count is incremented by the number of lines it took to write the totals, including the spacer lines. Next, the salesperson's total is added to the company total. Finally, the salesperson's

Algorithm	Flowchart

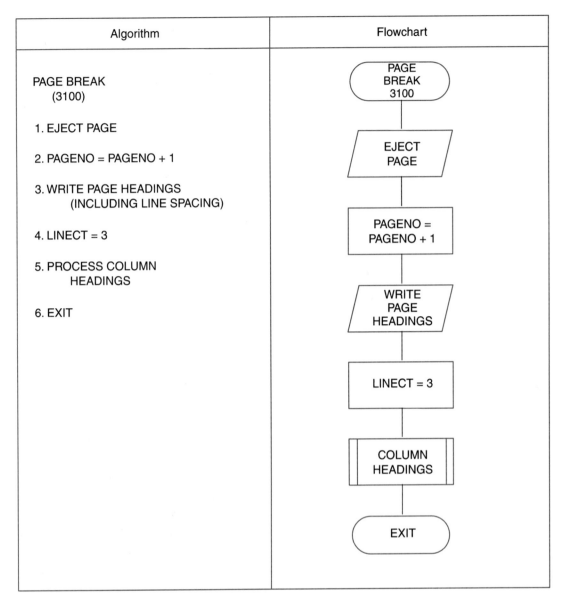

PAGE BREAK
 (3100)

1. EJECT PAGE

2. PAGENO = PAGENO + 1

3. WRITE PAGE HEADINGS
 (INCLUDING LINE SPACING)

4. LINECT = 3

5. PROCESS COLUMN
 HEADINGS

6. EXIT

Figure 15.2f Algorithm and Flowchart of PAGE BREAK Module for Sales Problem

total is set back to zero, and the salesperson's hold value (SNHOLD) is set equal to the new salesperson's number.

The Control-Break is the only new module required to process control-breaks. However, you must make changes in other modules. In the Init module, set the totals to zero and the hold value to the value of the control field for the first record. In the Calc module, test for the Control-Break. That is, test the present value of the control field against its hold value. When the two values are not equal, the Control-Break module will be processed. In the Wrap-up module, the control-break will be executed to process the totals for the last value of the control field.

Notice in Figure 15.3a the two changes in the Init module are to set the salesperson's total to zero, and the salesperson's hold value (SNHOLD) to the first person's number. In the Calc module (Figure 15.3b) the hold value is compared to the current

Algorithm	Flowchart

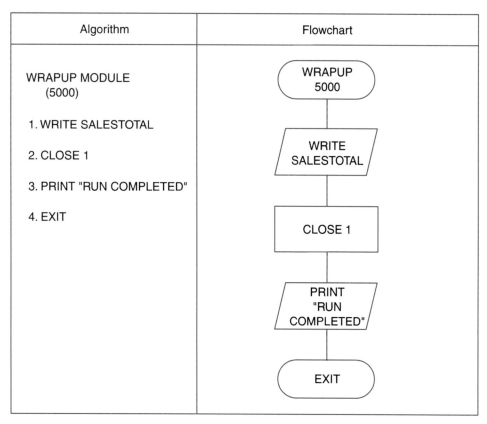

WRAPUP MODULE
 (5000)

1. WRITE SALESTOTAL

2. CLOSE 1

3. PRINT "RUN COMPLETED"

4. EXIT

Figure 15.2g Algorithm and Flowchart of WRAPUP Module for Sales Problem

salesperson number. When they are not equal, the record is the first record for the next salesperson and, therefore, the control-break has to be processed. The other change in the Calc module is that the salesperson's total is accumulated instead of the company's total. The company's total is accumulated in the Control-Break module using the salesperson's total. The last change is in the Wrap-up module in Figure 15.3d. The last salesperson's total must be written to the printer, and the company total must be updated before the completion of the processing; therefore, the control-break must be processed as the first instruction in the Wrap-up module.

Figure 15.3a–d shows all of the changes needed in the solution to the sales problem in order to include the subtotal for each salesperson. A single asterisk indicates where additional instructions were placed. The double asterisk indicates instructions that were changed.

Multiple Control-Breaks

Multiple control-breaks involve subtotals on more than one field. For example, the requirements of a problem might be to find the total sales for each region, for each district within each region, and for each salesperson within each district. In this case there would be three control fields—region, district, and salesperson. The file must be sorted using a concatenation of the fields on which there will be a control-break, starting with

Algorithm	Flowchart
INIT MODULE (1000) 1. OPEN 1, SALES 2. SALESTOTAL = 0 * 3. SALESMANTOTAL = 0 4. MAXLINECT = 60 5. PAGENO = 1 6. PROCESS READ * 7. SNHOLD = SN 8. PROCESS HEADINGS 9. EXIT * Additions to solution	INIT 1000 → OPEN 1, SALES → SALESTOTAL = 0 → SALESMANTOTAL = 0 → MAXLINECT = 60 → PAGENO = 1 → READ → * SNHOLD = SN → REPORT HEADINGS → EXIT

Figure 15.3a Revised INIT Module to add Control-Break to Sales Problem

the field that encompasses the largest grouping, and ending with the field that encompasses the smallest grouping. This file would be sorted the following way: regions in order within the company; districts in order within each region; salespersons in order within each district. The programmer would sort the file on the concatenation (the

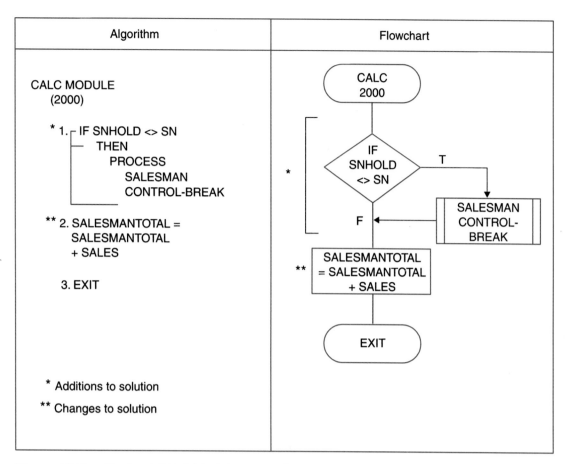

Algorithm	Flowchart
CALC MODULE (2000) * 1. IF SNHOLD <> SN THEN PROCESS SALESMAN CONTROL-BREAK ** 2. SALESMANTOTAL = SALESMANTOTAL + SALES 3. EXIT * Additions to solution ** Changes to solution	

Figure 15.3b Revised CALC Module to add Control-Break to Sales Problem

combining of strings by placing the first in front of the second) of the three fields, with the largest area of sales first, the second largest second, and the smallest third. In this case, the file would be ordered by region + district + salesperson.

All instructions processed in a single control-break also would be executed in each of the control-breaks in a multiple control-break. In the Init module, the programmer initializes all three totals to zero, and sets the hold values for all three control fields to the value of the first control field. In the Calc module, all control fields are tested, moving from largest to smallest. The programmer designs a Control-Break module for each control field, with the first instruction a process block for the next smaller group.

Again, using the sales file as an example, whenever the region changes, the district and the salesperson fields also must change. This sweeping change occurs because there is no overlap of districts between regions, or of salespersons between districts. The programmer sets up the solution so that the largest group is processed, which then processes the next largest group, and so forth, until the smallest group has been processed. Figure 15.4 shows the interactivity chart of the multiple control-break for the sales file discussed in this section.

Figure 15.5a–f shows the algorithms and flowcharts of a multiple control-break for the sales problem illustrated throughout this chapter. SNHOLD, DNHOLD, and RNHOLD are the hold values for the salesperson, the district, and the region, respectively. SN, DN, and RN are the current values of the salesperson's number, the district number, and the region number, respectively. Here again, the single asterisk indicates

Algorithm	Flowchart
SALESMAN CONTROL-BREAK (2300) 1. WRITE SALESMANTOTAL 2. LINECT = LINECT + 4 3. SALESTOTAL = SALESTOTAL + SALESMANTOTAL 4. SALESMANTOTAL = 0 5. SNHOLD = SN 6. EXIT	

Figure 15.3c New Salesman CONTROL-BREAK Module to add Control-Break to Sales Problem

where additional instructions were placed. The double asterisk indicates instructions that were changed.

Using Indicators for Program Control

As in other types of processing, programmers can use indicators in record processing to change the normal course of processing. A printed date of purchase or sales may be

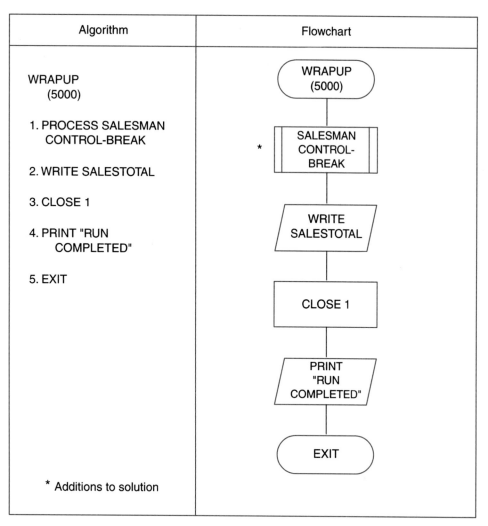

Algorithm	Flowchart
WRAPUP (5000) 1. PROCESS SALESMAN CONTROL-BREAK 2. WRITE SALESTOTAL 3. CLOSE 1 4. PRINT "RUN COMPLETED" 5. EXIT * Additions to solution	WRAPUP (5000) * SALESMAN CONTROL- BREAK WRITE SALESTOTAL CLOSE 1 PRINT "RUN COMPLETED" EXIT

Figure 15.3d Revised WRAPUP Module to add Control-Break to Sales Problem

required at the beginning of each page of a report and each time the date changes, but not with every record. The value of the control field may be printed only when it changes and at the start of a new page. Indicators are used to signal the computer regarding when to print the date or the value of the control field. Indicators also are used in data validation. Data validation is done by checking the data to see that it is correct or within proper range limits. When the computer finds an error, an indicator is set to TRUE to indicate that there has been an error. A programmer can then use this error indicator to signal the computer not to process the record that contains an error.

The sales example can be modified to show how a programmer can use an indicator to print the date of sales at the top of each page, as well as when the date of sales changes. The programmer sets an indicator to TRUE in both the Page-Break module and the Read Module to indicate when the date is to be printed. In the Print module, the indicator is tested; when it is on, the date of sales is printed. When it is off, the date of sales is not printed. Figure 15.6a–d illustrates the use of indicators in the sales problem used throughout this chapter.

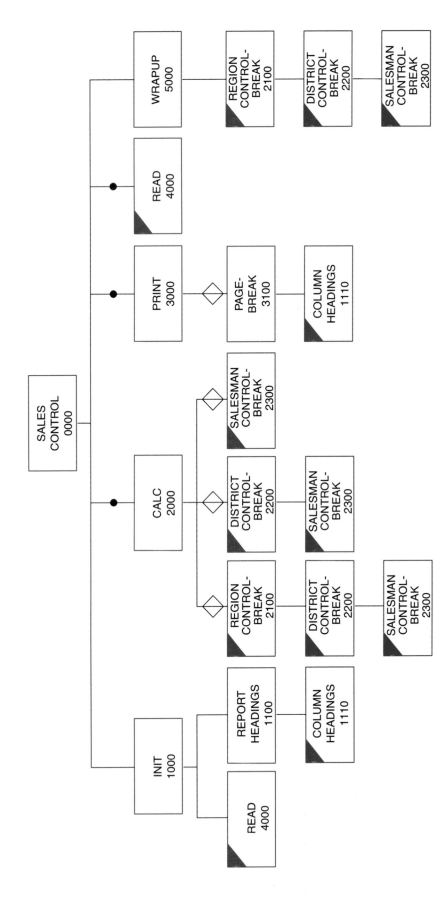

Figure 15.4 Interactivity (Structure) Chart for Multiple Control-Breaks

Algorithm	Flowchart

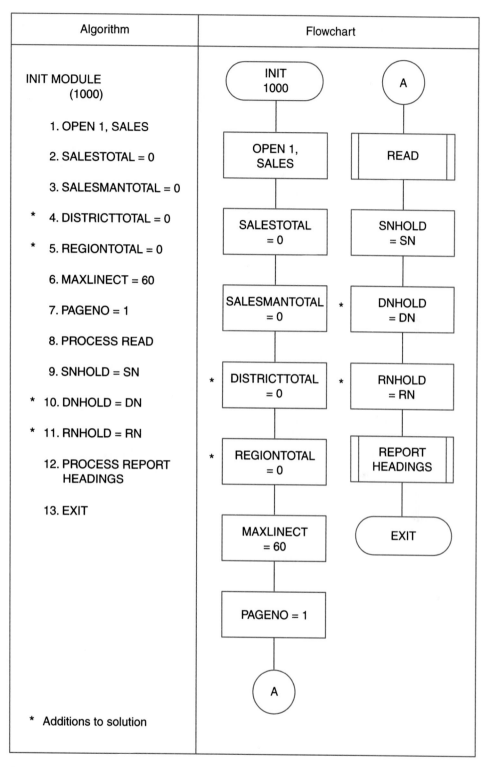

INIT MODULE
 (1000)

 1. OPEN 1, SALES

 2. SALESTOTAL = 0

 3. SALESMANTOTAL = 0

* 4. DISTRICTTOTAL = 0

* 5. REGIONTOTAL = 0

 6. MAXLINECT = 60

 7. PAGENO = 1

 8. PROCESS READ

 9. SNHOLD = SN

* 10. DNHOLD = DN

* 11. RNHOLD = RN

12. PROCESS REPORT
 HEADINGS

13. EXIT

* Additions to solution

Figure 15.5a Revised INIT Module to add Multiple Control-Breaks to Sales Problem

Algorithm	Flowchart

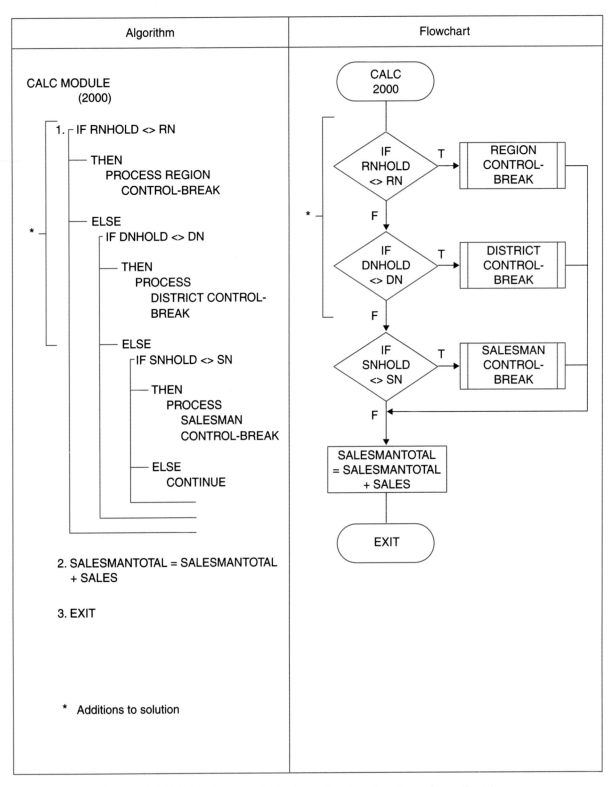

CALC MODULE
(2000)

1. IF RNHOLD <> RN

 THEN
 PROCESS REGION
 CONTROL-BREAK

 ELSE
 IF DNHOLD <> DN

 THEN
 PROCESS
 DISTRICT CONTROL-
 BREAK

 ELSE
 IF SNHOLD <> SN

 THEN
 PROCESS
 SALESMAN
 CONTROL-BREAK

 ELSE
 CONTINUE

2. SALESMANTOTAL = SALESMANTOTAL
 + SALES

3. EXIT

* Additions to solution

CALC
2000

IF
RNHOLD
<> RN

T → REGION
CONTROL-
BREAK

F

IF
DNHOLD
<> DN

T → DISTRICT
CONTROL-
BREAK

F

IF
SNHOLD
<> SN

T → SALESMAN
CONTROL-
BREAK

F

SALESMANTOTAL
= SALESMANTOTAL
+ SALES

EXIT

Figure 15.5b Revised CALC Module to add Multiple Control-Breaks to Sales Problem

Algorithm	Flowchart

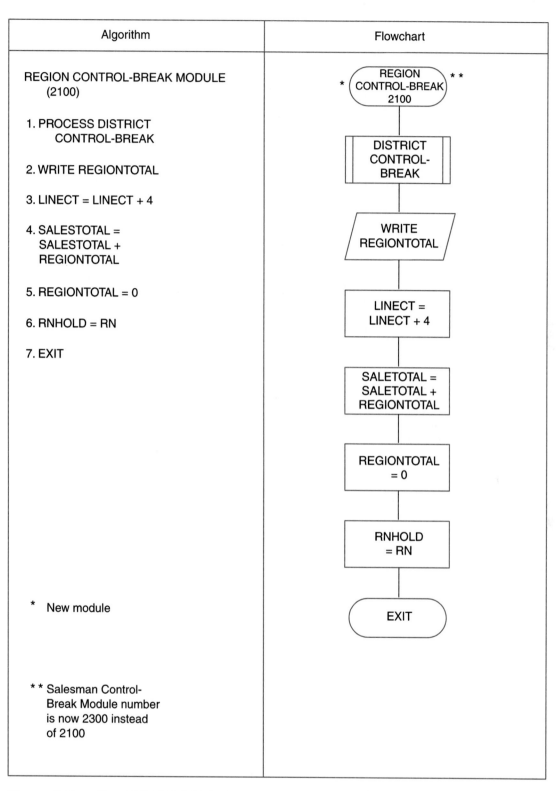

REGION CONTROL-BREAK MODULE
 (2100)

1. PROCESS DISTRICT
 CONTROL-BREAK

2. WRITE REGIONTOTAL

3. LINECT = LINECT + 4

4. SALESTOTAL =
 SALESTOTAL +
 REGIONTOTAL

5. REGIONTOTAL = 0

6. RNHOLD = RN

7. EXIT

* New module

** Salesman Control-
 Break Module number
 is now 2300 instead
 of 2100

Figure 15.5c New REGION CONTROL-BREAK Module to add Multiple Control-Breaks to Sales Problem

Algorithm	Flowchart
* DISTRICT CONTROL-BREAK (2200) 1. PROCESS SALESMAN CONTROL-BREAK 2. WRITE DISTRICTTOTAL 3. LINECT = LINECT + 4 4. REGIONTOTAL = REGIONTOTAL + DISTRICTTOTAL 5. DISTRICTTOTAL = 0 6. DNHOLD = DN 7. EXIT * New module ** The Salesman Control-Break now accumulates the district total instead of the company total.	

Figure 15.5d New DISTRICT CONTROL-BREAK Module to add Multiple Control-Breaks to Sales Problem

Error Handling

For data validation (finding errors in the data records), you need a new module to be processed from the Read module. Data validation is accomplished through the decision structure. Usually, straight-through logic is used because all data error conditions have

Algorithm	Flowchart
SALESMAN CONTROL-BREAK (2300) 1. WRITE SALESMANTOTAL 2. LINECT = LINECT + 4 3. DISTRICTTOTAL = DISTRICTTOTAL + SALESMANTOTAL 4. SALESMANTOTAL = 0 5. SNHOLD = SN 6. EXIT * This module is new	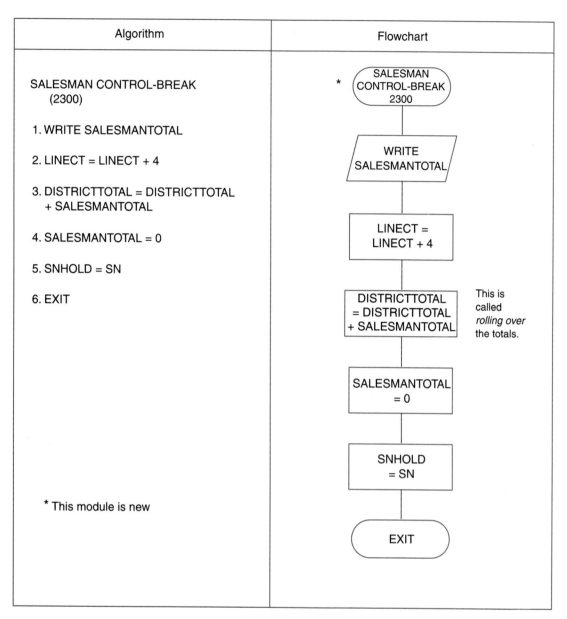

Figure 15.5e Revised SALESMAN CONTROL-BREAK Module to add Multiple Control-Breaks to Sales Problem

to be processed in order for the computer to find all errors in the record. The conditions for data validation are provided to the programmer by the problem. When the computer finds an error while processing the Error Validation module, a message is written to an error report, the error indicator is set equal to TRUE, and the processing of the Data Validation module or Error module is continued. If any errors are found, the record with errors is not processed, and the next record is read. This results in two reports—the regular report, which excludes the records with errors, and the error report (sometimes called an exceptions report), which includes them. For example, a Data Validation or Error module might check to see that the amount of a check is less than a given amount, or that an account number is all numbers, or that a name is all letters.

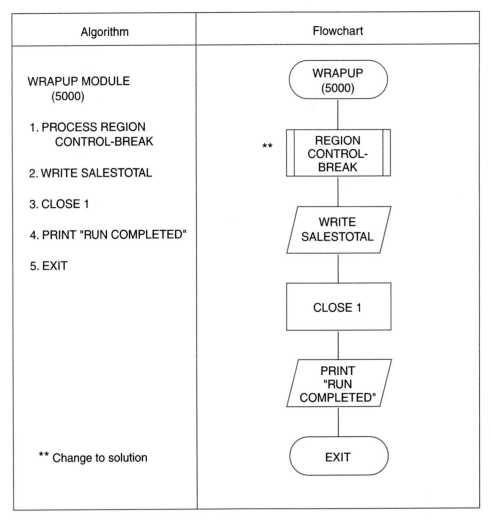

Algorithm	Flowchart
WRAPUP MODULE (5000) 1. PROCESS REGION CONTROL-BREAK 2. WRITE SALESTOTAL 3. CLOSE 1 4. PRINT "RUN COMPLETED" 5. EXIT ** Change to solution	WRAPUP (5000) → ** REGION CONTROL-BREAK → WRITE SALESTOTAL → CLOSE 1 → PRINT "RUN COMPLETED" → EXIT

Figure 15.5f Revised WRAPUP Module to add Multiple Control-Breaks to Sales Problem

After the computer reads the record in the Read module, it should process the Error module. In the programming, use a loop structure to read until a good record or an END-OF-FILE is found. Examples of the Read module and the Error module for the sales problem are found in Figure 15.7.

Often a large portion of the processing of a record is data validation. It is an important part of record processing because it helps eliminate problematic and costly errors.

Null Files

Another problem that arises when you are working with files is how to deal with the **null file,** or empty file. If a file is empty the computer prints an error message on the screen and processing halts. See Figure 15.8 to see the changes a programmer makes in the Init and the Wrap-up modules to accommodate a null file. The Init module is designed so a null file indicator is set to TRUE, an error message is printed, and all processing is skipped, except to close the files in the Wrap-up module if an END-OF-FILE is encountered after the primer READ.

Algorithm	Flowchart

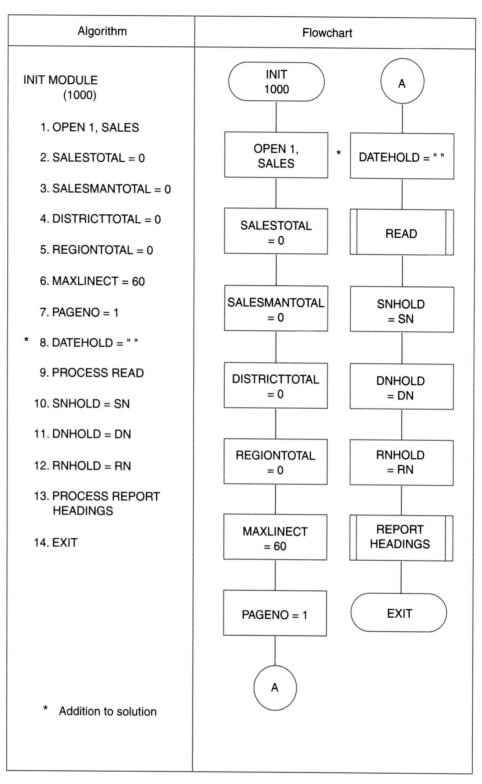

INIT MODULE
(1000)

1. OPEN 1, SALES

2. SALESTOTAL = 0

3. SALESMANTOTAL = 0

4. DISTRICTTOTAL = 0

5. REGIONTOTAL = 0

6. MAXLINECT = 60

7. PAGENO = 1

* 8. DATEHOLD = " "

9. PROCESS READ

10. SNHOLD = SN

11. DNHOLD = DN

12. RNHOLD = RN

13. PROCESS REPORT
HEADINGS

14. EXIT

* Addition to solution

Figure 15.6a Modified INIT Module to add the Indicator to the Sales Problem

Algorithm	Flowchart

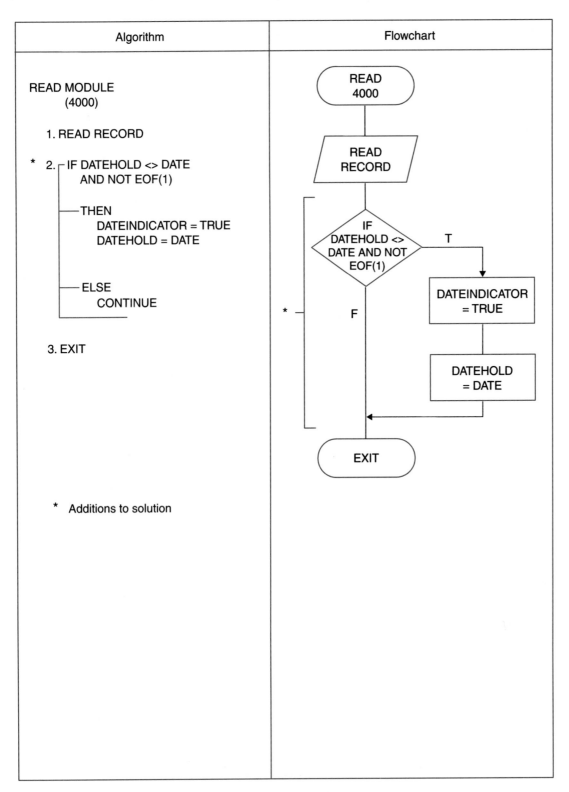

READ MODULE
 (4000)

 1. READ RECORD

* 2. ┌ IF DATEHOLD <> DATE
 AND NOT EOF(1)

 ├─ THEN
 DATEINDICATOR = TRUE
 DATEHOLD = DATE

 ├─ ELSE
 CONTINUE

 3. EXIT

 * Additions to solution

Figure 15.6b Modified READ Module to add the Indicator to the Sales Problem

Algorithm	Flowchart
PRINT MODULE (3000) 1. ⌐ IF LINECT >= MAXLINECT └─ THEN PROCESS PAGE-BREAK * 2. ⌐ IF DATE INDICATOR └─ THEN WRITE DATE W/O {CR} DATEINDICATOR = FALSE 3. WRITE RECORD INFORMATION 4. LINECT = LINECT + 1 5. EXIT * Additions to solution	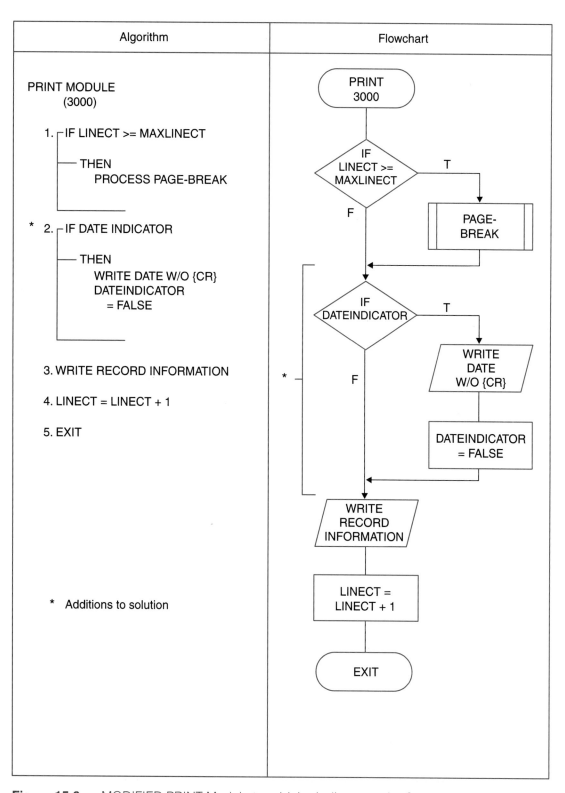

Figure 15.6c MODIFIED PRINT Module to add the Indicator to the Sales Problem

Algorithm	Flowchart
PAGE-BREAK MODULE (3100) 1. EJECT PAGE 2. PAGENO = PAGENO + 1 3. WRITE PAGE HEADINGS 4. LINECT = 3 * 5. DATEINDICATOR = TRUE 6. PROCESS COLUMN HEADINGS 7. EXIT * Additions to solution	

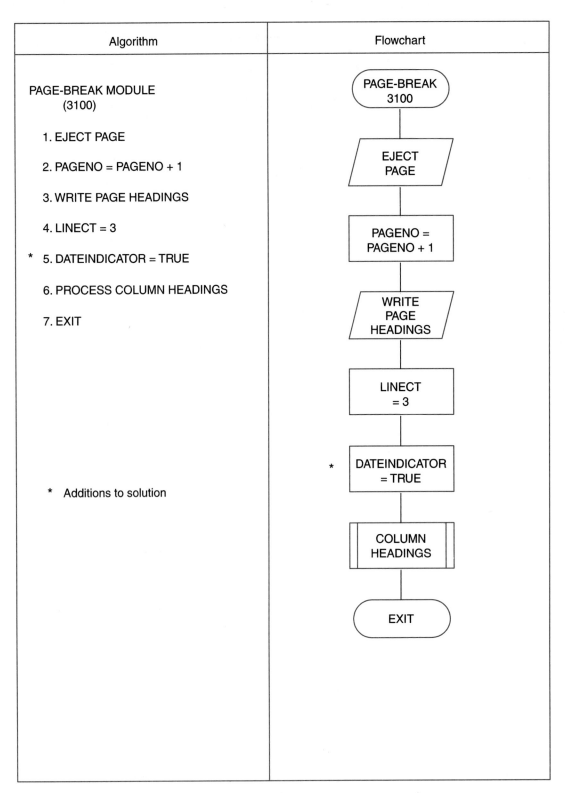

Figure 15.6d Modified PAGE-BREAK Module to add the Indicator to the Sales Problem

Algorithm	Flowchart
READ MODULE (4000) 1. REPEAT READ RECORD ERRORINDICATOR = FALSE PROCESS ERROR UNTIL ERROR INDICATOR = FALSE OR EOF(1) 2. IF NOT EOF(1) AND DATEHOLD <> DATE THEN DATEINDICATOR = TRUE DATEHOLD = DATE 3. EXIT	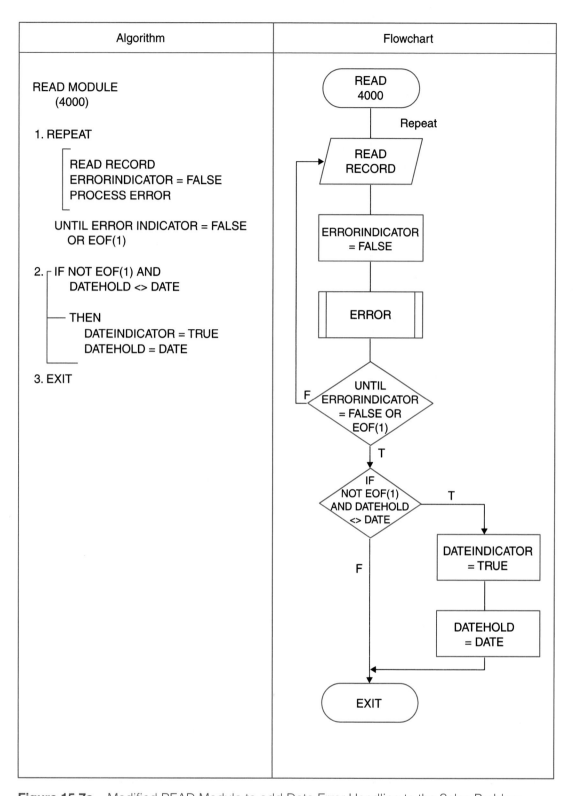

Figure 15.7a Modified READ Module to add Data Error Handling to the Sales Problem

Algorithm	Flowchart
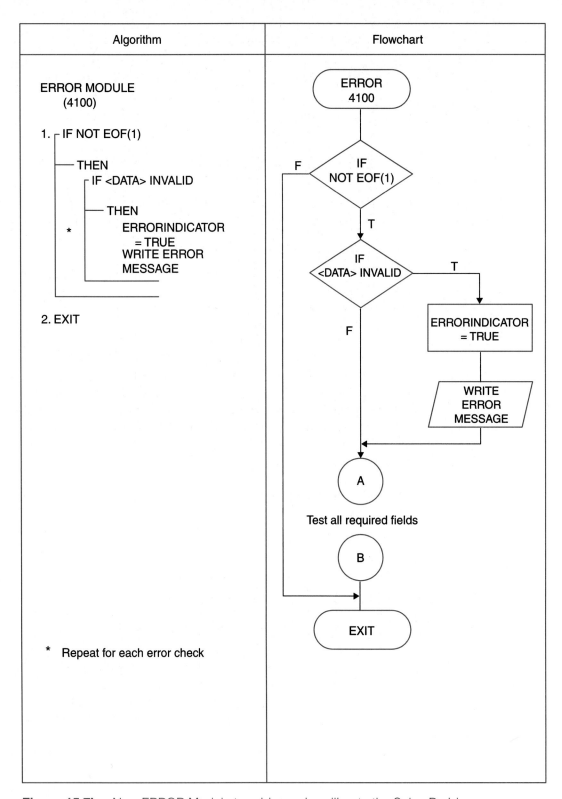	

ERROR MODULE
(4100)

1. ⌐ IF NOT EOF(1)

 ├── THEN
 ⌐ IF <DATA> INVALID

 ├── THEN
 * ERRORINDICATOR
 = TRUE
 WRITE ERROR
 MESSAGE

2. EXIT

* Repeat for each error check

Figure 15.7b New ERROR Module to add error handling to the Sales Problem

Algorithm	Flowchart

Algorithm

INIT MODULE
 (1000)

1. OPEN 1, SALES

2. SALESTOTAL = 0

3. SALESMANTOTAL = 0

4. DISTRICTTOTAL = 0

5. REGIONTOTAL = 0

6. MAXLINECT = 60

7. PAGENO = 1

8. DATEHOLD = " "

9. PROCESS READ

10. ⌐ IF EOF(1)

 — THEN
 PRINT ERROR
 MESSAGE

 NULLINDICATOR
 = TRUE

 — ELSE
 SNHOLD = SN
 DNHOLD = DN
 RNHOLD = RN
 PROCESS REPORT
 HEADINGS

11. EXIT

Figure 15.8a Modified INIT Module to handle a Null File in the Sales Problem

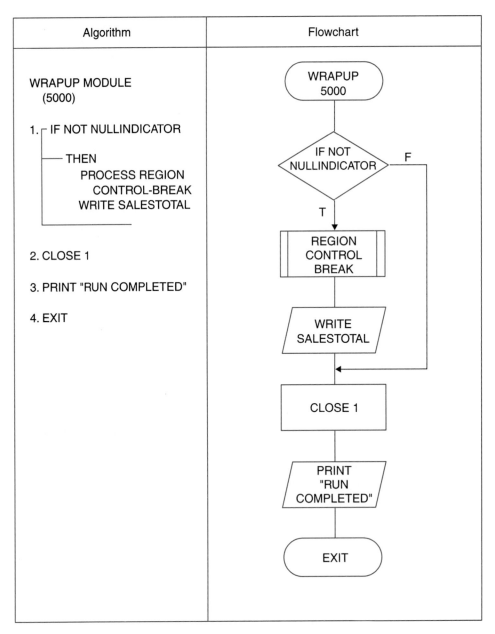

Algorithm	Flowchart

Algorithm

WRAPUP MODULE
 (5000)

1. ┌ IF NOT NULLINDICATOR

 ├── THEN
 PROCESS REGION
 CONTROL-BREAK
 WRITE SALESTOTAL

2. CLOSE 1

3. PRINT "RUN COMPLETED"

4. EXIT

Flowchart

WRAPUP 5000

IF NOT NULLINDICATOR — F

T

REGION CONTROL BREAK

WRITE SALESTOTAL

CLOSE 1

PRINT "RUN COMPLETED"

EXIT

Figure 15.8b Modified WRAPUP Module to handle a Null File in the Sales Problem

Summary

The concepts of record processing, including control-breaks, indicators, and data validation, are important to both programming and applications. Control-breaks are important because many business reports call for subtotals or other intermediate results, such as the number of customers a salesperson contacted or the number of items in an inventory.

A good programmer designs output reports with the user in mind. Report headings, data placement, and general organization are very important.

A programmer handles data errors, null files, or changes in processing with the use of indicators. A finished solution should be as correct and as foolproof as possible. Data-entry errors are made often. Even if 99% of entry records are correct, that means there is still one error in every one hundred entries. With thousands of pieces of data entered daily, there are many errors. Indicators are used any time the programmer needs to indicate to the computer that another set of instructions is to be used.

For computer users in business, sequential-access processing techniques such as those covered in this chapter are keys to productivity.

New Terms

control-break	null file
multiple control-break	page-break

Questions

1. Name the items to consider when designing a record. Explain each.
2. What is a page-break?
3. What are control-breaks?
4. What is a null file?
5. Why is it important to consider a null file when processing files?

Problems

Using the input records and systems flowcharts developed in Chapter 14, develop a solution for the following problems. Your solution should include a problem analysis chart, an IPO chart, a data dictionary, an interactivity chart, algorithms and flowcharts for each module, and an output report design.

1. Develop a solution to calculate the gross pay for the employees of a company. The input records should include the following: name, address, Social Security number, pay type (H = hourly, S = salary), rate of pay, and hours worked. The output report should include all input data plus the gross pay.

2. A sales manager wants a program that lists all of the company's customers according to the salesperson assigned to each. She would also like the number of the salesperson's customers to be totaled at the end of the list. The name of the salesperson should be printed only at the top of the page or when the processing encounters a new salesperson. The input file contains the following data: the customer's name, address, and telephone number, and the salesperson's name.

3. Develop a solution for an inventory problem. The input record should include the following: the inventory number, the department number, the store number, the item description, the quantity, the wholesale price, the markup percentage, and the vendor. The output report should include the input data plus the retail price, the

(Problem 3 continued on next page)

total value of the item in stock (quantity * price), and the total wholesale and retail values of the stock for each department and for each store. The company totals should be printed at the end of the report.

4. A sales manager would like to be able to track the clients of each of her salespeople. She would like to know when each customer was last contacted, the outcome of the contact (not interested, sale made, call back later, no contact made), the sale amount if a sale was made, and the name and number of the salesperson. She would like the list of clients to be in chronological order according to the last contact time by salesperson. She would also like a total of sales per day per salesperson. Develop the solution to the manager's problem. In what order should the file be sorted?

 What would you have to do if the manager also wanted a client list in alphabetical order?

Chapter 16

Sequential-Access File Updating

Overview

Creating Files
The Master File
Transaction Files
Activity Files
Backup Files
Updating the Master File Using a Transaction File
A Useful Alternative Method

Objectives

When you have finished this chapter, you should be able to:

1. Design a solution to update sequential files.
2. Explain the use of a master file.
3. Explain the use of a transaction file.
4. Explain the use of an activity file.
5. Explain the use of a backup file.

Businesses are dynamic, and so are their data-processing requirements. Therefore, the data files used in business must be easily updated. Chapter 15 explained how to process a sequential-access file, including data validation, but it did not explain how to update one. This chapter explains and illustrates the technique for updating a sequential file using a transaction file.

Updating a file means adding, deleting, or changing records to keep the information in the file current.

1. *Adding records.* When new data are generated, new records have to be added to a file. When a new employee starts work, data about that person must be

added to the personnel file; when a sale has been made, the data from the sales receipt must be added to the sales file, and so on.

2. *Deleting records.* There must be a way to delete obsolete data from a file. When a record is deleted, the user chooses one of several options about what to do with it. These options include:
 a. the record may be discarded, never to be seen again;
 b. the record may be placed in an archival file, so that the information can be retrieved at a later date, if needed;
 c. the record may be placed in an activity file to show when it was deleted.

3. *Changing records.* Data in a record must be changed from time to time for a variety of reasons. The data may be incorrect, or they may have changed. When data have changed, the date and the changes themselves can go into an activity file for future inquiries.

Files can be processed in two ways. The first method is **batch processing,** in which data collected over a period of time are processed all at once without human intervention. This is the most efficient for the computer because it doesn't have to wait for a person to respond. However, it is not as efficient for people since they have to wait until a batch run is made. Because of the relatively long batch turnaround time, data that are currently being used are frequently not up to date. The more efficient type of processing for people is **interactive processing,** in which data are entered and updated at the same time. In this case, the data in the file are as current as the data being entered at any given time. The person communicating with the computer is interacting with it through commands or menus. The use of commands and menus to update random-access files is discussed in Chapter 17.

Since sequential files are processed in order, it is inefficient to update them interactively. Therefore, sequential files are best updated by batch processing. All of the changes are put into a transaction file, which is a separate file that is periodically used to update all the files at one time. This update may be done daily, weekly, or as often as company policies require.

Creating Files

You have to create a file before you can use it. That is, you have to design the structure of the file and communicate it to the computer. This structure includes the layout of the record and the kind of file—sequential or random access. The files discussed in this chapter, as well as all other files, are designed by the methods described in Chapter 14. Use the record data dictionary and layout form (Appendix D, Figure D.6) to design the structure of the record. When you create a file, you designate the fields of each record and specify the attributes of each field, such as the name of the field, the length of the field, the data type, and the number of decimals for numeric data. All records have different data but the same attributes. Then you put together the record layout to specify the way the record will look.

When you create a new file from an old one, all previous data in the file are destroyed. Therefore, create a new file over an old one only when it is necessary, and when you wish to delete the information in the old file.

The Master File

The **master data file** is the file in which all the data needed for processing are kept current. Common types of master files include inventory files, payroll files, and mailing list files. The files used in the sequential-access solutions in Chapter 15 are master data files. The data are arranged in order of processing, and they are kept in the same order during processing and after the updates are completed. In this chapter, the terminology *old master file* and *new master file* are used. The former means the master file before the update, and the latter expression refers to the master file after the update.

Transaction Files

Transaction files contain all the data needed to update the master file. You construct a transaction file by entering records that contain new data for the master file. The entries are made as new data arise and then are sorted according to the requirements of the master file. The records in the transaction file must be in the same order as that in the master file since they will be used to update the master file records sequentially. That is, if the master file is sorted in account number order, then the transaction file must be sorted in account number order. Or if the master file is sorted according to the last name, then the transaction file must be sorted accordingly.

Activity Files

All completed file updates are listed in an **activity file** (audit trail file). Each time a user executes the update program, he or she will append the activity file (add new entries to the end of the file) to show what has happened to the master file. The activity file is the only file in a company from which the records are never deleted. It is important because it is the only file that can tell auditors what has happened to any particular record.

Backup Files

Backup files contain the records as they were before the last file update. It is important to keep backup files in case something happens to the current master file. Machine or human errors that destroy part or all of a file are fairly common occurrences. If the current master file is destroyed, the user can copy the backup file into the old master file, rerun the update program with the last transaction file, and then start the processing again. There are two ways to make a backup file: one is to rename the old master file (containing the data that existed before the most recent update) as a backup; the other is to copy a file—old, new, master, transaction, or activity—onto another auxiliary storage medium, either disk or tape.

Updating the Master File
Using a Transaction File

Figure 16.1 shows a generalized systems flowchart to update a sequential-access file. Figure 16.2a–d shows a specific solution for updating a file. You design an update solution by using the loop structure and the decision structure. Be cautious to avoid common programming errors, such as allowing the addition of a record with a key field that already exists. The key field, remember, is the field that has a unique piece of data that is used to order the master file and the transaction file. This error can be caused by incorrectly entering a key in the transaction file. Entering an incorrect key can cause another error: trying to delete or change a nonexistent record. Therefore, part of the solution should contain checks so that any transaction records containing such errors are not used to update the master file but instead are written to an error output report. Check data for errors by using an IF/THEN/ELSE instruction. When an error is found, that transaction record will not be processed but will be written to the error output record. See Chapter 15 for more information about error checks.

[An alternative solution to the problem of updating a sequential file is presented in the final section of this chapter and illustrated in Figure 16.4a–d (pp. 402–405). See the discussion there.]

After the data in a transaction record have been used to update the file, a new transaction record is read into internal memory. Likewise, after a master file record has been updated, or found to have no changes, a new master file record is read into internal memory. Through the use of transaction codes—codes that indicate to the computer what type of transaction is to take place, whether adding the record, deleting the record, or changing the record—and the comparison of the transaction record to the master file record, the file is updated record by record.

The master file might contain payroll data, inventory data, or sales data. The technique of updating is the same for all sequential files and does not depend on the type of data in a given file. The example in Figure 16.2, therefore, indicates the type of files to be read in and used rather than using specific data. The following notes explain each of the steps in Figure 16.2a–d.

Notes For Figure 16.2

Definitions
- MF—old master file
- TF—transaction file
- AF—activity file
- HF—new master file
- TCODE—transaction code (a field in the transaction record)
 "A"—add a record
 "D"—delete a record
 "C"—change a record
- MFKEY—master file key field
- TFKEY—transaction key field
- ERRORCODE—code that indicates an error has been found: a 1 indicates an attempt to change or delete a nonexistent record; a 2 indicates that a record with a duplicate key is being added.

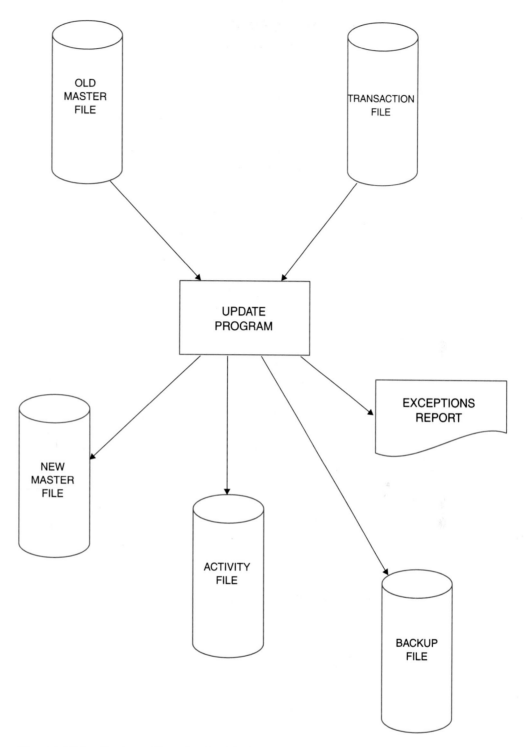

Figure 16.1 Systems Flowchart to Update a Sequential-Access File

The numbers of the notes correspond to the circled numbers in the flowchart.

1. This loop continues as long as there are records in both the old master file and the transaction file. When the END-OF-FILE (EOF) of either file is reached, the processing continues at 7.

UPDATE FILE MODULE

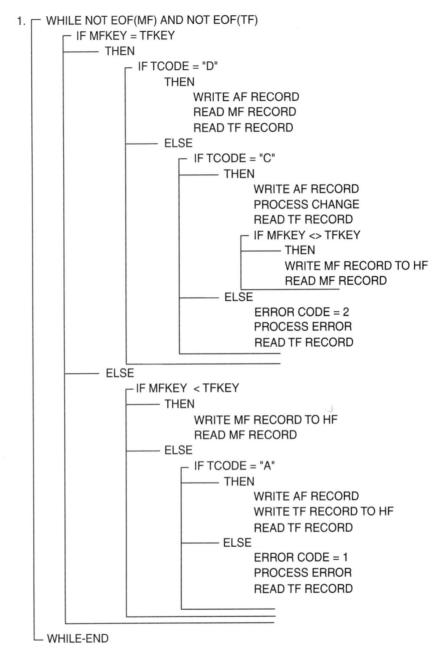

```
1. ┌ WHILE NOT EOF(MF) AND NOT EOF(TF)
   │    ┌ IF MFKEY = TFKEY
   │    ├──── THEN
   │    │         ┌ IF TCODE = "D"
   │    │         │    THEN
   │    │         │        WRITE AF RECORD
   │    │         │        READ MF RECORD
   │    │         │        READ TF RECORD
   │    │         ├──── ELSE
   │    │         │         ┌ IF TCODE = "C"
   │    │         │         ├──── THEN
   │    │         │         │        WRITE AF RECORD
   │    │         │         │        PROCESS CHANGE
   │    │         │         │        READ TF RECORD
   │    │         │         │        ┌ IF MFKEY <> TFKEY
   │    │         │         │        ├── THEN
   │    │         │         │        │      WRITE MF RECORD TO HF
   │    │         │         │        │      READ MF RECORD
   │    │         │         ├──── ELSE
   │    │         │         │        ERROR CODE = 2
   │    │         │         │        PROCESS ERROR
   │    │         │         │        READ TF RECORD
   │    ├──── ELSE
   │    │         ┌ IF MFKEY < TFKEY
   │    │         ├──── THEN
   │    │         │        WRITE MF RECORD TO HF
   │    │         │        READ MF RECORD
   │    │         ├──── ELSE
   │    │         │         ┌ IF TCODE = "A"
   │    │         │         ├──── THEN
   │    │         │         │        WRITE AF RECORD
   │    │         │         │        WRITE TF RECORD TO HF
   │    │         │         │        READ TF RECORD
   │    │         │         ├──── ELSE
   │    │         │         │        ERROR CODE = 1
   │    │         │         │        PROCESS ERROR
   │    │         │         │        READ TF RECORD
   └ WHILE-END
```

Figure 16.2a Sequential-File Updating Through a Transaction File—Algorithm

2. When the old master file key field is equal to the transaction file key field, the only two processes that occur are changing or deleting records. When the two fields are not equal, either the old master file record is to remain the same, or the transaction file record is added to the master file. There may be many records that will not have any updates made to them.

3. When the transaction code is equal to "D", then the information is written to the activity file, and another master file record and another transaction file

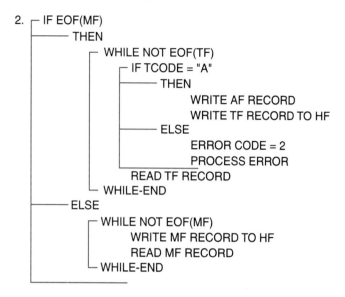

```
2.  ┌ IF EOF(MF)
    ├──── THEN
    │         ┌ WHILE NOT EOF(TF)
    │         │    ┌ IF TCODE = "A"
    │         │    ├──── THEN
    │         │    │          WRITE AF RECORD
    │         │    │          WRITE TF RECORD TO HF
    │         │    ├──── ELSE
    │         │    │          ERROR CODE = 2
    │         │    └          PROCESS ERROR
    │         │      READ TF RECORD
    │         └ WHILE-END
    ├──── ELSE
    │         ┌ WHILE NOT EOF(MF)
    │         │    WRITE MF RECORD TO HF
    │         │    READ MF RECORD
    │         └ WHILE-END
    └
```

Figure 16.2b Sequential-File Updating Through a Transaction File—Algorithm

3. CLOSE FILES (MF, TF, HF, AF)

4. REPLACE BACKUP FILE WITH MF

5. REPLACE MF WITH HF

6. EXIT

record are read. Since the record was not written to the new master file, it has been deleted.

4. When the transaction code is equal to "C", then the information is written to the activity file, the master record is changed internally in the computer's memory (not in external memory), and then the changed master file record is written to the new master file. Another transaction file record is then read.

 This transaction file key is checked against the current master file key. If they are equal, processing continues without reading in another master file record. If they are not equal, then no more updating needs to be done to the current master file record, and another one is read.

5. If the transaction file and master file keys are equal and the transaction code does not equal either "D" or "C", then there is an error in the transaction record. When an error occurs, an error message is written to an error report, and another transaction record is read.

6. When the master file key is less than the transaction file key, then nothing has to be done to that record. The master file record is written to the new master file, and another master file record is read.

Flowchart

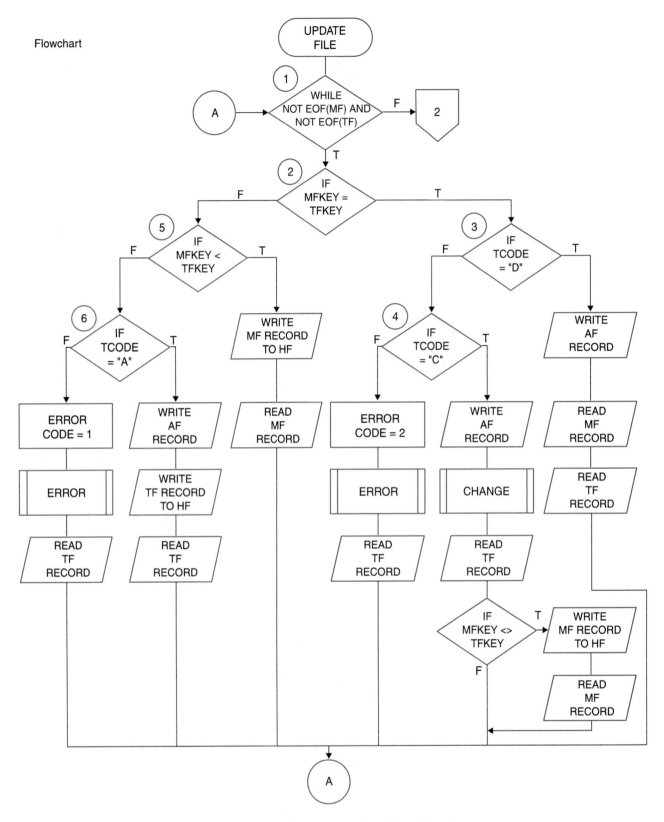

Figure 16.2c Sequential-File Updating Through a Transaction File—Flowchart

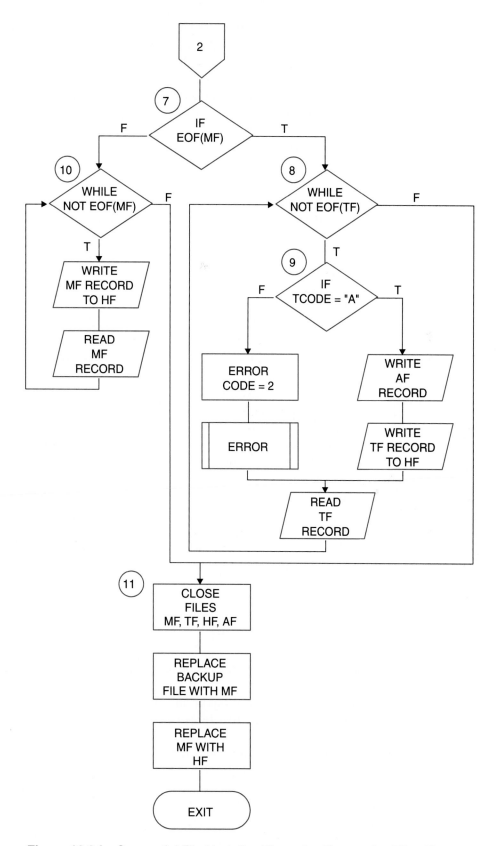

Figure 16.2d Sequential-File Updating Through a Transaction File—Flowchart

7. When the master file key is greater than the transaction file key, then the transaction file should be added to the master file. When the transaction code is equal to an "A", then the information is written to the activity file, the transaction record is written to the new master file, and another transaction record is read. If the master file key is greater than the transaction file key and the transaction code does not equal "A", then there is an error in the transaction record. An error message is then written to the error report, and another transaction record is read.

8. When an EOF occurs in either the old master file or the transaction file, the preceding loop stops, and the program processes the remaining records on the file that has not encountered an EOF. That is, if the EOF is encountered on the old master file, then the remaining transaction file records are processed; if the EOF is encountered on the transaction file, then the remaining old master file records are processed.

9. The transaction file records are processed until the EOF on the transaction file has been reached.

10. The remaining records on the transaction file must be records to be added to the master file. When the transaction code is equal to "A", then the information is written to the activity file, the transaction record is written to the new master file, and another transaction file record is read. If the transaction code is not equal to "A", then there is an error in the transaction file record. When an error occurs, an error message is written to the error report, and another transaction file record is read.

11. The master file records are processed until the EOF on the old master file is encountered. Since there are no more transactions to be done on the master file, these records are simply written to the new master file and a new record is read.

12. After the EOF has been encountered on both the old master file and the transaction file, all files are closed. Then the data in the backup file are replaced with the data in the old master file, and the data in the old master file are replaced with the data in the new master file. It is then ready to be used for the next update processing.

Putting It All Together

Figure 16.3a–j demonstrates a complete solution for an update to an inventory file. The Control module (Figure 16.3b) is very simple: the Init is processed, then the Update module, and finally the Wrap-up module. The Init module of Figure 16.3c opens the four files and reads the first record in the old master file (the inventory file) and the transaction file.

The Update module (Figure 16.3d) contains three loops. The first loop continues as long as there are records in both files. Either the second or third loop completes the process of updating the inventory file. In the first loop, one module (Equal) is processed when the transaction file key (TINVNO) is equal to the master file key (INVNO) for the current record; another module (Notequal) is processed when they are not equal. In the second loop, when the end-of-file has been reached for the old master file, the remaining transaction records should all be add records and are added in the ADD module. If the transaction code is not equal to "A", then the ERROR module is processed. In the third

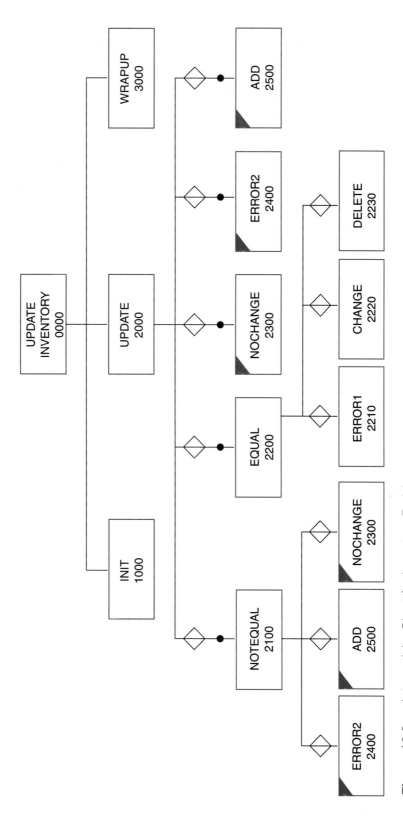

Figure 16.3a Interactivity Chart for Inventory Problem

Algorithm	Flowchart
UPDATE INVENTORY CONTROL 0000 1. PROCESS INIT 2. PROCESS UPDATE 3. PROCESS WRAPUP 4. END	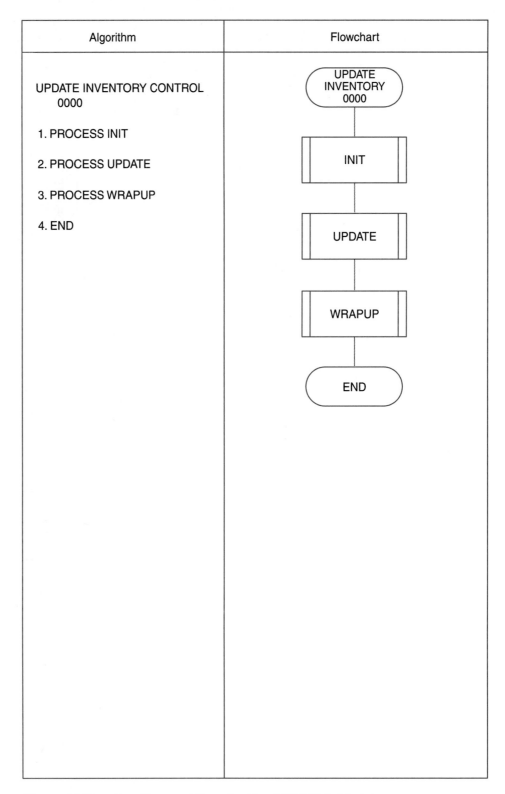

Figure 16.3b Algorithm and Flowchart for CONTROL Module

Algorithm	Flowchart
INIT MODULE 1000 1. OPEN 1, INVENTORY FOR INPUT 2. OPEN 2, TRANSACTION FOR INPUT 3. OPEN 3, ACTIVITY FOR APPEND 4. OPEN 4, TEMPORARY FOR OUTPUT 5. READ (1) RECORD 6. READ (2) RECORD 7. EXIT	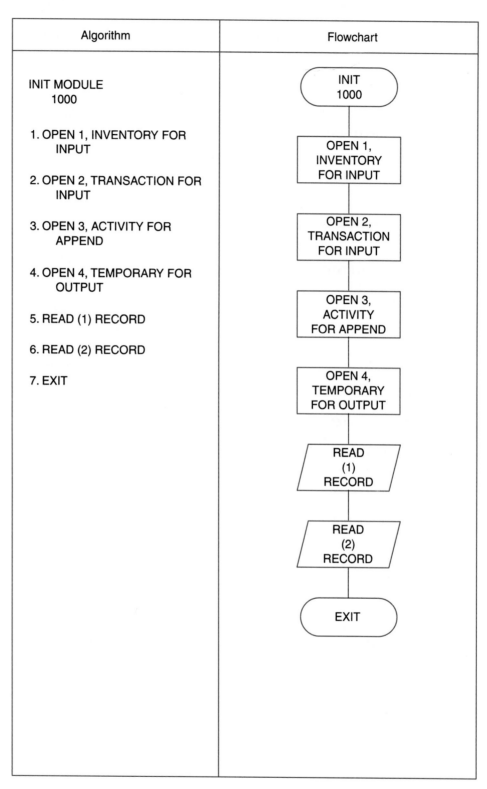

Figure 16.3c Algorithm and Flowchart for INIT Module

Algorithm	Flowchart
UPDATE MODULE 2000 1. ┌ WHILE NOT EOF(1) AND NOT EOF(2) ┌ IF TINVNO = INVNO ├ THEN PROCESS EQUAL ├ ELSE PROCESS NOTEQUAL └ WHILE-END 2. ┌ IF EOF(1) ├ THEN ┌ WHILE NOT EOF(2) ┌ IF TCODE = "A" ├ THEN PROCESS ADD ├ ELSE PROCESS ERROR2 └ WHILE-END ├ ELSE ┌ WHILE NOT EOF(1) PROCESS NOCHANGE └ WHILE-END 3. EXIT	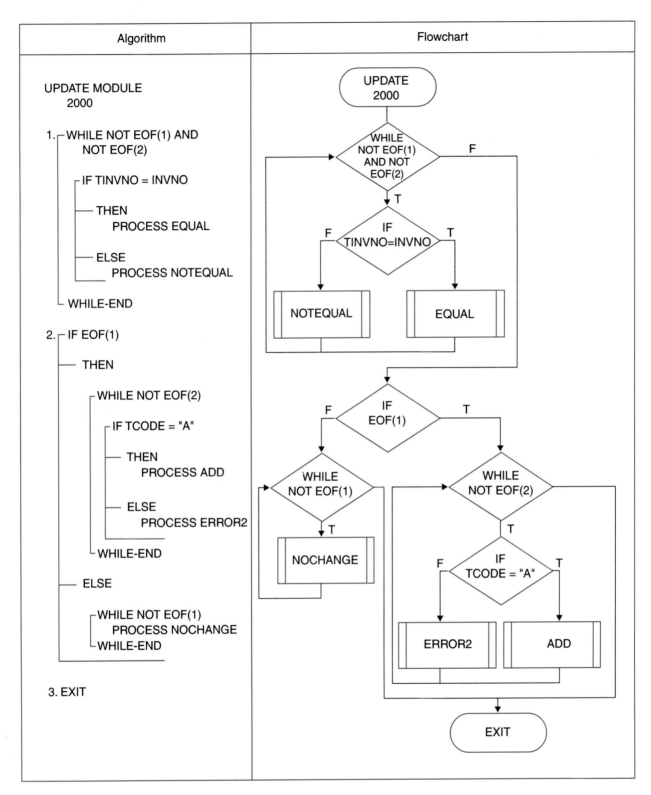

Figure 16.3d Algorithm and Flowchart for UPDATE Module

Algorithm	Flowchart

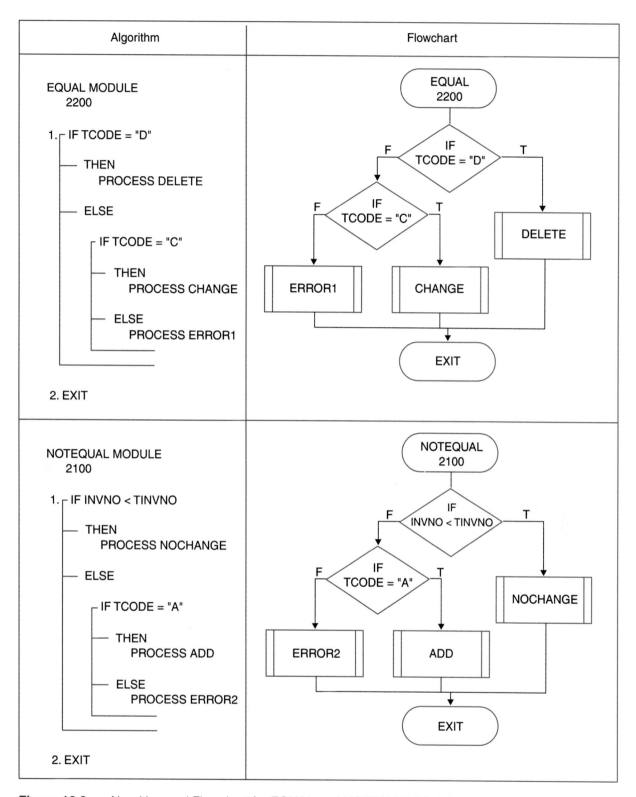

EQUAL MODULE
2200

1. ⌐ IF TCODE = "D"

— THEN
 PROCESS DELETE

— ELSE

 ⌐ IF TCODE = "C"

 — THEN
 PROCESS CHANGE

 — ELSE
 PROCESS ERROR1

2. EXIT

NOTEQUAL MODULE
2100

1. ⌐ IF INVNO < TINVNO

— THEN
 PROCESS NOCHANGE

— ELSE

 ⌐ IF TCODE = "A"

 — THEN
 PROCESS ADD

 — ELSE
 PROCESS ERROR2

2. EXIT

Figure 16.3e Algorithm and Flowchart for EQUAL and NOTEQUAL Modules

Algorithm	Flowchart
NOCHANGE MODULE 2300 1. WRITE (4) RECORD 2. READ (1) RECORD 3. EXIT	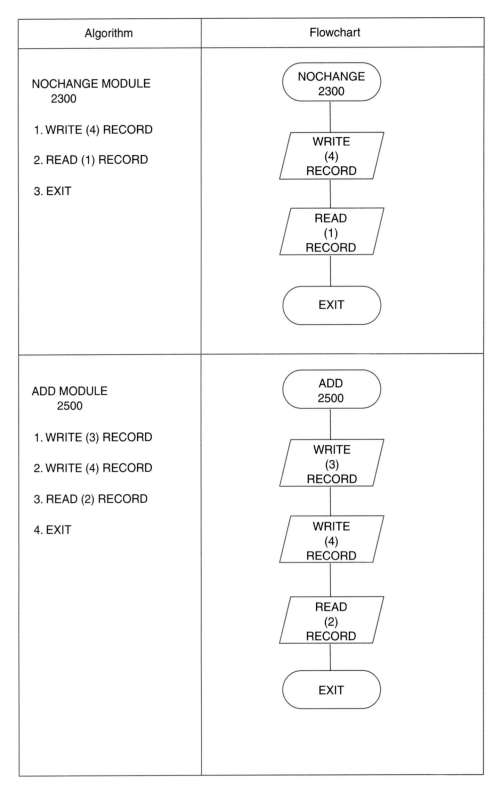
ADD MODULE 2500 1. WRITE (3) RECORD 2. WRITE (4) RECORD 3. READ (2) RECORD 4. EXIT	

Figure 16.3f Algorithm and Flowchart for NOCHANGE and ADD Modules

Algorithm	Flowchart
ERROR2 MODULE 2400 1. WRITE ERROR MESSAGE (NONEXISTENT INVNO) 2. READ (2) RECORD 3. EXIT	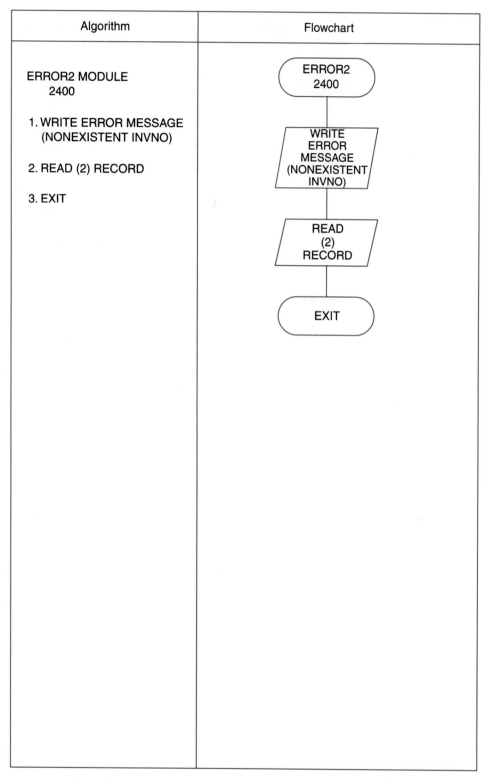

Figure 16.3g Algorithm and Flowchart for ERROR2 Module

Algorithm	Flowchart

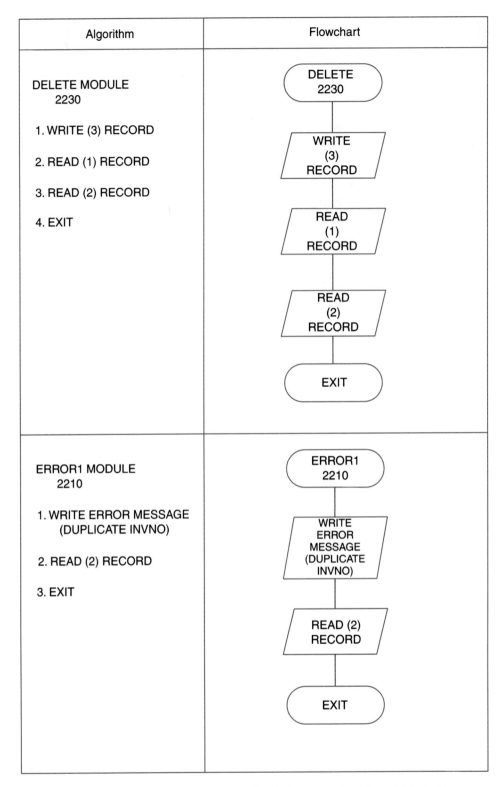

DELETE MODULE
2230

1. WRITE (3) RECORD

2. READ (1) RECORD

3. READ (2) RECORD

4. EXIT

ERROR1 MODULE
2210

1. WRITE ERROR MESSAGE
(DUPLICATE INVNO)

2. READ (2) RECORD

3. EXIT

Figure 16.3h Algorithm and Flowchart for DELETE and ERROR1 Modules

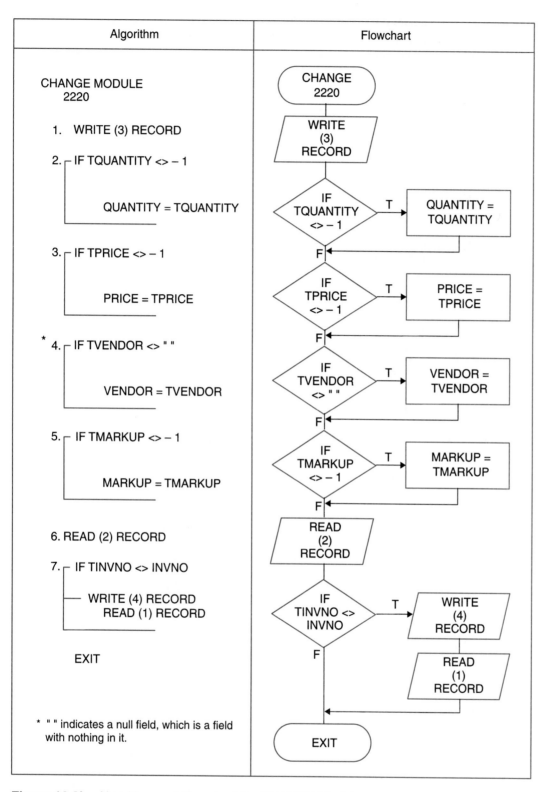

Algorithm	Flowchart
CHANGE MODULE 2220 1. WRITE (3) RECORD 2. ┌ IF TQUANTITY <> – 1 │ │ QUANTITY = TQUANTITY └ 3. ┌ IF TPRICE <> – 1 │ │ PRICE = TPRICE └ * 4. ┌ IF TVENDOR <> " " │ │ VENDOR = TVENDOR └ 5. ┌ IF TMARKUP <> – 1 │ │ MARKUP = TMARKUP └ 6. READ (2) RECORD 7. ┌ IF TINVNO <> INVNO │ ├─ WRITE (4) RECORD │ READ (1) RECORD └ EXIT * " " indicates a null field, which is a field with nothing in it.	

Figure 16.3i Algorithm and Flowchart for CHANGE Module

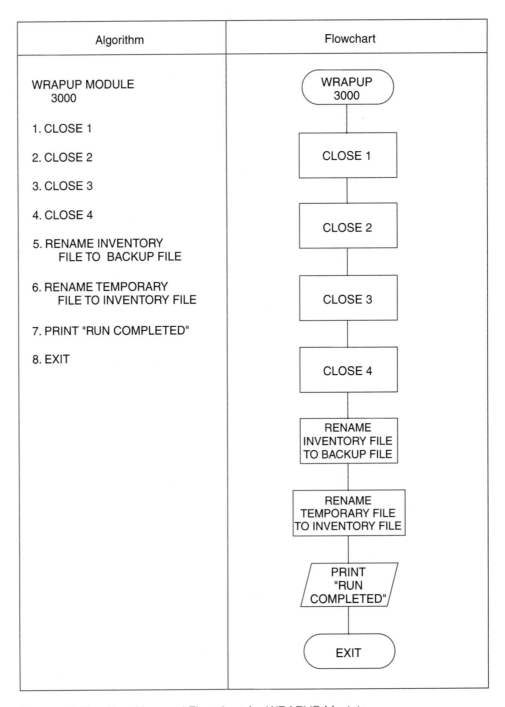

Algorithm	Flowchart
WRAPUP MODULE 3000 1. CLOSE 1 2. CLOSE 2 3. CLOSE 3 4. CLOSE 4 5. RENAME INVENTORY FILE TO BACKUP FILE 6. RENAME TEMPORARY FILE TO INVENTORY FILE 7. PRINT "RUN COMPLETED" 8. EXIT	WRAPUP 3000 → CLOSE 1 → CLOSE 2 → CLOSE 3 → CLOSE 4 → RENAME INVENTORY FILE TO BACKUP FILE → RENAME TEMPORARY FILE TO INVENTORY FILE → PRINT "RUN COMPLETED" → EXIT

Figure 16.3j Algorithm and Flowchart for WRAPUP Module

loop, the END-OF-FILE for the transaction file has been reached and there are no more updates to the master file. The remaining master file records are copied to the new master file in the Nochange module. Either the second or third loop is processed, not both.

In the Equal module (Figure 16.3e), the processes that can take place are to change the record (transaction code = "C") or to delete the record (transaction code = "D"). If the transaction code equals anything else, it is an error and the Error1 module is processed. In the Notequal module, when the master file key is less than the transaction file key, there are no updates to that master file record. When it is not less than the transaction file key,

then the transaction code should be an "A" and the transaction record is added to the master file. If it is not an "A", then there is an error and the Error2 module is processed.

The Nochange module (Figure 16.3f) writes the data to the new master file and reads the next record from the old master file. The Add module writes the transaction record to the activity file and the new master file and then reads the next record from the transaction file.

The Error2 module (Figure 16.3g) writes an error message to the output error report and reads the next transaction file record. If there is an error in the transaction file, that record should not be processed. The Delete module (Figure 16.3h) writes the transaction file record to the activity file and reads the next record from the old master file and the transaction file. The Error1 module writes an error message to the output error report and reads the next transaction file record.

The Change module (Figure 16.3i) is the most involved in that each field must be checked to see if the data in that field need to be changed. In this solution, -1 or " " (a null field—a field with nothing in it) indicates there is to be no change in that field. First, the transaction record is written to the activity file. Then the master file record is changed. The updated master file record is written to the new master file. The next transaction file record is read and checked to see if it is still equal to the master file record. If it is equal, there are more changes to be made to the master file record; if not, the next master file record is read.

In the Wrap-up module (Figure 16.3j), the files are closed, the old master file is renamed as the backup file, the new master file is renamed as the inventory file, and a message is printed to the screen that the run is completed.

A Useful Alternative Method

Figure 16.4a–d shows an alternative solution to the sequential file update problem. The main difference between the alternative solutions is how the update is driven. In the other examples in this chapter, the update is driven by the key. All updates, including adding records, changing records, and deleting records, are completed at the same time in key order. In the alternate method, the update is driven by the type of update; that is, all adds are completed at one time, as are the deletions and changes. When using the alternative method the transaction file must be sorted using the transaction code as the primary key and the primary key of the master file as the secondary key.

Summary

File updating is an essential part of working with files. The process of updating files includes adding, deleting, and changing records. Computer users use batch processing to update sequential files since random processing of sequential records is time-consuming on the computer. Updating a sequential file involves the following files: a transaction file, which contains all the updates; an activity file, which contains a record of all processes completed on the file; a backup file, which is the master file before the last update; a new master file, the updated master file; and, of course, the old master file, which contains the records to be updated. Sequential file updating involves updating the master file with the additions, deletions, and changes listed in the transaction file. After an update is completed, a new transaction file is created, and after a period of time, the user processes the master and the transaction files together again.

```
1. OPEN TF, AF
2. READ TF RECORD
3. HCODE = TCODE
4. WHILE NOT EOF(TF)
      OPEN MF, HF
      READ MF RECORD
      WHILE NOT EOF(MF) AND NOT EOF(TF) AND TCODE = HCODE
         IF TCODE = "A"
            THEN
               IF MFKEY = TFKEY
                  THEN
                     PROCESS ERROR 1
                     READ TF RECORD
                  ELSE
                     IF MFKEY < TFKEY
                        THEN
                           WRITE MF RECORD TO HF
                           READ MF RECORD
                        ELSE
                           WRITE AF RECORD
                           WRITE TF RECORD TO HF
                           READ TF RECORD
            ELSE
               IF TCODE = "D"
                  THEN
                     IF MFKEY = TFKEY
                        THEN
                           WRITE AF RECORD
                           READ MF RECORD
                           READ TF RECORD
                        ELSE
                           IF MFKEY < TFKEY
                              THEN
                                 WRITE MF RECORD TO HF
                                 READ MF RECORD
                              ELSE
                                 PROCESS ERROR 2
                                 READ TF RECORD
                  ELSE
                     IF MFKEY = TFKEY
                        THEN
                           WRITE AF RECORD
                           PROCESS CHANGE
                           WRITE MF RECORD TO HF
                           READ TF RECORD
                           IF MFKEY <> TFKEY
                              THEN
                                 READ MF RECORD
                              ELSE
                                 CONTINUE
                        ELSE
                           IF MFKEY < TFKEY
                              THEN
                                 WRITE MF RECORD TO HF
                                 READ MF RECORD
                              ELSE
                                 PROCESS ERRORS
                                 READ TF RECORD
      WHILE-END
      WHILE NOT EOF(MF)
         WRITE MF RECORD TO HF
         READ MF RECORD
      WHILE-END
      CLOSE FILE (MF, HF)
      REPLACE BACKUP FILE WITH MF
      REPLACE MF WITH HF
      WHILE-END
5. CLOSE FILE (TF, AF)
6. EXIT
```

Figure 16.4a Algorithm for Alternative Method of Sequential-Access File Update

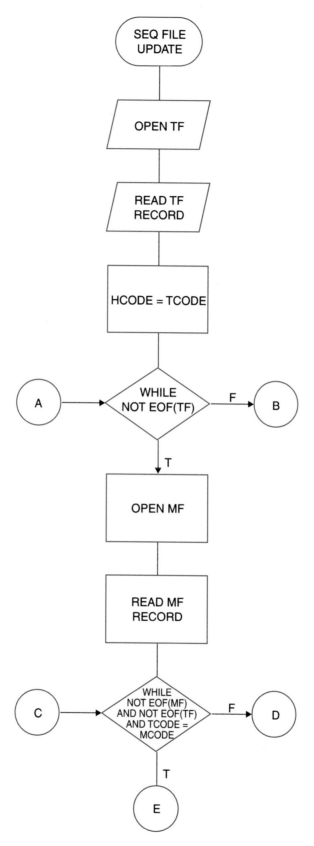

Figure 16.4b Flowchart for Alternative Method of Sequential-Access File Update

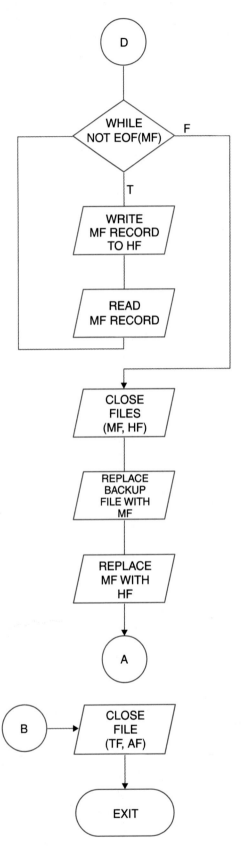

Figure 16.4c Flowchart for Alternative Method of Sequential-Access File Update

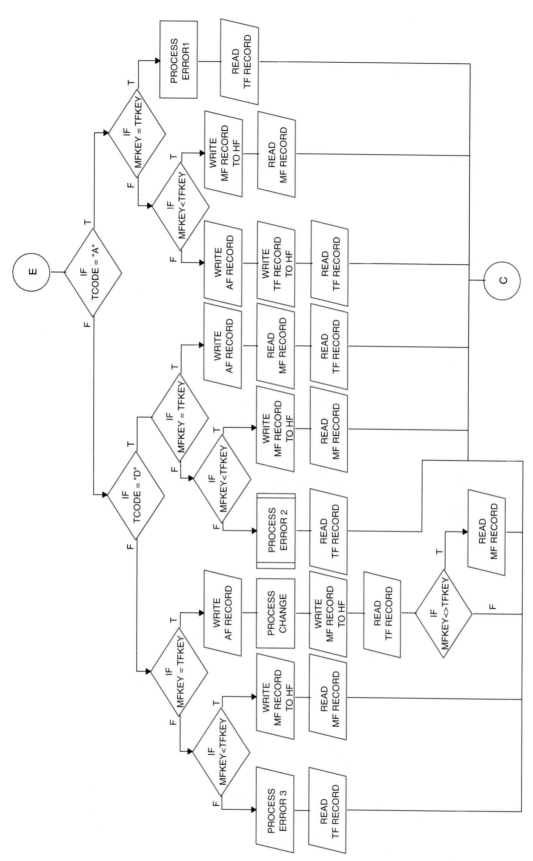

Figure 16.4d Flowchart for Alternative Method of Sequential-Access File Update

New Terms

activity file

backup file

batch processing

interactive processing

master file

transaction file

Questions

1. What processes are included in updating a file?
2. State the functions of the following files:
 a. old master file
 b. transaction file
 c. activity file
 d. new master file
 e. backup file

Problems

1. Develop a solution and data dictionary to update a client file given the following master and transaction file records: (The master file is in client number order.)

Master File Record

client #	name	telephone	address	city-state	zip code	last call	amount of last order

Transaction File Record

client #	tcode	name	telephone	address	city-state	zip code	last call	amount of last order

2. Develop a solution and data dictionary to update a book file for a library. After each update is complete, the librarian would like a copy of all new listings. Design the master file record and the transaction file record. The data in the master file record should include:
 a. reference number
 b. author
 c. title
 d. publisher
 e. ISBN
 f. copyright date
 g. number of copies
 h. subject
3. Develop a solution to update the file used in Problem 1 in Chapter 15, on page 379.
4. Develop a solution to update the file used in Problem 3 in Chapter 15, on page 379.

Chapter 17

Random-Access File Processing and Updating

Overview

Index Files
Updating Random-Access Files Using an Interactive Process

Objectives

When you have finished this chapter, you should be able to:

1. Explain the concept of index files.
2. Design a solution that requires updating of random-access files through the use of interactive processing.

Many computer applications require that the user be able to inquire about specific data in a file at any time or require that the data be current at all times. Random-access files enable the user to retrieve the most current information available on the computer. Chapter 14 discussed the different ways that random-access files could be updated, either interactively or sequentially, as the need required.

Menu-driven programs for random-access files allow the user to update, make an inquiry, or print a report. A record is accessed by giving the record number to the computer, which then retrieves it from the auxiliary storage device. The major problems with this type of retrieval are that the user normally does not know the record number and that the records normally are not physically in the required order but, instead, are in the order in which they were originally entered. Therefore, the programmer must use some method to permit accessing the desired record using the proper key. There are many ways to access the desired record as fast as possible and in some type of order. As mentioned in Chapter 14, the method taught in this text is that which uses index files.

Index Files

When using indexed files, there are one or more index files for each master file. An **index file** consists of two fields: The first is the key and the second is the corresponding record number. The index file is small enough to fit into the internal memory of the computer in order to speed up the retrieval of a random-access record. The index file is then sorted using the key field. This file is used to process the master file.

Figure 17.1 demonstrates the use of index files. The master file that contains all the data needed for a client file is stored on an external storage device. To develop an index file, such as the NAMEINDEX file, the file is first created with two fields. The first field is the name field, and the second is the record number of that name. This NAMEINDEX file is then sorted by one of many sorting techniques. The file is then used to process the MASTERFILE in alphabetical order of the names. Within the solution, the first record number in the MASTERFILE to be used is the value of the RECORDNO field in the

	MASTER FILE				INDEX ON NAME NAMEINDEX FILE			INDEX ON NUMBER NUMBERINDEX FILE	
RECORD NO.	NAME	NUMBER	OTHER DATA		NAME	RECORDNO		NUMBER	RECORDNO
1	DUNLAP	210		1	ADAMS	7		25	7
2	JOHNSON	100		2	BENNET	4		40	18
3	WHITE	125		3	BLACK	8		50	9
4	BENNET	275		4	BROWN	16		75	11
5	GREEN	225		5	COOPER	9		100	2
6	SMITH	200		6	DUNLAP	1		125	3
7	ADAMS	25		7	GREEN	5		150	14
8	BLACK	400		8	GREY	15		175	12
9	COOPER	50		9	HOLDING	13		200	6
10	OLSON	350		10	JOHNSON	2		210	1
11	WILLIAMS	75		11	LITTLE	14		225	5
12	NELSON	175		12	NELSON	12		250	15
13	HOLDING	335		13	OLSON	10		275	4
14	LITTLE	150		14	SMART	18		300	18
15	GREY	250		15	SMITH	6		325	16
16	BROWN	325		16	STONE	17		335	13
17	STONE	300		17	WHITE	3		350	10
18	SMART	40		18	WILLIAMS	11		400	8

Figure 17.1 Index Files

NAMEINDEX file, which is 7. Therefore, the seventh record is the first to be used, because it is the first record in alphabetical order. The second record to be used in the NAMEINDEX file is the fourth record in the MASTERFILE; the third is the eighth record; and so on. If at another time the MASTERFILE is to be accessed by NUMBER, then the NUMBERINDEX is used in a similar manner. Thus, by creating the NAMEINDEX and NUMBERINDEX files, the MASTERFILE does not have to be sorted again every time the requirements of the problem demand that the file be used in a different order.

Another use for the index file is to find a single record. This is done through either a *sequential search* of the file or through another search method, such as the *binary search*. (The method for a binary search is illustrated in Chapter 9.) The sequential search is the easiest, but it is usually the most time consuming; the key field of the index file is checked, record by record, until the correct record is found. Then the record is retrieved from the master file by using its record number from the index file. Figure 17.2 demonstrates the sequential method of searching for a record. In Figure 17.2, the NAMEINDEX file has been read into two parallel arrays called NAME and RECNO. The search takes place by checking the SEARCHNAME against the NAME array element by element until they match. After the loop has concluded, the element number of the array is checked to see if it is greater than the number of records in the MASTERFILE. If it is, then the record was not found. If the element number is less than or equal to the number of records in the MASTERFILE, then the record is retrieved from the MASTERFILE by using the corresponding record number in the RECNO array to find the correct record in the MASTERFILE.

Updating Random-Access Files
Using an Interactive Process

A menu-driven program is often used to update and process random files. The control menu lists the options for the user. The options usually include the following:

1. Add records
2. Delete records
3. Change records
4. Query
5. Print records
6. Quit

The processing always loops back to the control menu after completing one of the tasks such as adding a record. The only way out of the processing is for the user to use the quit option. There may be submenus within any of the listed options. For example, the Report module may have a submenu of various types of reports so that the user can select the type desired. In the add, delete, and change-a-record modules, processing continues until there are no more records to be processed. Then the program returns the user to the main menu. The program uses an index file to search for a particular record in the file.

The sections of Figure 17.3a–m show the development of a random-file update and processing problem for a mailing list. A company needs a program to maintain a mailing list for advertising purposes. The company collects data for the list from the mail and from people who request advertising material. The company needs a program to update the list and to print labels and a directory.

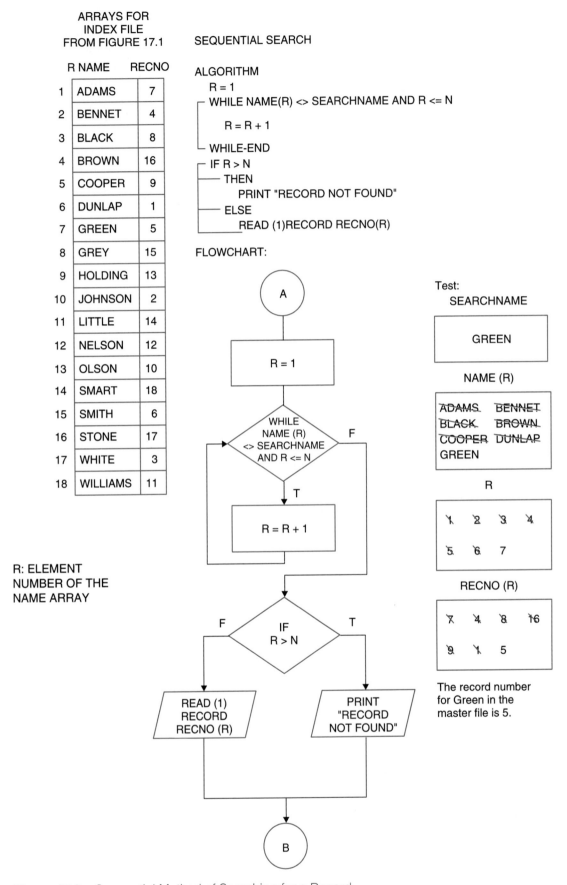

ARRAYS FOR
INDEX FILE
FROM FIGURE 17.1

	R NAME	RECNO
1	ADAMS	7
2	BENNET	4
3	BLACK	8
4	BROWN	16
5	COOPER	9
6	DUNLAP	1
7	GREEN	5
8	GREY	15
9	HOLDING	13
10	JOHNSON	2
11	LITTLE	14
12	NELSON	12
13	OLSON	10
14	SMART	18
15	SMITH	6
16	STONE	17
17	WHITE	3
18	WILLIAMS	11

R: ELEMENT
NUMBER OF THE
NAME ARRAY

SEQUENTIAL SEARCH

ALGORITHM
 R = 1
 WHILE NAME(R) <> SEARCHNAME AND R <= N

 R = R + 1

 WHILE-END
 IF R > N
 THEN
 PRINT "RECORD NOT FOUND"
 ELSE
 READ (1)RECORD RECNO(R)

FLOWCHART:

Test:
 SEARCHNAME

 GREEN

NAME (R)

ADAMS BENNET
BLACK BROWN
COOPER DUNLAP
GREEN

R

1 2 3 4
5 6 7

RECNO (R)

7 4 8 16
9 1 5

The record number
for Green in the
master file is 5.

Figure 17.2 Sequential Method of Searching for a Record

Record Data Dictionary

File Name: MAILLIST

Field Number	Item Description	Field Name	Field Length	Data Type	# of Decimals	Comments
1	Last Name	LNAME	15	C		Indexed on this field
2	First Name	FNAME	15	C		
3	Address	ADDRESS	20	C		
4	City	CITY	18	C		
5	State	STATE	2	C		
6	Zip Code	ZIPCODE	10	C		Will be used as a Secondary Key

Record Layout

1 15	16 30	31 50	51 68	69 70	71 80
LNAME	FNAME	ADDRESS	CITY	STATE	ZIPCODE

Figure 17.3a Menu-Driven Random-Access File Update and Reports Program: Data Dictionary and Record Layout

Record Data Dictionary

File Name: INDEXFILE

Field Number	Item Description	Field Name	Field Length	Data Type	# of Decimals	Comments
1	Last Name	NAME	15	C		Ordered on this field
2	Record Number	RECNO	4	I		

Record Layout

1 15	16 19	
NAME	RECNO	

Figure 17.3b Record Data Dictionary and Record Layout for INDEXFILE

411

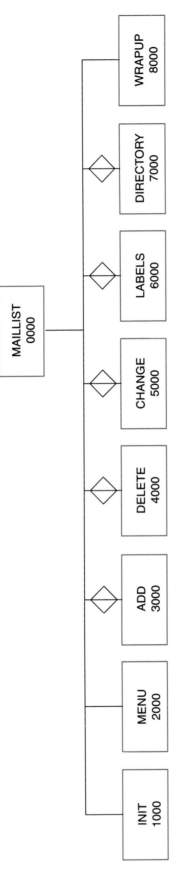

Figure 17.3c Interactivity Chart

Algorithm	Flowchart

Algorithm:

CONTROL MODULE
0000

1. PROCESS INIT

2. PROCESS MENU

3. WHILE CHOICE <> 6

 CASE OF CHOICE

 =1: PROCESS ADD

 =2: PROCESS DELETE

 =3: PROCESS CHANGE

 =4: PROCESS LABELS

 =5: PROCESS DIRECTORY

 END-OF-CASE

 PROCESS MENU

 WHILE-END

4. PROCESS WRAPUP

5. END

Figure 17.3d Algorithm and Flowchart for CONTROL Module

Algorithm	Flowchart
INIT MODULE 1000 1. PRINT INSTRUCTIONS 2. OPEN 1, MAILLIST 3. OPEN 2, INDEXFILE 4. READ (2), NAME (1), RECNO (1) 5. N = 1 6.⌐ WHILE NOT EOF(2) N = N + 1 READ (2), NAME (N) RECNO (N) └ WHILE-END 7. CLOSE 2 8. EXIT NAME = Name array RECNO = Array of record numbers N = Number of records in the data file and index file	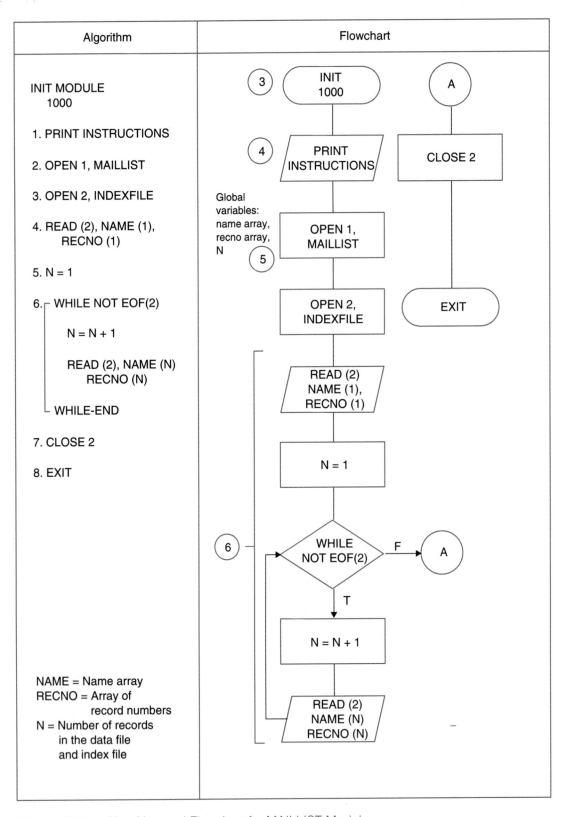

Global variables: name array, recno array, N

Figure 17.3e Algorithm and Flowchart for MAILLIST Module

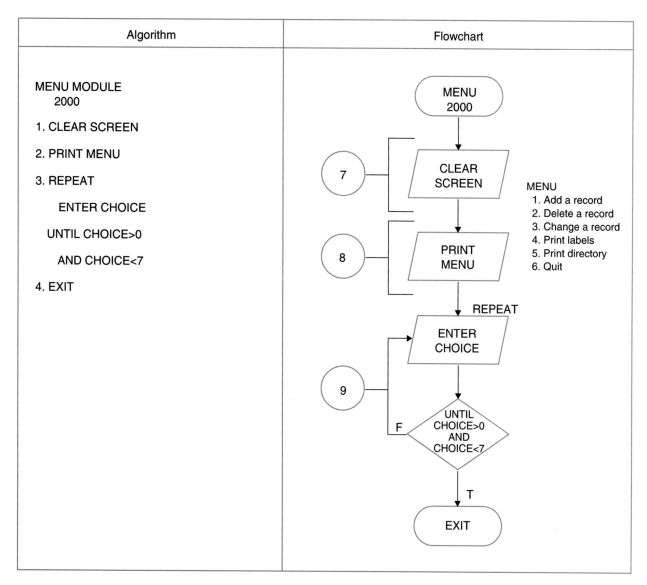

Algorithm	Flowchart
MENU MODULE 2000 1. CLEAR SCREEN 2. PRINT MENU 3. REPEAT ENTER CHOICE UNTIL CHOICE>0 AND CHOICE<7 4. EXIT	MENU 2000 7 — CLEAR SCREEN 8 — PRINT MENU MENU 1. Add a record 2. Delete a record 3. Change a record 4. Print labels 5. Print directory 6. Quit REPEAT ENTER CHOICE 9 F — UNTIL CHOICE>0 AND CHOICE<7 T EXIT

Figure 17.3f Algorithm and Flowchart for MENU Module

Notes on Figure 17.3a–m

The following notes explain the solution to the mailing list problem illustrated in Figure 17.3a–m. Items 1 and 2, which describe the field data and the layouts of the two records necessary to the solution, are illustrated in Figures 17.3a and 17.3b. The items numbered 3 through 51 correspond to the circled numbers in Figures 17.3d through 17.3m.

1. This is the MAILLIST file field data dictionary and record layout. The key field is the last name. There are 80 characters per record.
2. This is the INDEXFILE file record layout. There are 19 characters per record.
3. This setup module for the program is equivalent to the Init module. This is the first module that is executed since these are the first tasks that need to be completed. Once the program executes these instructions, it never returns to this module.
4. This instruction prints instructions for the user.

Algorithm	Flowchart

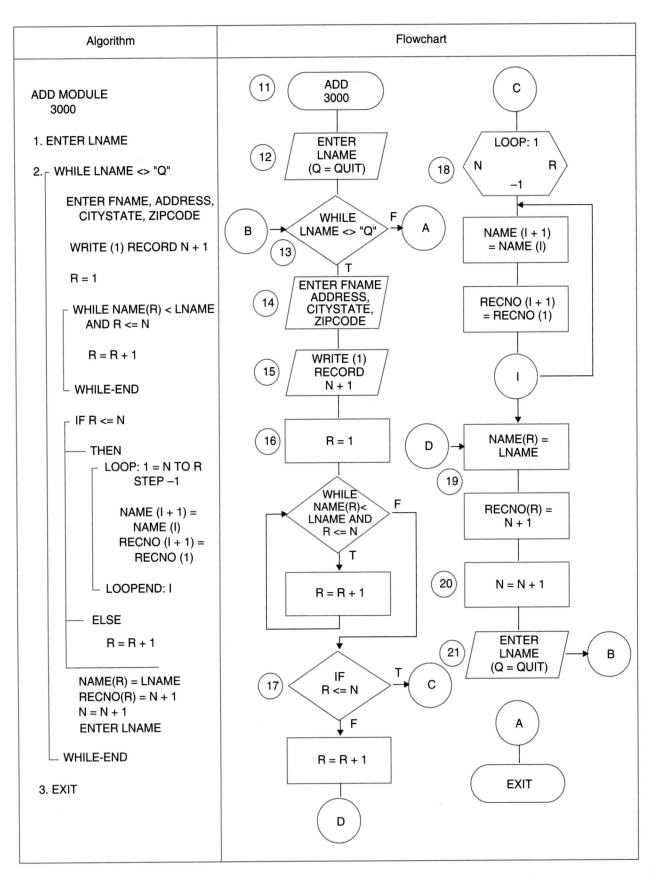

ADD MODULE
 3000

1. ENTER LNAME

2. ⌐ WHILE LNAME <> "Q"

 ENTER FNAME, ADDRESS,
 CITYSTATE, ZIPCODE

 WRITE (1) RECORD N + 1

 R = 1

 ⌐ WHILE NAME(R) < LNAME
 | AND R <= N

 | R = R + 1

 └ WHILE-END

 ⌐ IF R <= N

 | ── THEN
 | ⌐ LOOP: 1 = N TO R
 | | STEP −1
 | |
 | | NAME (I + 1) =
 | | NAME (I)
 | | RECNO (I + 1) =
 | | RECNO (1)
 | |
 | └ LOOPEND: I
 |
 | ── ELSE
 | R = R + 1
 |
 | NAME(R) = LNAME
 | RECNO(R) = N + 1
 | N = N + 1
 | ENTER LNAME
 |
 └ WHILE-END

3. EXIT

Figure 17.3g Algorithm and Flowchart for ADD Module

Algorithm	Flowchart

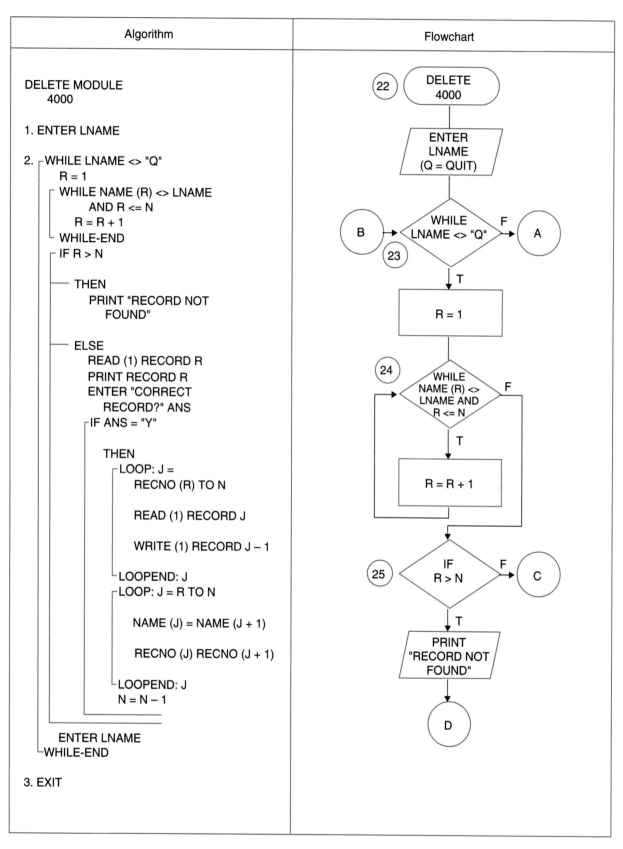

Algorithm:

DELETE MODULE
 4000

1. ENTER LNAME

2. WHILE LNAME <> "Q"
 R = 1
 WHILE NAME (R) <> LNAME
 AND R <= N
 R = R + 1
 WHILE-END
 IF R > N

 THEN
 PRINT "RECORD NOT
 FOUND"

 ELSE
 READ (1) RECORD R
 PRINT RECORD R
 ENTER "CORRECT
 RECORD?" ANS
 IF ANS = "Y"

 THEN
 LOOP: J =
 RECNO (R) TO N

 READ (1) RECORD J

 WRITE (1) RECORD J − 1

 LOOPEND: J
 LOOP: J = R TO N

 NAME (J) = NAME (J + 1)

 RECNO (J) RECNO (J + 1)

 LOOPEND: J
 N = N − 1

 ENTER LNAME
 WHILE-END

3. EXIT

Flowchart:

22 — DELETE 4000

ENTER LNAME (Q = QUIT)

B → WHILE LNAME <> "Q" — F → A
23
T

R = 1

24 — WHILE NAME (R) <> LNAME AND R <= N — F
T

R = R + 1

25 — IF R > N — F → C
T

PRINT "RECORD NOT FOUND"

D

Figure 17.3h Algorithm and Flowchart for DELETE Module

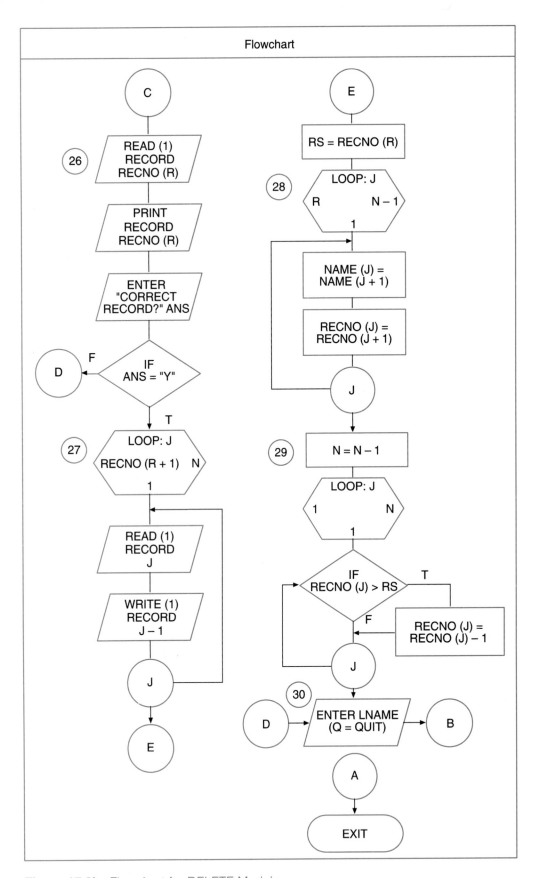

Figure 17.3i Flowchart for DELETE Module

Algorithm	Flowchart
CHANGE MODULE 5000 1. ENTER LNAME 2. ┌ WHILE LNAME <> "Q" R = 1 ┌ WHILE NAME (R) <> LNAME AND R <= N R = R + 1 └ WHILE-END ┌ IF R > N ├─ THEN PRINT "RECORD NOT FOUND" ├─ ELSE READ (1) RECORD R PRINT RECORD R ENTER "CORRECT RECORD" ANS ┌ IF ANS = "Y" ├─ THEN ENTER FNAME, ADDRESS, CITYSTATE, ZIPCODE WRITE (1) RECORD R ENTER LNAME └ WHILE-END 3. EXIT	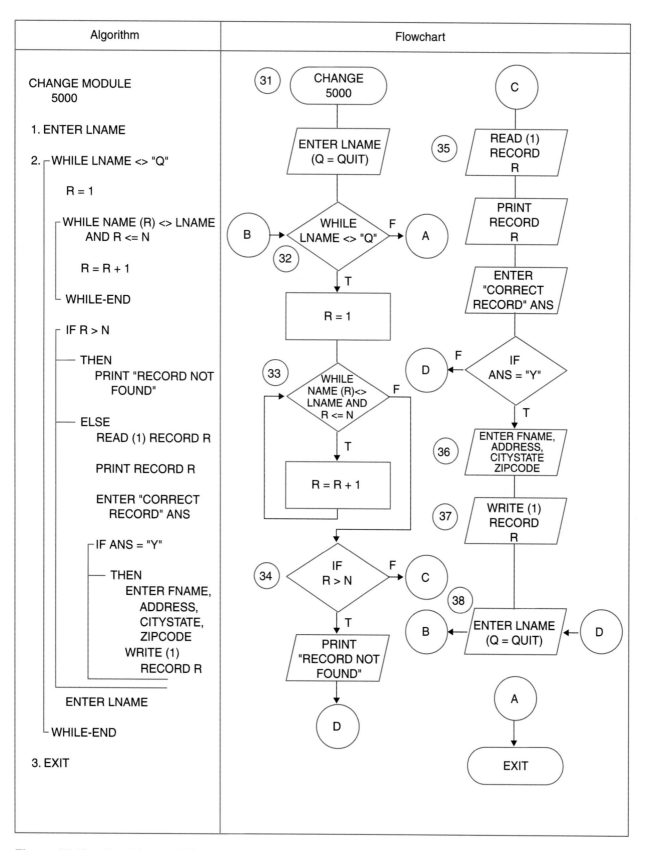

Figure 17.3j Algorithm and Flowchart for CHANGE Module

Algorithm	Flowchart

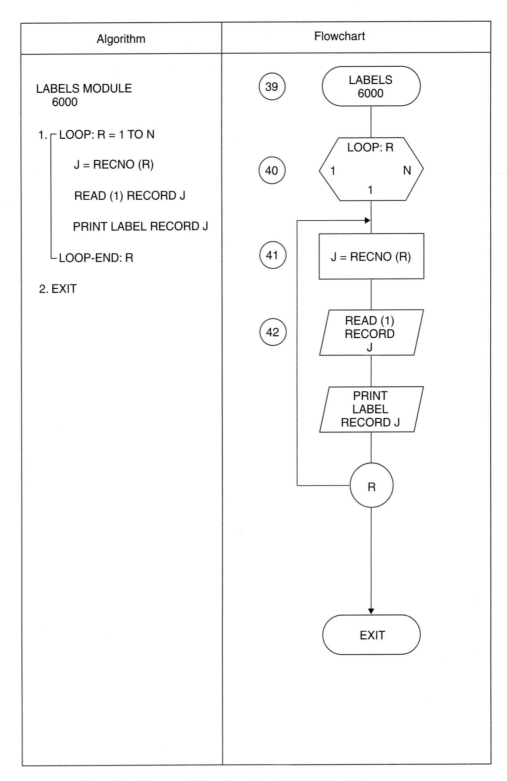

LABELS MODULE
 6000

1. ┌ LOOP: R = 1 TO N

 J = RECNO (R)

 READ (1) RECORD J

 PRINT LABEL RECORD J

 └ LOOP-END: R

2. EXIT

(39) LABELS 6000

(40) LOOP: R — 1 — N — 1

(41) J = RECNO (R)

(42) READ (1) RECORD J

PRINT LABEL RECORD J

R

EXIT

Figure 17.3k Algorithm and Flowchart for LABELS Module

Algorithm	Flowchart
DIRECTORY MODULE 7000 1. ⌐LOOP: R = 1 TO N J = RECNO (R) READ (1) RECORD J PRINT RECORD J └LOOP-END: R 2. EXIT	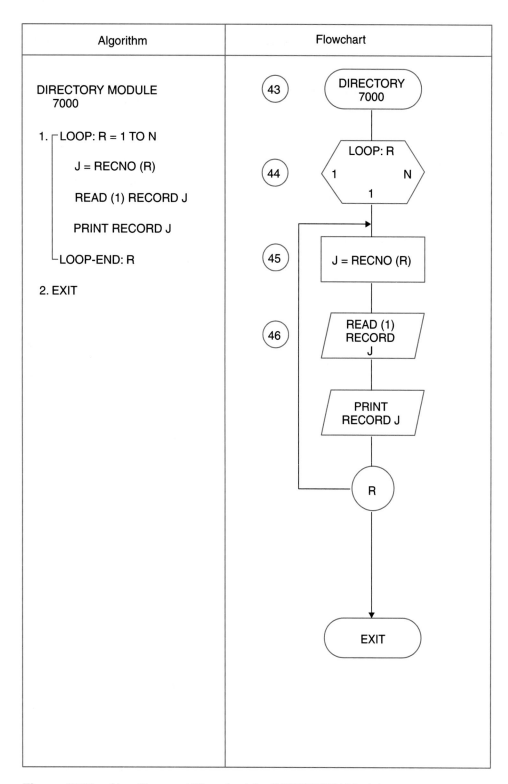

Figure 17.3I Algorithm and Flowchart for DIRECTORY Module

Algorithm	Flowchart

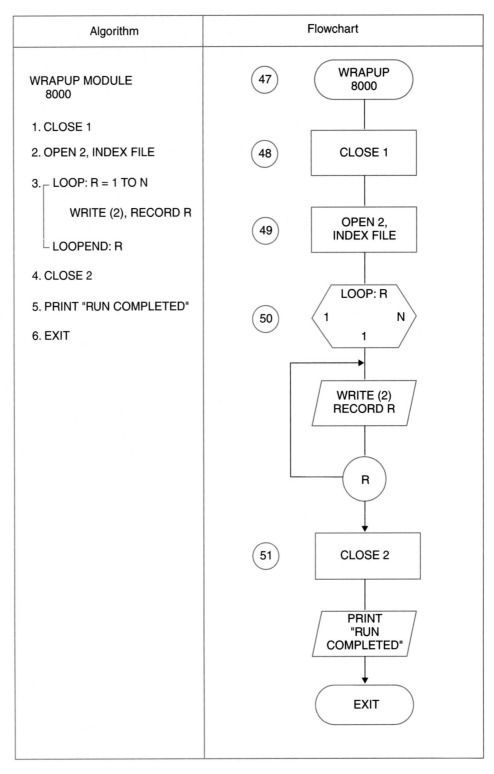

WRAPUP MODULE
 8000

1. CLOSE 1

2. OPEN 2, INDEX FILE

3. ┌ LOOP: R = 1 TO N

 WRITE (2), RECORD R

 └ LOOPEND: R

4. CLOSE 2

5. PRINT "RUN COMPLETED"

6. EXIT

47 — WRAPUP 8000

48 — CLOSE 1

49 — OPEN 2, INDEX FILE

50 — LOOP: R 1 N 1

WRITE (2) RECORD R

R

51 — CLOSE 2

PRINT "RUN COMPLETED"

EXIT

Figure 17.3m Algorithm and Flowchart for WRAPUP Module

5. These instructions open both files.
6. These instructions copy the index file into parallel arrays. The index file is then closed since it will not be used again until the Wrap-up module.
7. This instruction clears the screen.
8. This instruction prints the main menu.
9. This loop enters the user's choice from the main menu. It loops until the entered number is within the range of choices on the menu.
10. This CASE OF instruction, along with the user's choice from the menu, chains one of the modules.
11. This module adds a record.
12. This instruction enters the last name of the first record to be added to the file.
13. This instruction loops as long as the last name is not equal to "Q". This allows multiple records to be added at the same time without going through the menu in the driver each time.
14. This instruction enters the remaining data for each record.
15. This instruction appends the record to the file (adds the record to the end of the file).
16. This instruction searches to find the place in the index file to insert the key and the record number.
17. When FALSE, this decision indicates the new record should be appended to the end of the index file.
18. This loop moves all records down one record in the file to leave room for the new record to be inserted.
19. These two instructions insert the values of the key field and the record number of the new record into the parallel arrays (the index arrays).
20. This counter increases the number of records by one.
21. This instruction allows the user to enter the last name of the next record to be added to the file. The user enters a Q when there are no more records to be added.
22. This is the module to delete a record.
23. This instruction loops until there are no more records to be deleted.
24. This loop searches to find the record to be deleted.
25. When this condition is TRUE, the record could not be found. The processing then continues at the end of the loop by entering another name.
26. These instructions read and print the record and then ask the user whether it is the correct record. If it is not, then the processing continues at the end of the loop by entering a new name.
27. This loop moves the records up. In doing so the requested record is deleted.
28. This loop moves the elements up in the index arrays.
29. This counter decreases the number of records by one.
30. This instruction allows the user to enter the next record to be deleted. If there are no more records to be deleted, the user enters a Q.
31. This is the module to change data in a record.
32. This instruction loops until there are no more records to change.
33. This loop searches for the record to change.
34. If this condition is TRUE, the record could not be found. Processing then continues at the end of the loop by entering a new name.
35. These instructions read and print the record and then ask the user whether it is the correct record. If it is not, then the processing continues at the end of the loop by entering a new name.
36. This instruction enters the changed data.

37. This instruction writes the record back to its original location in the file.
38. This instruction allows the user to enter the name of the next record to be changed. If there are no more records to be changed, the user enters a Q.
39. This is the module to print labels.
40. This loop prints the labels for all records.
41. J is the pointer that points to the next record to be processed. RECNO(R) is the corresponding record number in the index arrays.
42. These instructions read the record to be processed and print the label of the record in the memory.
43. This is the module to print the directory.
44. This loop prints the information for all records.
45. J is the pointer that points to the next record to be processed. RECNO(R) is the corresponding record number in the index arrays.
46. These instructions read the record to be processed and print the data on the record for the directory.
47. This module processes all the necessary tasks after the files have been updated and processed.
48. This instruction closes the MAILLIST data file.
49. This instruction opens the INDEXFILE.
50. This loop copies the data in the index arrays to the INDEXFILE.
51. This instruction closes the INDEXFILE.

Summary

Random-access files are becoming more and more important as businesses demand more up-to-date information. Random-access files are used any time data might be needed at a moment's notice and/or when data must be as current as possible. The program must be interactive because the user needs to converse regularly with the computer through the keyboard and monitor.

The student programmer has to learn some specialized techniques and concepts to work with random-access files.

Many random-access programs allow the user to both update and process files through the use of a menu. The menu usually consists of options to add, delete, and change a record, make a query, print reports, and quit. The program loops until the user is ready to quit.

New Term

index file

Questions

1. What is the difference between a random-access file and a sequential-access file?
2. Can you efficiently process a sequential-access file randomly? Explain.
3. Can you process a random-access file sequentially? Explain.
4. Define *index*. What is the purpose of an index?

Problems

Develop solutions for the following problems. Your solution should include a problem analysis chart, an IPO chart, a structure chart, a data dictionary, output report designs, and the algorithms and flowcharts for all modules.

1. **a.** Use the random-access file developed in Problem 5 in Chapter 14 for an inventory system for a TV or radio station. (LP records, cassettes, CDs, and VCRs are all used by the station.) How would this file be ordered? Which field(s) would be used for searching? Which fields might need to be updated? What reports would be needed?

 b. Develop a solution to update and print the records for the file in Problem 1a.

2. Complete Problems 1a and 1b for the following:
 a. a book inventory
 b. the students and their grades for a course
 c. a checking account for an individual

3. An airline has information on all of its flights from departure to arrival time. The company needs to update information constantly about the number of seats sold, the number of seats remaining, the time of departure, the estimated time of arrival, the departure airport, and the destination airport. The file must be updated whenever a reservation is made or canceled, or when a flight begins or ends. The airline has to be able to inquire about a particular flight and about all flights to one particular airport within a certain time frame. The company also needs a report on all completed flights. Develop the solution to this problem.

Unit Five

Supplementary Exercises

Develop a complete solution, including a data dictionary, to the following problems:

1. The financial director of a private school would like a tracking system for the students at the school.
 a. Design the file. The information on the student includes: name, home address and telephone, age, sex, class status, parent's name (one parent), dormitory room, tuition, room fees, board fees, book fees, computer fees, miscellaneous fees, date of payment on each charge.
 b. Design the reports needed:
 Student directory
 Charges
 Outstanding fees
 List of students by dormitory
 List of students by class
 c. Develop a solution using sequential-access files to output the needed reports.
 d. Develop an update program.
2. Develop Problem 1 using random-access files.

UNIT SIX

PROBLEM SOLVING FOR APPLICATION SOFTWARE

Chapter 18

Introduction to Application Software

Overview

Types of Horizontal Software
Steps to Learning a New Software Package
Software Compatibility Problems
File Organization

Objectives

When you have finished this chapter, you should be able to:

1. List and define the major types of horizontal software.
2. List and explain the steps needed to learn a new application software package.
3. List and explain some of the compatibility problems associated with application software.

Microcomputers started a revolution in the computer industry during the late 1970s. Until that time, computers were used only by large businesses because of the initial and operating costs of a computer. Microcomputers made computers affordable for almost anyone, but many people who were interested in using them were not interested in programming them.

The microcomputer industry responded to this new consumer need by developing software programs for the general public. The first software was **vertical software.** Vertical software is purpose specific; that is, it is for dental accounting, retail inventory, recreational games, educational programs, and so forth. In the early 1980s the software industry started to sell **horizontal software.** Horizontal software is broader in purpose and capable of solving many kinds of problems.

Programmers use application software for creating reports, external documentation, budgets, and other documents pertaining to the programs they are developing. It is important for the programmer to learn to use these packages efficiently to increase their

productivity. Document processing is important to you as a programmer because you may want to include a database report, a graphic, a spreadsheet, or another type of document in your reports.

Types of Horizontal Software

The main types of horizontal software in use today are word processing, graphics, desktop publishing, data-management, and spreadsheet programs. Other types of horizontal software to develop applications for end users include Web page development and multimedia development. However, we will not discuss these packages in this text. Both of these types of packages require the use of good screen design techniques. Just about anyone can develop a Web page; however, it requires graphic design training to develop attractive Web pages.

The use of graphics with all productivity packages is increasing. Graphics can be obtained from a variety of sources, including buying clip art, photo clips, and drawing your own. There are also packages that allow you to create animation and capture sound and video. These are part of multimedia. The use of multimedia in applications is also increasing in use. Multimedia is defined as combining text, graphics, video, animation, and sound in developing an interactive application. Chapters 12 and 13 present topics on the design and development of graphics user interfaces for use in the development of multimedia applications. The following chapters will not include these topics. They will, however, include the use of still graphics in connection with word processing, desktop publishing, and spreadsheets.

Word processing enables the user to edit text electronically. Desktop publishing is the combination of word processing and computer graphic arts. This type of software at the present is usually a page layout program; however, the trend is to combine word processing and desktop publishing into one package. Data management allows the user to enter data easily, organize it, and create reports. Spreadsheets enable a user to calculate worksheets with a computer instead of a calculator and to recalculate easily whenever the data change. These tasks are performed in business every day. Just as a person does not have to be a jeweler to use a watch or a mechanic to drive a car, users no longer need to know how to program, in the traditional sense, to perform these tasks on a computer. However, computer users do still need to know how to operate the computer and how to use the software. The end user can use all productivity packages without knowing anything about programming. However, many horizontal productivity software packages have built-in programming languages. The word *programming* may not be used, but that is really what the end user is doing. Word processors and spreadsheets have macros that the end user can write. A macro is a short set of instructions or keystrokes that can be saved for use whenever needed. Database management systems have a more extensive built-in language. So the end users must have some knowledge of programming if they are going to use the software package to the extent it is intended. Chapters 19, 20, and 11 explain the type of languages used for each type of application.

Word processing, elementary graphic concepts, and desktop publishing topics are presented in Chapter 19. Spreadsheets and elementary presentation graphics topics are presented in Chapter 20. Refer to Chapter 11 for relational database management systems topics. These chapters give an overview of how these application software packages can improve your productivity.

Computers and horizontal software packages are designed to help the user make better decisions and to increase productivity, efficiency, and accuracy. For this reason,

horizontal software packages are called **productivity packages.** If after learning to use a horizontal software package your productivity has not increased, then you are not using the package correctly, you are not using the correct package, or you should not be using a computer for that particular application. It is very important to choose the correct tool to complete the job.

Steps to Learning a New Software Package

A software package contains the following parts:

1. One or more diskettes or a CD-ROM containing the programs, computer-based tutorials, and any other files necessary to use the package.
2. One or more manuals. These manuals provide instructions for the use of the software, manual-based tutorials for learning the package, installation instructions, and other information for the end user.

It is advisable to be systematic when you are learning a new software package. Here are some recommended steps to learning new software.

1. Look through the manuals to get an overview of the sections for future reference.
2. Find the section in one of the manuals on how to install the program *for your computer.* Computer systems vary. Software packages, however, are universal, so they have to be adapted to a specific computer. When you **install a software package,** you match it to the type of printer, the type of monitor, and the type of disk drive in your computer setup. Follow the directions in the manual very closely. If the program comes on diskettes, the first step usually is to copy all of the diskettes onto other diskettes and to put the originals away for safekeeping. The second step is to run an install or setup program—a program that helps the user install the software package. Most installation programs are very user friendly. If new hardware is bought, then software to be used with the new hardware may need to be reinstalled.
3. Run the computer-based **tutorials.** Tutorials consist of demonstrations and instructions for the productivity packages. Start the tutorial and work through the lessons. Do not try to remember everything the first time, but try to get an understanding of the capabilities of the package.
4. Examine the manuals and get a general idea of their contents. Find all the indexes. Often there will be more than one index in a manual. You can find an answer to most questions by using the indexes and the table of contents.
5. Look through any tutorials in the manuals. Complete the tutorial exercises as needed. After each tutorial, complete the same task with other data. Run the computer-based tutorial as often as you need to.
6. Complete one of the types of tasks for which the package was designed, just to get a feel for the package. When problems occur, read the manual to find a solution. Use the online help as much as possible. Today's application packages often have almost the complete manual available for use under the help menus. Some software packages offer online tutorials for the end user to use without exiting the package. These are extensive and provide timely reminders of how to accomplish something. Keep a notebook of solutions to problems, and repeat the steps that caused the problem, to better understand what happened. If there is a problem that you cannot solve, ask other users

about it or call the company hotline. Most of the larger software houses offer help to those who register software with them.

7. Add several features every time you use the package for a new problem. Do not be content with using only a few features of a full-featured package.

Productivity packages are designed to help. They can be frustrating at first because they are complex. However, it doesn't take too long to become familiar with them. Once you are comfortable with them, the capabilities and speed of the packages will amaze you. Once you overcome initial fears and frustrations, the programs are fun and exciting.

Software Compatibility Problems

Depending on the computer system and software packages you are using, you may have compatibility problems when you try to import files (transfer data files from another program to the one you are using) or to export files (transfer data files from the program you are using to another program). An example of compatibility would be to create a graphic in a paint package and import it to a desktop publishing package. The saved file may or may not be compatible with the software package doing the importing. If it is not, then you have to use a utility to convert the data from the format in which it is saved to a format that is acceptable to the importing software package.

Most word-processing packages import and export text as ASCII or RTF (Rich Text Format) files without any problem. However, ASCII files are saved without any of the special print features produced in the word-processing package. Many application software packages have lists of formats that the package will accept on opening a document. Likewise, there are many formats in which the end user can save a document. This includes other versions of the same package. Most manufacturers of software make sure their new versions are **upward compatible.** This means that documents produced in previous versions of the software are compatible with the new version. The term **downward compatible** means that a document produced by a new version is compatible in a previous version of the software.

Integrated software combines multiple types of full-featured horizontal software into one package. The advantage of using integrated software is the ease of importing and exporting data from one type to another type of software. The disadvantage is the decrease in the number of features of each package.

File Organization

You will be creating many data files during the time you are working on the computer. It is important that these files are organized so you can find them again without a lot of hunting. Program files will automatically be placed in folders when they are installed. However, it is up to you to organize data files into folders. You need to design the organization of your files in a way that you know where files are located. If you were using a file cabinet, you would not put every paper in the same folder. You would organize the documents into folders labeled according to what has been put into the folder. You do the same thing on the computer.

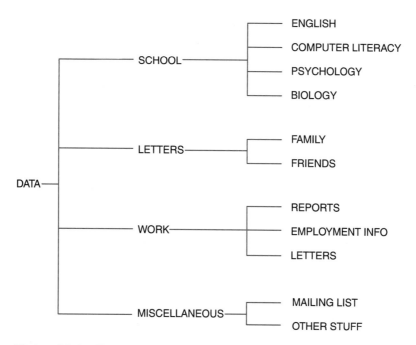

Figure 18.1 Example Data File Organization System

Containers in which you put computer files are called **folders.** You will be organizing folders within folders. The main folder should be labeled accordingly—such as MY DOCUMENTS, or DATA, or NORM'S DATA and MAUREEN'S DATA if you have several people using the computer. Within these major folders, you will need to discover the best way to organize your own documents, such as a folder called SCHOOLWORK that contains your schoolwork, another folder called PERSONAL LETTERS, another called MISCELLANEOUS, and so on. Within your schoolwork folder, you might have a folder for each class. It is very frustrating to know you have created a file for a report due in English tomorrow and cannot find where you saved the file. If you save the paper in your ENGLISH folder within the SCHOOLWORK folder, you can find it immediately. Figure 18.1 gives an example of how one person might organize his or her data into folders.

Summary

With the advent of the microcomputer came the need for software. Part of this need has been met through the development of such productivity packages as word processing for text editing, graphic packages, desktop-publishing packages for combining word-processing and graphics capabilities, data-management systems for data organization, and spreadsheets for automatic calculation of worksheets. These packages require the user to learn how to use a computer and to learn how to enter instructions to solve problems, but they do not require any programming in the traditional sense. Problem solving for applications requires different skills than for languages.

The steps to learning a new package include installing the software for a particular computer, using the tutorials, examining the manuals, and using the package for practice problems.

New Terms

downward compatible

folders

horizontal software

install a software package

productivity package

tutorials

upward compatible

vertical software

Questions

1. Define the term *productivity package*.
2. Name at least three types of productivity packages.
3. State the steps to learning a new productivity package.
4. What is meant by "installing your software"?
5. What is meant by *upward compatibility?*
6. What is meant by *downward compatibility?*
7. Why are upward and downward compatibility important to the end user?

Problem

1. Design a data file organization system for yourself.
 a. List the types of data files you expect to develop.
 b. Design your data file organization system so you can easily find these files. Be sure to name the folders according to what you will save in them.

Chapter 19

Problem Solving for Word Processing and Desktop Publishing

Overview

Word Processing
 Functions of a Word Processor
Graphics
Desktop Publishing
Some Design Concepts
Word-Processing and Desktop-Publishing Planning
Types of Documents

Objectives

When you have finished this chapter, you should be able to:

1. Define the concepts of a well-designed document.
2. List and describe the types of documents produced by a word processor and a desktop publisher.
3. Define the basic features of a word processor.
4. Define the basic features of a desktop publisher.
5. Develop a planning guide for the page layout of a document.

Word processing is one of the key programs that you as a programmer will use. You probably have been using word processing for some time already. It is the package that you will use to write your reports, external documentation, communications with colleagues and clients, and many other documents. You may want to use a desktop publisher. However, most documents you will be creating can easily be completed using a word processor. This chapter will review some of the basic functions of a word processor and a desktop publisher and present design concepts for making a professional document.

 Word processing and desktop publishing are discussed here, in a problem-solving text, because some problem solving is involved when a user designs and produces a

document. A **document** is the text and graphics placed in one file. **Word-processing** software enables the user to design, edit, and view a document before it is printed. A **desktop publisher** is a page layout program. It will take text from a word processor and graphics from a graphics program and combine them into a professional quality document. Some word processors have limited desktop-publishing capabilities, and some desktop publishers have limited word-processing capabilities; if at all possible, you should use the software best designed for your particular needs.

Word processing is the most widely used productivity application. Word processing is coming very close to replacing the typewriter both at home and in the office. It is usually the first package implemented when a computer system is purchased. Most people learn to use a word processor in a relatively short time, but it takes longer to use it efficiently. With a word processor, text no longer has to be typed and retyped every time there is a correction or revision. The text is stored electronically on a separate file and is accessed through the word-processing program. The extent of what a word-processing package can do depends on the functions provided in the package.

Word Processing

The primary goal of a word processor is to increase productivity by editing text electronically. Editing is accomplished through such basic functions as cursor control for moving around in the document, formatting pages, correcting text, copying and moving text from one place to another, checking spelling, style, and grammar, enhancing characters, and merging and separating documents. The functions found in a package vary from few to many, depending on the package. Some functions are automatically performed by the computer; others are performed manually by the user. The user executes functions through commands given directly to the computer, by choosing the desired functions from a menu, or by the click of a mouse button.

Functions of a Word Processor

A word processor has many functions to increase productivity. These functions can be separated into categories. The first is text formatting. Text formatting allows you to select how the text will look on the page. The second is text editing or correction. This allows you to change the text after it has been entered into the computer. The third is file manipulation. This allows you to save, print, export, and perform other activities with your text. The fourth is tools that allow you to add columns, tables, graphics, sort data, check spelling, and so on. The fifth function is the ability to build macros to allow you to combine multiple tasks into one task. The sixth is the view function, which allows you to view your document in multiple ways. The last function is communication tools that allow you to access information from outside resources. You need to learn to use these functions efficiently along with the help menus to increase your productivity when creating a document. When you are planning a document you need to keep these functions in mind to decide how you will use them to create the document in the most efficient way.

Text formatting. Ask yourself the following questions when you are considering the function of text formatting:

- What type of fonts will you use? There are many types of fonts. Most fonts are used for headlines, special effects, and so on. There are only a few fonts that are used with body text. You want to choose a font for the body text that is readable and that corresponds to the style of the document. If the document is a formal

document, you might want to use Times Roman; however, if the document is informal and has a more modern style, you might want to use Arial. Check the fonts on your word processor and evaluate which font you would use for which purpose. Since many word processors only give you the names of the fonts and do not show how each font looks, it is useful to make a document to refer to that shows the appearance of each font.

- What size text will you use? The size of the text depends on what it will be used for. The body text of most documents will use 10 or 12 point sizes. However, you may want to use larger fonts for headings, for cover pages, for flyers, or other purposes. You may want to use smaller sizes for superscripts, subscripts, or on drawings. Choose the size that fits the document. You may want to use 10 point for a long document or 12 point for a hard to read document. Text in columns should be sized so that $1^1/4$ to $1^1/2$ alphabets fit on one row of text. If the text is larger than this, there will be too many hyphenated words, the text will be harder to read, and aesthetically it looks bad. If it is smaller that this, the text looks too crowded and is too hard to read.

- Will you use color to emphasize your text? If your document will not be copied, you may want to use color to emphasize some important text.

- What will be your margins? Your document will dictate your margins. If you plan to bind your document, be sure to have the left, or inside margins, wider than the right, or outside margins, to leave room for the binding. This is usually an extra quarter of an inch unless the binding is extra wide. If the document is a little to long or short for the number of pages you desire for your document, adjust the margins a little so the document will fit.

- What type of line and paragraph spacing will you use? Your document may be single, double, or $1^1/2$ spaced. You may want extra space between your paragraphs, as in a letter. Let the computer do the spacing for you. It saves time and energy.

- Will you use numbering or bullets for lists? If so, what type of numbering or bullets will you use? If you have lists, or if you are creating an outline, you will want to use numbering or bullets. If you use bullets, be sure to choose an appropriate bullet for the list and the document. A tree or some other symbol used for a bullet in a list of tree species, for example, may be appropriate for some documents but not others, depending on the formality of the document. You may want to use a check mark, an asterisk, a dot, a dash, or some other appropriate symbol or you may want to have a numbered list. Again let the computer put in the numbers and the bullets, as the computer will return to the correct spot for the next line.

- What type of alignment will you use? The four types of alignment are left, right, center, and full justification. Left alignment means that the characters are aligned on the left and the right side is jagged. Right alignment is the opposite, the characters are aligned on the right with the left side jagged. Center alignment means that all lines are aligned in the center of the margins. Full justification means that both the right and the left sides are aligned with the extra space distributed throughout the line. Traditional documents should be full justified. However, the trend today is to use left alignment, because it is considered be more aesthetic as the extra spaces bother the eye. This is a matter of opinion. This trend has come about because of desktop publishing.

- Will you use borders or shading with any of your text? Borders and shading should be used to highlight featured text or other text elements that are separate from the body text. Use these tools sparingly, as too much use can be distracting to the body text.

- What will be the tab settings? What type of tabs will you use? You need to set the tabs and the tab leader, if one is used, at the beginning of the document, and at any place in the document where tabs change. This should be part of your planning document. Also, you need to decide what type of tabs you will be using: left, right, center, or decimal. Refer to Figure 19.2 on page 445 to see how different types of tabs look.
- Will you use drop caps anywhere? Drop caps are large first letters in a paragraph. They are used at the beginning of a chapter or a section of a large document such as a book. They are used very seldom, but when needed can add to the appearance of the text.
- Will you use boldface, italicized, or underlined text? Boldface, italicized, or underlined text is used to emphasize the text. Again you need to use these enhancements sparingly.
- Will you use other text enhancing such as strikethrough, embossed, superscript, subscript, and so on?
- Will you be using footers and headers? If so, what needs to be included? Footers are one or more lines of text repeated at the bottom of all pages. Headers are the same but at the top of the page. These are placed outside of the margins, so when you set up your margins make sure you leave enough room for your headers and footers. The page number usually goes in either the header or the footer. You may also include the date and other text. The placement of the page number depends on the document. You may want the page number centered, or at the right, or at the outside corners if the document is to be printed two sided.

Text Editing. The second function is text editing or text correction. You will use these tools after you have entered your text into the computer. One of the most important of these tools for you to learn is the use of cut or copy and paste. This allows you to either cut a selected section out of your document or copy it to the clipboard. The clipboard is a holding place in the computer's memory where cut or copied material (text or graphics) is placed until it is recalled. Only one item may be placed on the clipboard at one time. If you cut or copy another section, the first is deleted.

When you want to place the text or graphic you have cut or copied, you will use the paste tool. The paste tool only copies the data to the document. You may paste the data as many times as needed. You may also go to another program and paste the data into that program. This is an easy way to take data from a document in one program to a document in another program without many compatibility problems. The data on the clipboard will be deleted only if other data is cut or copied to the clipboard or if the computer is shut off.

Other tools for text editing include the backspace key and the delete key. These keys delete text to the left and right of the cursor; the backspace key deletes to the left and the delete key deletes to the right. The insert key toggles between inserting characters and typing over characters to make corrections in the text. The find and replace tool allows you to find a specific word and replace it with another word. If you want to replace text with the opportunity to look at each occurrence throughout your document, you should use the find and replace tool.

File manipulation. The third function is file manipulation. You will want to save your document to be used at a later time, or you may want to save it under a different name. You may want to preview your document before printing. You will probably want to print your document at some point. When you finish working on your document, you will need to close it. At a later time, you will want to open it to use it again. You will

need to start new documents. You may want to save your document in different formats so other programs will be able to recognize it. These are all file manipulation tools.

Tools. The fourth function is productivity tools. The spell checker and grammar checker are the tools that you will use most in this category. The spell checker checks to see if the word is in its dictionary. If the spell checker finds the word, it assumes the word to be correct. Most word-processor dictionaries contain approximately 55,000 words. You will find words that are not in this dictionary; therefore, you may want to start building your own dictionary.

The grammar checker follows a set of rules. If you break any of these rules, the checker will alert you that an error has occurred. You may want these to tools to check your work as you type and even to do a certain amount of correction automatically. These are options that you will use as you are entering the text.

The thesaurus allows you to choose a synonym for a word. Other tools in this category are columns and tables. You would use these tools to easily set up columns in a newsletter, or set up a table of data with calculations, or any other document that would require the use of columns and tables. Mail merge is another tool in this category. Mail merge allows you to merge data from a database into a document. This tool will be discussed more completely later in this chapter.

Macros. The fifth function is the ability to build **macros** to allow you to combine multiple tasks into one task. Macros are either recorded as you perform the tasks or they are written by following specific instructions and rules. Macros are an advanced function in word processors and will take practice to really use them effectively.

View. The view function allows you to view your document in multiple ways. You may want to view the document in an normal way with the screen as your electronic paper, or you may want to view it as a page layout in order to see the graphics as well as the text. To add graphics, you usually must be in the page layout view in the word processor. Many word processors allow you to develop a Web page and export the document as an HTML formatted document. Find the view that you like to work in and that is best for the document that you are creating.

Communication. The last function is communication tools. This function allows data to be imported or accessed from the World Wide Web or another program such as a spreadsheet or a graphics program. This includes the use of OLE (Object Linking and Embedding), which allows the linking of data between the word processor and another program such as a database or a spreadsheet. More about OLE will be discussed in Chapter 21.

The functions you need to think about when planning a document are the following:

- Text formatting tools
- Productivity tools, especially columns, tables, sorting, and graphics
- Macros
- Communication tools

Graphics

Graphics to be used with word processors or with a desktop publisher enhance the aesthetics and clarify the meaning of the information presented in the document. There are several sources of graphics. One of the easiest sources of artwork is **clip art.** Clip art packages contain pictures that can be imported into the document on which you are working. Photo collections and clip art collections can be found on disk or CD-ROM.

Another means of importing artwork is through the use of a scanner. A scanner allows you to take a paper picture and digitize it into the computer. A further source of art is to draw it yourself through the use of a paint or draw package. Any of these methods is an appropriate means of producing artwork for your document. In this chapter, we will not be concerned about where the art came from but how to use it. The only problem with importing artwork is compatibility. Certain graphics packages are not compatible with many word processors and desktop-publishing packages on the market. If the packages are not compatible, then you will need to go through the process of converting to a compatible format. Directions for doing so are found in the manuals for the software.

Desktop Publishing

Desktop-publishing application software packages are page layout software packages. They are designed to take text from a word processor and graphics from various sources and easily place and manipulate them on a page to be aesthetically pleasing. The text from the word processor is unformatted. The formatting is accomplished in the desktop publisher. The graphic images come from clip art, photo art, a scanner, or original art. Some text can be entered using the desktop publisher; however, a desktop publisher is not designed to be as efficient as a word processor at entering text. Some artwork can be created in the desktop publisher, but it has very limited graphics capabilities. Again, the desktop publisher is not a draw or a paint program. Desktop publishers allow the user to flow text easily around graphics, to use guidelines to place text and graphics accurately on the page, to set up a master page, and to add lines, circles, and boxes on a page along with many other things. If there is plain text only on a page, then a word processor is probably the best tool. If there are graphics to be placed on the page; lines, circles, or boxes to be used; or fancy headings, fonts, and so on, to be used, then the best tool is a desktop publisher. We will not be concerned with the specific package in this text; instead we are concerned with the plan of the document.

Some Design Concepts

A word-processing program and a desktop publisher considerably simplify planning the appearance of the text of a document, the first step in its production. Planning includes **formatting** the pages, paragraphs, lines, and words; that is, designing page length, margins, paragraph indentation, line spacing, and so forth. A document planning guide—a form that shows the layout of a document page with margins, footers, and headers—can help in planning the format of the page. However, a sketch showing the format is just as helpful. Figure 19.1 presents an example of a page planning guide (see also Figure D.9, Appendix D). The planner makes notes on the guide to use later when actually creating the document.

Although preplanning may seem like a waste of time, a few minutes spent doing it saves many minutes of shaping the document by trial and error on the computer. Certainly, there will be changes to the document in the final stages, but if the document is well planned, the changes will be minimal.

Planning a well-designed document requires a knowledge of page design concepts. These include the placement of graphics; the use of white space; the use of lines, boxes, and circles; the placement and alignment of text; the use of effective fonts and point sizes; and the understanding of the movement of the eye on the page. It is also

Header: _____

Left Margin: _____ Top Margin: _____ Right Margin: _____

Tabs:

Bottom Margin: _____

Footer: _____

	Other Information	
Paper Size: _____ Page Orientation: _____ Justification: _____	Graphics To Be Used:	Reminders: Mark Guidelines Mark Page Number Placement

Figure 19.1 Page Planning Guide for Word Processing

important to consider the purpose of the document. After all, a memo should have a different design from an advertising flyer.

One of the most important considerations in designing a document is to understand how the eye naturally scans a page without any redirection. The eye naturally falls in the upper right-hand part of the page, then follows the page in an S-shaped path. Therefore, it starts at the right, moves to the upper left of the page, then to the bottom right and finally to the bottom left. If another part of the page is to be seen first, then part of the design needs to redirect the eye to the specified part. This can be done with graphics, lines, circles, color, or any other eye-catching mechanism. When redirecting the eye, though, make sure you provide redirection to stay on the page. The use of an arrow is a good tool; however, it can redirect the eye right off the edge of the page. Hence, we offer *Rule Number One:* Keep the eye on the page through the use of redirection methods until you want the reader's eye to go to another page.

The use of fonts and point sizes can increase the aesthetics and readability of a document unless too many are used. Font refers to how the characters look. There are many font styles. Points refer to the size of the characters. Point sizes generally go from 9 points (smallest) to 128 points (largest), although this is also determined by the software package. Generally, you should use no more than two different fonts and four different point sizes on a page, although there are always exceptions. *Rule Number Two:* Use no more than two fonts and four point sizes on a page. When more are used, the document becomes cluttered, confusing, and unreadable.

Lines, boxes, and circles should also be used in moderation. The use of a box surrounding the document can pull the information together. However, sometimes it makes the document very confining. Use boxes, lines, and circles to emphasize elements that are very important, or to confine a graphic or an idea to a particular area. Too many boxes, lines, and circles will only confuse the ideas you intended to emphasize. Boxes should be aligned with the text. Alignment provides an invisible line connecting the elements together; therefore, alignment is very important. When elements are aligned, the document looks neater and better. *Rule Number Three:* Use properly aligned boxes, lines, and circles sparingly.

The placement of graphics on the page should enhance the attractiveness of the document. The graphics should pertain to the document. You should not use clip art of a boat when the article is talking about the desert, nor would you place a palm tree on a memo. The graphic should be relevant or it should not be there. Do not clutter the page with graphics. Use them sparingly. You should keep your document as simple as possible, while still getting the idea across to your readers. You should use prepackaged photo art or clip art whenever possible. It is easy to overestimate the time it takes to make your own artwork or to take your own photos. As with boxes, alignment is very important. A graphic should be aligned with text and other graphics on the page. *Rule Number Four:* Use relatively few graphics to enhance the aesthetics of the document, and increase communication of the document.

The use of white space increases readability. Still, do not put too much white space on a page. Leading is the space between the lines. There should be enough space to increase the readability, but not so much that the eye does not flow naturally. On a flyer or a brochure, or on a business card, use white space to attract the eye. When you are creating a flyer, there should be as little text as possible, without sacrificing pertinent information. White space should enhance the attractiveness of the flyer. *Rule Number Five:* Be conscious of the use of white space, using enough to enhance readability but not so much as to lose continuity.

The placement of the text is a very important aspect of designing the document. After all, the reason the document is being created is so that it can be read. The placement

of the text should make it as readable as possible. It should flow around the graphics in a way that will increase interest in the subject. Consider the justification (alignment of the text). There are four different types of justification: aligned left, aligned right, justified right and left, and aligned center (ragged outside). Text should be aligned with the graphics on the page. This adds continuity to the document along with good eye flow. Kerning can add to the aesthetics of the document. Be careful of hyphenation. Do not hyphenate over a graphic or to another page, unless absolutely necessary. *Rule Number Six:* Carefully place the text to increase readability, interest, and aesthetics.

Consider the type of document you are creating. Be aware of the purpose of the document. A flyer needs, first, to attract attention, and, second, to be easily read. A memo should be very simple in format. A newsletter should have more graphics intermixed with the text than a poem. *Rule Number Seven:* Design the document with its purpose in mind.

Color is expensive to add to your document, although it does add to the attractiveness of the page. Color should be added to enhance and attract the eyes of the reader to pertinent information. Many times color is added to a document, but the color does not allow the reader's eye to flow to the rest of the page. Using colored paper is a way to cut the cost and yet add some color. *Rule Number Eight:* Use color to enhance the elements of a document and to attract the eye to the document.

All documents should communicate a message. When your document is finished, test its effectiveness by answering the question: Does this document communicate the intended message? If it doesn't, you need to modify the document until the message is clearly communicated. *Rule Number Nine:* Make sure the document is effective in communicating the message.

1. Keep the reader's eye on the page through the use of redirection methods until you want it to go to another page.
2. Use no more than two fonts and four point sizes on a page.
3. Use boxes, lines, and circles sparingly.
4. Use relatively few graphics to enhance the aesthetics of the document and only use graphics that are relevant.
5. Be conscious of the use of white space, using enough to add readability, but not so much as to lose continuity.
6. Carefully place the text to increase readability, interest, and aesthetics.
7. Design the document with its purpose in mind.
8. Use color to enhance the elements of a document and to attract the eye to the document.
9. Make sure the document is effective in communicating the message.

Above all, use common sense and good judgment to see if you like to look at the document. If it is pleasing to your eye (please be honest), then it probably will be pleasing to others.

Word-Processing and Desktop-Publishing Planning

Planning a document is important. Just as in all other applications on the computer, there are certain steps you should complete in planning your document. The following section lists the steps to accomplish a well-designed document. No matter what the purpose of the document, it should be done well and in the least amount of time. Remember the rules of good design in every step of the planning process. Use the page planning

guide shown in Figure 19.1 to help in the planning of your document. This guide is also found in Appendix D in Figure D.9.

Step 1: Determine your audience. This is important since the design will differ from audience to audience. The audience determines the terminology used, the form of the document, and the overall appearance. An advertisement would be designed differently to advertise a party for the office staff versus advertising for a public band concert.

Step 2: Determine the purpose of the document. You need to know why this document is being developed before you can design it. A memo will have a different design than a report.

Step 3: Determine the page setup. The page setup includes margins, headers, footers, page numbering, justification, size of paper, and orientation of paper. The margins include the top, bottom, left, and right margins. If the document is to be bound, the margin needs to be wider on the left or if the pages are to be copied back to back, then you will need to have a right page and a left page. In this case, if this is to be a bound document, there will be inside margins and outside margins, the inside margins being larger than the outside margins. A header is one line of text placed in the top margin and printed on every page. This might include the name of the chapter, the name of the book, the page numbers, or the date. A footer is basically the same as the header except that it is in the bottom margin. The page numbers are normally placed as part of the header or the footer, depending on the desired placement. Justification or alignment of the text is important. Left alignment means the text is aligned on the left; right alignment means the text is aligned on the right; justification means the text is aligned on both the left and right; and center justification means the text is centered. The size of paper includes letter size, legal size, envelope size, or any custom paper size. The orientation of the paper includes portrait (tall) or landscape (wide). Portrait is printed on the page so the left and right margins are on the long sides of the paper. Landscape is printed on the page so the left and right margins are on the narrow sides of the paper.

Step 4: Determine where the guidelines are to be placed. There are three types of guides. The first type is margins. Included are the top, bottom, right, left, header, and footer margins. See step 3 for more information about margins. The second is column guides. These designate the separation between columns on a page. The third type of guides is ruler guides or blue line guides. These are lines used to align text and graphics on the page. They can be horizontal or vertical. All guides are visible on the monitor only and are not printed.

Step 5: Determine the type and placement of all graphics. This will include artwork, pictures, clip art, scanned images, photo images, mastheads for newsletters, and the captions that accompany the graphics.

Step 6: Determine the placement of the text and what the text will include. The text needs to be written and the placement of the text on the page needs to be determined. The format of tables and columns of data is determined at this time. This would include any mathematical formulas that need to be entered and the determination of tabs. There are five types of tabs: right tab—text is aligned to the right of the column; left tab—text is aligned to the left of the column; center tab—text is centered in the column; decimal tab—a right-aligned tab with the decimal points aligned; and tab leader—a right tab that includes a dotted line placed between the last text character on the left and the beginning character of the text in the column. See Figure 19.2 for example of tabs. Also, at this time, determine any files that need to be inserted from other applications such as a report from a spreadsheet program or a report from a database management system.

Step 7: Determine the type styles of all text. Figure 19.3 shows some of the different type styles available. The font and size are the first to be determined. The font should be chosen with the audience and purpose in mind. You probably would not use a

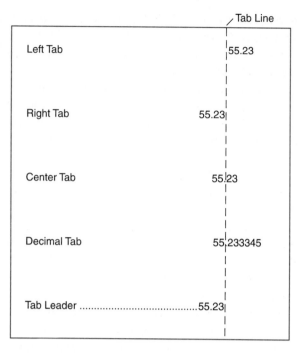

Figure 19.2 Types of Tabs

Bold type Style
Italic type style
Outline type style
Shadow type style
~~Strike through type style~~
<u>Underline type style</u>
Examples of fonts:
 Font
 Font
 Font
 Font
 Font
Examples of sizes:
 10 point
 12 point
 18 point
 24 point

Figure 19.3 Examples of Type Styles, Fonts, and Sizes

cursive font for a very formal report. Many fonts are available on the market today. Check to see what fonts are provided for your use. It is very useful to have examples of the fonts and the sizes printed out and kept in your notebooks. Some words need to be emphasized. To do this, you would boldface the characters or italicize them. Characters used in mastheads of newsletters many times need to be emphasized by outlining the characters or putting a shadow on them. In a document that is being revised, the characters that are being deleted are often struck out instead of being deleted. In this case you

would use the strikethrough type style. Some characters need to be raised or lowered as superscripts or subscripts. Others need to be placed with one character on another character, as in the case of an accent over a letter. You also need to determine whether or not you want words to be hyphenated.

Step 8: Determine any other special needs of the document. These special needs might include development of an index or a table of contents; special enhancement using boxes, circles, or lines; added color; or anything else you would like to place on this document.

Types of Documents

A word processor can be used to create various types of documents. It can be used by a writer to compose a book, by a secretary to type and polish a letter or memo, by a teacher to create assignments or tests, or by a student to finish a paper or report. Different people bring different needs to a word processor.

A report, an article, or a book is a lengthy textual document that needs very few enhancements, such as boldface or underlining. It is usually composed on the computer, edited, and then printed in its final form. The planning required for this type of document is an outline and page formatting.

A word processor increases productivity with this type of document simply because it enables the user to design, edit, and print the document electronically. The user can simplify footnoting by placing footnotes immediately following the paragraph in which they appear, and then after the document is completed, moving them to the bottom of the correct page. The footnotes should initially be typed exactly as they will appear. When you footnote in this manner, the footnotes at first take up space in the body of the report. During typing, the writer can edit and move paragraphs without having to keep track of the footnotes—they move with the paragraphs—or without having to guess how many lines to leave at the bottom of the page. Some word processing packages have a footnote feature.

The memo is a document intended for distribution. The form is generally standard. A memo can be distributed in print or as a message sent on E-mail. The memo is stored electronically and can be easily updated later for redistribution. This type of document is short, and the enhancements are few. They might include boldface, underlining, and indentation. With the word processor, increased productivity again lies in the speed of editing and printing.

Templates, or boilerplates, are empty forms or shells that can be used repeatedly with different data. Templates include forms for letters, schedules, or any other forms needed in the workplace. A newsletter can be set up as a template. Figure 19.4 shows an example of a newsletter template. The template for a newsletter should include the masthead, guidelines, and any other elements included in every newsletter. The example in Figure 19.4 includes the masthead, the box around the news, and the place for the editor's column.

Increase in productivity lies with the use of the completed template. The template will be completed once and then used many times. For example, a form for a schedule can be completed in just a few minutes. The form is then saved in a file of its own. When the user needs to fill it in, she or he copies the form into a new file. The Search or the Search-and-Replace function can help the user find the data needed to fill in the form. When using a word processor, there are two modes of correcting text. One is the

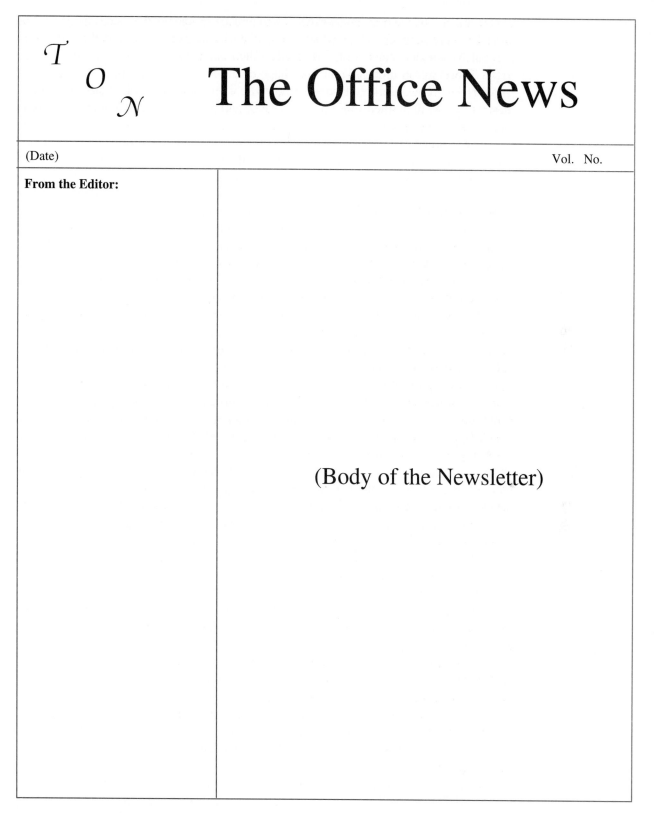

Figure 19.4 Example of a Newsletter Template

replacement mode, which enables the user to replace characters that need changing. The other mode is the insert mode, which allows the user to insert characters that have been left out. When characters are inserted, all of the remaining characters are moved to the right. Therefore, in order to keep the shape of the form when filling in data, the user should turn off the insert mode, or else the vertical lines and structure of the form will be destroyed. The final filled-in form is then saved and printed. In the planning stages, only the blank form is designed.

An instructor can create lesson plans, assignments, and tests for one term and then easily modify them for the next term. Although the word processor does not significantly increase productivity for the instructor the first semester, it does each time the course is offered thereafter. In addition, word processing improves an instructor's productivity by helping in the production of all of the types of documents mentioned earlier.

Advertising employs many types of documents. A flyer is a one-page sheet advertising an event. A brochure is more elaborate and more costly. It is usually six to twelve pages and is folded for sending in the mail. Some desktop publishers have page templates set up for brochures, so all you have to do is to add text and graphics. Business cards can be designed. Stationery, signs, and all kinds of forms can be designed for businesses.

Resumes are another type of document that can be designed for use with a word processor or desktop publisher. Figure 19.5 shows one form of a resume. You would need to fill in the information. The main sections of the document include your name, address, and phone number; objectives of employment; education; experience; other information; and references. The section headings should be enhanced. The information under each heading should be indented. The information should be as concise as possible without leaving out any pertinent information. The resume should be no longer than two pages. Productivity is increased by the use of electronic editing and the ease of modification of the resume.

Manuals can be created very easily through the use of the master page setup. A master page has elements that are found on all pages, including margins, guidelines, headers, footers, page numbers, and so forth. The text is written in a word processor, and then can be placed on the pages at a relatively rapid speed. The manual can then be modified without a great deal of difficulty when revisions are necessary.

<div align="center">

Name
Street Address
City, State Zip Code
Telephone

</div>

Objectives:

 (Short narrative about your employment objectives.)

Education:

 (List all schools you have attended and the degrees you hold.)

Experience:

 (List all employment experience.)

References:

 (List all references or state "Available upon request.")

Figure 19.5 Example of a Resume Template

Documents that require file merging can be accomplished on a word processor. There are two types of merges. The first involves merging two or more documents into the same file. The user will find it faster to develop small files and then merge them in another file later than to create the large document in one file. As a text gets larger, the computer takes longer to transfer from one part of the document to another. Therefore, this type of merge saves time for the user and the computer.

The second type of merge is merging data with text. In this type, often used for form letters, the document is in one file called the shell or the primary file, with variables placed wherever a word or words will differ from one printed copy to another. A second file called the data file or secondary file contains the data that will be entered as the values of the variables. One record from the data file contains the information for one printed copy of the completed document. When merging the two files, the computer matches the data in the data file to the variable name in the text file to print the copy of the document. Therefore, it is important that the two files be developed together. Once the files have been developed, the merge function can be executed, and, assuming the two files have been set up correctly, the printed document will differ for each record in the data file. See Figures 19.6a–b for a sample of the primary and secondary files. Productivity is increased since one form letter and one data file can produce many individualized letters.

There are many other types of documents, ranging from poems to books with many graphics, that can be desktop published. In doing these, and all other documents, you need to be aware of the capabilities of your software and hardware. There are limitations with all packages, and you need to create your documents with the limitations, as well as the special features of your package, in mind.

Primary File: Fields will be indicated by an F followed by the field number

F1
F2
F3, F4

Dear F5:

I am writing to inquire about an entry level programming job with your company.

Having just graduated from University of the North Coast, in Computer Information Systems, I am eager to begin my career. Attached is my resume. I have summarized my qualifications and my experience in my resume. I believe I could be a valuable asset to your company.

I can be reached at 707-545-4321. I would welcome a chance to discuss employment possibilities with you. Please let me know if there is a job opening.

I am looking forward to hearing from you.

Sincerely,

John Johnson
5555 H Street
Arcata, CA 95521

Enclosure: Resume

Figure 19.6a Mail Merge—Primary File

Field: 1	2	3	4	5
Dr. John Smith	545 Wood St.	New Town	CA 95367	Dr. Smith
Mr. Roy Jones	764 Tree St.	Old City	WA 98700	Mr. Jones
Ms. Jane Brown	888 Oak St.	Middle Town	OR 96787	Ms. Brown

Figure 19.6b Mail Merge—Secondary File

Many word processors have the ability to create macros. Macros are short sets of keystrokes and/or instructions to accomplish tasks. These macros are then saved with the document to be used whenever the document is used. An example of a macro would be the keystrokes to type a name that is used many times throughout the document. This improves productivity by using one keystroke to execute many keystrokes.

Summary

Word processing is one of the four major productivity packages available to the computer user. It allows text to be edited electronically. The features of word-processing packages vary greatly, so a user should choose one that is best for the types of documents to be created.

Desktop publishing is a composition program, which merges text and graphics onto one page. Follow the rules for good page design when designing any type of document. The steps to produce a well-designed document are:

- *Step 1:* Determine your audience.
- *Step 2:* Determine the purpose of the document.
- *Step 3:* Determine the page setup.
- *Step 4:* Determine where the guidelines are to be placed.
- *Step 5:* Determine the type and placement of any graphics.
- *Step 6:* Determine the placement of the text and what the text will include.
- *Step 7:* Determine the type styles of all text.
- *Step 8:* Determine any other special needs of the document.

Many types of documents can be produced by a word processor and a desktop publisher, including reports, newsletters, form letters, manuals, flyers, and so forth.

New Terms

clip art	macro
desktop publishing	templates
document	word processing
format	

Questions

1. What is word processing?
2. What is a document?

3. What are some of the functions of a word processor?

4. What are some of the types of documents?

5. Why use a word processor?

6. In what ways are problem-solving techniques important in word processing?

7. Describe how the six steps of problem solving described in Chapter 1 would be used in word processing.

8. What is the difference between a word processor and a desktop publisher?

Problems

Develop a document plan using the document planning guide, rules for good design, and the eight steps to produce a well-designed document for the following problems:

1. Select a poem and develop a plan to put it on a word processor or a desktop publisher.

2. Develop an hourly schedule for a week for yourself. Then, using a planning guide, develop the design of that document that would use the fewest possible keystrokes. Never retype the same line twice; that is what the copy function is for. Remember, there are 66 lines per page and 80 characters per line.

3. Develop a letter that has at least seven variables at different locations in the text, and develop the data file for the letter. Specify at least three records of data.

4. Answer the following questions about a flyer you received in the mail, in the newspaper, or at a store (answer the questions with the eight rules of document design in mind):
 a. What do you see first on the page? Why?
 b. What do you see next? Why?
 c. Is there anything on the page that you missed at first glance?
 d. How many fonts were used?
 e. How many different point sizes were used?
 f. How many colors were used?
 g. How much text was used?
 h. Did you know what the flyer was advertising at first glance?
 i. How effective do you think this flyer is? How would you change the flyer to make it more effective?

5. Design your own flyer to advertise a product or service that interests you.

6. When designing a page for a report, or a flyer, how can you accomplish the following?
 a. keeping the eye on the page
 b. an uncluttered look
 c. an aesthetically attractive format (one that is pleasing to the eye)

7. Design a brochure advertising an interesting new product.

8. Analyze a brochure you received. Answer the same questions as in Problem 4.

Chapter 20

Problem Solving for Spreadsheets

Overview

Spreadsheet Basics

Design of a Spreadsheet
Steps to Design a Spreadsheet

Macros
Steps to Complete the Spreadsheet on the Computer

Graphs and Presentation Graphics

Objectives

When you have finished this chapter, you should be able to:

1. Explain the basic principles of a spreadsheet.
2. List and explain the steps to design a spreadsheet.
3. Develop and evaluate expressions using functions, constants, and cell references.
4. Develop a planning guide for a spreadsheet that will be entered into the computer at a later date.

Programmers use spreadsheets for project management, budgets, cost analysis, and so on. A **spreadsheet package** allows the user to design a worksheet so that the computer will do the calculations and continually display the results in matrix form. The package simplifies accounting, cost analysis, forecasting, grades, and many other tasks that require mathematical calculations. Spreadsheets increase productivity because the computer does the calculation and recalculation of the data.

The time a worker saves by using a spreadsheet does not necessarily lie in the initial calculation but in the recalculations. The **worksheet** is a display on the monitor of answers to the calculations, along with the data to use in the calculations and the labels that tell the user what the numbers on the screen represent. It takes time to initially set up the worksheet and to test it to make sure the answers are correct. The testing is important

because the chance of serious error is greater with a spreadsheet program than with any of the other productivity packages. If there is an error in a word-processing document or in a piece of data in a database management system, the impact on the results is usually minimal. However, if there is an error in an equation or in assumed data in a spreadsheet, the results could be catastrophic. Once the initial calculations are completed on the worksheet, it's quick and easy to recalculate any of the figures.

Spreadsheet Basics

A spreadsheet package consists of several parts. The first part is the worksheet itself. On the monitor, all or part of the worksheet appears as a grid of boxes arranged in rows and columns. Worksheet boxes, called **cells,** are designated by row and column identifiers. A cell is where the row meets the column. A cell may contain one or more characters. Cursor movement in the worksheet is from cell to cell rather than from character to character. The worksheet displays mathematical information and results in a form similar to manual worksheets.

A second part of a spreadsheet package, also found on the screen, is the Control Panel and status line. This area is located above the worksheet. There are three standard lines in the status area. The first is the menu bar. The second is the tool bars. These are icon substitutions for menu items. The third is the input line, where the computer communicates with the user. The status line is an information line and is found at the bottom of the screen. It displays such information as whether the computer is calculating, what type of data is in the cell where the cursor is, and whether there is an error.

Another part of a spreadsheet package is the contents of the cells. For each displayed worksheet cell, another cell in the computer memory contains user-established instructions to the computer regarding how and what to put into the worksheet cell for display on the screen. Displayed worksheet cells show only the results of instructions, not the instructions themselves. The contents of the cell in which the cursor is presently found are shown on the status line in the status area. The type of data making up the contents of the cell is shown on the information line in the status area. The contents of a cell can be one of two data types. The first is a label, which is character or string data type. Labels allow the user to identify rows and columns. The second type is a value, which is a numeric data type. Calculations, called formulas, and constants are considered values in a spreadsheet. Formulas are expressions (see Chapter 2).

A fourth part of the spreadsheet package is the **commands.** Commands help the user set up and use the spreadsheet. This is accomplished through the use of menus and tool bars. Pull-down menus allow the user to choose the needed command. Some spreadsheet software packages have tool bars. The click of the mouse on an icon in the tool bar will accomplish a task normally found in one of the menus. Many of the commands have keyboard equivalents. Some of the commands include cut and paste, save and retrieve, assigning cell attributes, graph, data query, and many more. Many of the commands require the use of a **range** or block of cells. A range or a block of cells consists of a single cell, a consecutive group of vertical cells, a consecutive group of horizontal cells, or a rectangular block of cells. In reference to Figure 20.1, A1 would be a single cell range, A1 to A3 would be a vertical range of cells, A1 to D1 would be a horizontal range of cells, and A1 to D3 would be a block range of cells. The number and types of commands depend on the spreadsheet package.

The cell is the central part of the spreadsheet package. Each cell is designated by column letter and a row number (see Figure 20.1). Rows are numbered in order, starting at the top of the spreadsheet with number 1. Columns are designated with letters starting

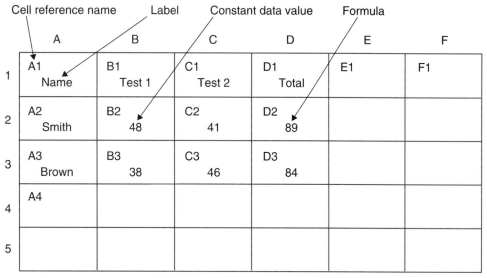

Cell reference name Label Constant data value Formula

	A	B	C	D	E	F
1	A1 Name	B1 Test 1	C1 Test 2	D1 Total	E1	F1
2	A2 Smith	B2 48	C2 41	D2 89		
3	A3 Brown	B3 38	C3 46	D3 84		
4	A4					
5						

CELL CONTENTS

A1: Name C1: Test 2
A2: Smith C2: 41
A3: Brown C3: 46
B1: Test 1 D1: Total
B2: 48 D2: +B2+C2
B3: 38 D3: +B3+C3

Figure 20.1 A Worksheet and the Cell Contents

with A on the left-hand side of the spreadsheet. The name of the cell is the column letter followed by the row number; for example, the cell found in Column C and row 5 is called C5. Some spreadsheet programs have three-dimensional spreadsheets. Pages are designated by letters, numbers, or names. A cell found in page B, column C, row 5 would be called B:C5. Pages, cells, and blocks of cells can be named relevant names through the use of a command.

Many options are available to the user to specify how the data in a cell will be displayed. The first option is what is to be entered into the cell. Three things can be entered: labels, which are string data; values, which are numeric; and formulas, which are mathematical expressions. Another option is the width of the cells in one column (the number of characters in each cell). The width can be specified as something other than the default value. (A **default value** is a value that is built into the program; the computer uses this value unless the user specifies otherwise.) When a cell contains values or formulas and the width of the cell is insufficient, the program will tell the user that there is an error. When the content of a cell is a label, and the column is not wide enough, it will spill over into the next cell. The cell width must be the same for all cells in a column. The command to change the default value for width allows the user to change the width of the column but not of an individual cell.

A third option is that the user can specify the format of the cell. The various formats include right or left justification or centering of labels, as well as value formats for fixed decimals, scientific notation, currency that includes the dollar sign ($), percentages that include the percent sign (%), and the insertion of commas and other symbols for values and formulas. Figure 20.2 shows examples of value formats.

The user can change cell contents through an editing process. Contents can be erased, moved to another cell, or copied to another cell or cells. When a user copies the contents of one cell to other cells, the cell names in the formulas change. These are designated relative cell references. A formula in cell D2 designated as B2 + C2 will copy

Value Format	Number Before Formatting	Number After Formatting
General	12234.543	12234.543
Number (with comma and 2 decimal places)	12234.543	12,234.54
Currency (2 decimal places)	12234.543	$12,234.54
Percentage (0 decimal places)	0.25	25%
Fraction	0.25	1/4
Scientific (2 decminal places)	12234.543	1.22E+04
Date	36204	February 13, 1999

Figure 20.2 Examples of Cell Formats

Number of Customers for XYZ Company

	Day 1	Day 2	Subtotals	% of Total
Store 1	23	15	38	#VALUE!
Store 2	35	26	61	#DIV/0!
Store 3	42	34	76	#DIV/0!
Total			SUM(D4:D6)	

Copying without using absolute references in percentage calculation.

Number of Customers for XYZ Company

	Day 1	Day 2	Subtotals	% of Total
Store 1	23	15	+B4+C4	+D4/D7
Store 2	35	26	+B5+C5	+D5/D8
Store 3	42	34	+B6+C6	+D6/D9
Total			SUM(D4:D6)	

Copying without using absolute references in percentage calculation.

Number of Customers for XYZ Company

	Day 1	Day 2	Subtotals	% of Total
Store 1	23	15	38	21.71%
Store 2	35	26	61	34.86%
Store 3	42	34	76	43.43%
Total			175	

Copying using absolute references in percentage calculation.

Number of Customers for XYZ Company

	Day 1	Day 2	Subtotals	% of Total
Store 1	23	15	+B4+C4	+D4/D7
Store 2	35	26	+B5+C5	+D5/D7
Store 3	42	34	+B6+C6	+D6/D7
Total			SUM(D4:D6)	

Copying using absolute references in percentage calculation.

Figure 20.3 Absolute Reference Examples

to cell D3 as B3 + C3. If one of the cells in the formula should not change, then it must be designated an absolute cell reference. This designation is usually accomplished by the insertion of a $ before the column letter and row number. See Figure 20.3 for an example.

A variety of other features are available to users of spreadsheets. Some commands can be executed either globally or through a range of values, such as changing column widths or formatting cells. Other commands are designated as either global or range, but not both. Global commands address the entire worksheet; *range commands* address only a designated part, such as a single cell, part of a row of cells, part of a column of cells, or a block of cells. Other commands address particular tasks to be accomplished.

Cell protection is a feature of many spreadsheets. It guards against the accidental destruction of a cell's contents by blocking the cursor out of the cell, which means the user is locked out. Also, a row or a column can be hidden from view. This guards against someone seeing sensitive information. In either case, the user has to specify the cells to be protected and/or hidden.

Recalculation, another spreadsheet feature, can be done in several ways. *Manual* recalculation means that it will be done only when the user specifies it. *Automatic* recalculation means that the worksheet will be recalculated every time a value or a formula in a cell is changed. *Natural* recalculation is done in a natural order, similar to the way a person would do it. This means that the computer decides which cells should be calculated first according to the cells the formulas use. *Row* recalculation is completed row by row, beginning with the upper left cell, working across the row, and then moving

down the spreadsheet to the next row. *Column* recalculation occurs column by column, beginning with the upper left cell and moving down the column, and then moving across the spreadsheet to the next column. When the spreadsheet is large, it saves time to recalculate manually so the recalculation is done only periodically rather than every time any cell is filled or changed. With a small spreadsheet, the time for recalculation is relatively short, but on a larger spreadsheet the time can be considerable.

Sometimes the user has to insert or delete a row or a column. Often the user makes insertions or deletions to improve the appearance of a worksheet. The spreadsheet automatically updates any cells that reference the altered cells.

The row labels in the left-hand column and the column labels in the top row usually define the row and the columns. When the user scrolls the worksheet down or to the right, these labels disappear to the top or to the left. They can be fixed to make them stay on the screen so the user can see them at all times, including absolute references.

The window feature on some spreadsheets enables the user to view and derive information from two sections of the worksheet at the same time.

All or part of a worksheet and/or the cell contents can be printed. Depending on the package, other options are available, such as headers, footers, variations on margins, and scaling.

File operations are an important part of the commands of a spreadsheet package. The user can save a worksheet to an auxiliary storage device under many different names if desired. The program saves only the cell contents; when the file is retrieved, the worksheet on the screen is recalculated. Through other file operations, files can be combined, parts of worksheets saved, directory listings printed on the screen, and data disk drives changed.

Some spreadsheet packages include the ability to graph the data and create interactive slide shows. Also, some packages include data management features, such as finding a piece of data, sorting a range of cells, and easily filling in tables.

Many different spreadsheet packages are available. The more features a package has, the more flexible it is, but also the harder it is to learn. Even though packages with fewer features may be easier to learn, they are less flexible and can become frustrating because they lack certain useful features.

Design of a Spreadsheet

A spreadsheet is an electronic calculator. Since a spreadsheet is primarily a mathematical tool, many people have problems with the mathematics end of working with a spreadsheet. It is important to be able to glean the facts out of a problem, whether you are solving a textbook problem or a real-life problem. The PAC (problem analysis chart) introduced in Chapter 3 will help. Word problems are a problem for some people. However, if you can eliminate the words while keeping the facts, you no longer have a word problem. The PAC will help you do this. Use the PAC to separate the given data, the required results, and the needed formulas from the problem. This is the first step in the problem-solving stage of designing a spreadsheet.

Steps to Design a Spreadsheet

Just as in all other aspects of computer work, it is important to plan what you are going to do before sitting down to start work on the computer. The following are the steps you should follow to complete your planning of a spreadsheet. Remember as you plan that this is just a starting point. You will modify this plan as you test out your solution.

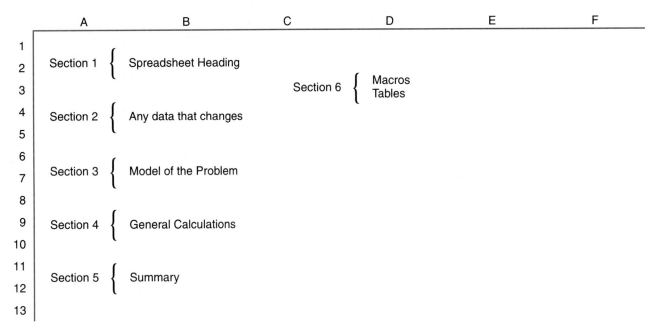

Figure 20.4 General Design of Most Spreadsheets

Step 1: Use the PAC to glean the given data, the required results, and the needed formulas. As you are reading through the problem, write down all facts in the appropriate box. You will use this to develop the cell contents in later steps.

Step 2: Create the overall design of the spreadsheet. Figure 20.4 shows a general design of most spreadsheets used for business analysis. The headings are at the top. The headings include such things as the company name, the title of the spreadsheet, the date, the time, and any other pertinent information about the spreadsheet. The second section includes all changes that are made from calculation to calculation. This would include such things as percentage increase or decrease, inflation rates, and the like. This third section is the model. This would include such things as number of employees, salaries, number of students, and so forth. The next section is where the general calculations take place. This would include such things as costs, revenues, or grade calculations. The next section is the summary section. This would include such things as the gross profit, tax, net profit, or average test scores. The last section is placed out of the area of the other sections. This section includes the macros and any tables required for the general calculations, such as tax tables.

Step 3: Place the labels on the worksheet. Using a piece of paper, or the form found in Appendix D (Figure D.1), draw horizontal and vertical lines to simulate a worksheet. Make sure the columns are wide enough to write your formulas. Next label the columns with letters (A, B, C, . . .) and rows with numbers (1, 2, 3 . . .). Next designate the column widths. Check the default values since you may want to use these column widths. Now is the time to start designating the cell contents. Starting with the first section of your overall plan, fill in the labels in the proper cells. The labels will designate the overall worksheet heading, the column headings, the row headings, plus any other general labels that need to be entered. You should designate the format of each cell (right, left, or center justified).

Step 4: Enter the constant data, which are any data that will not change often. This includes data such as quantity, wholesale price, names, base salaries, deductions, and scores. You should designate the format for each cell (fixed, currency, percent, date, etc.).

Step 5: Enter the formulas (the mathematical expressions). Formulas are used to calculate the displayed value of a cell. Cells are referenced by the row and column in which they are found. The contents of cells that contain formulas are expressions, whereas the value of the expression is found in the worksheet cell on the screen. This value can be used in any other cell by using the cell reference name. Mathematical operators are used to perform the calculations. In some spreadsheet programs, the user can name a cell or a group of cells by giving it a variable name. This is called a **range name.** This chapter will not use range names.

Formulas use mathematical operators in the calculations. They also can use functions. Many functions are specific to spreadsheets. The following are some of the functions you might find in a spreadsheet package:

Mathematical Functions:

INT(*x*)	Returns the integer part of *x*
MOD(*x, y*)	Returns the whole number remainder of *x/y*
ROUND(*x, n*)	Returns the number *x* rounded to *n* decimal places
SQRT(*x*)	Returns the square root of *x*

Other Functions:

IF(condition, *x, y*)	Returns *x* if condition is true, *y* if false
CHOOSE(*x, v*0, *v*1, *v*2, *v*3, . . .)	Returns the value of the expression in the *x*th position in the list $v_0 - v_n$
VLOOKUP(*x, range, column*)	Looks up the value *x* in the table at *range* (range where the table is located) and then returns the value found in *column* (column number in the table starting with the left-most column as zero)
DATE(*year, month, day*)	The Julian calendar date of *year, month, day*
NOW	The Julian calendar date of today

Financial Functions:

CTERM(*int, fv, pv*)	Returns the number of periods for the investment of *pv* to grow to *fv*, earning the rate of *int*
DDB(*cost, salvage, life, period*)	Returns the double-declining depreciation of an asset, during *period*, given *cost, salvage value*, and the *life* of the asset
FV(*pmt, int, term*)	Returns the future value of *term* number of payments, earning *int* rate
IRR(*guess, range*)	Returns the internal rate of return of the cash flows in *range*, with a guess needed for calculating purposes
NPV(*int, range*)	Returns the net present value of a series of cash flows in *range*, discounted at *int* rate
PMT(*prin, int, term*)	Returns the payment needed to pay off the loan of *prin*, at the *int* interest rate, over *term* number of periods
PV(*pmt, int, term*)	Returns the present value of *term* number of *pmt* payments, discounted at the *int* interest rate

RATE(*fv, pv, term*)	Returns the rate needed for *pv* to grow to *fv* over *term* number of periods
SLN(*cost, salvage, life*)	Returns the straight-line depreciation of an asset, given *cost, salvage value,* and the *life* of the asset
SYD(*cost, salvage, life, period*)	Returns the sum-of-the-years digits depreciation of an asset, during *period,* given *cost, salvage value,* and the *life* of the asset
TERM(*pmt, int, fv*)	Returns the term needed for periodic payments of *pmt* to grow to *fv* with an interest rate of *int*

Statistical Functions:

AVG(*range*)	Returns the average of the values in *range*
COUNT(*range*)	Returns the number of values in *range*
COUNTIF(*range, condition*)	Returns the number of *x* values in the range
MAX(*range*)	Returns the highest value in *range*
MIN(*range*)	Returns the lowest value in *range*
STD(*range*)	Returns the standard deviation of the values in *range*
SUM(*range*)	Returns the summation of the values in *range*
VAR(*range*)	Returns the variance of the values in *range*

See Figure 20.5a–b for examples of functions.

Data							
34.00		0.00	33.00	0.10	4.00	23.00	23.00
54.00		200.00	45.00	0.20	300.00	45.00	45.00
56.00		300.00	67.00	0.90	3.00		0.00
76.00		500.00	54.00	0.40		67.00	67.00
Statistical Functions					**Other Functions**		
SUM(range)	220.00				IF(condition, true, false)	3,000.00	
AVERAGE(range)	55.00						
	45.00				CHOOSE(x, v1, v2, v3, v4, v5)	50.00	
	33.75						
					VLOOKUP(x, range, column)	0.90	
MAXIMUM(range)	76.00						
MINIMUM(range)	34.00				**Date, Time Functions**		
					TODAY()	36,353.00	
COUNT(range)	4.00						
	3.00				NOW()	36,353.71	
	4.00						
					DATE(year, month, day)	36,081.00	
COUNTIF(range, condition)	3.00				TIME(hour, minute, second)	0.34	
					Financial Functions		
					PMT(rate, nper, pv)		
						63.74	
					FV(rate, nper, pmt)	11,615.56	
					PV(rate, nper, pmt)	9,413.07	

Figure 20.5a Spreadsheet Functions as Displayed

Data							
34.00		0.00	33.00	0.10	4.00	23.00	23.00
54.00		200.00	45.00	0.20	300.00	45.00	45.00
56.00		300.00	67.00	0.90	3.00		0.00
76.00		500.00	54.00	0.40		67.00	67.00

Statistical Functions		Other Functions	
SUM(range)	SUM(A4:A7)	IF(condition, true, false)	IF(A5<A4,2000,3000)
AVERAGE(range)	AVERAGE(A4:A7)		
	AVERAGE(G4:G7)	CHOOSE(x, v1, v2, v3, v4, v5)	CHOOSE(F4,10,30,20,50,40)
	AVERAGE(H4:H7)		
		VLOOKUP(x, range, column)	VLOOKUP(F5,C4:E7,F6)
MAXIMUM(range)	MAX(A4:A7)		
MINIMUM(range)	MIN(A4:A7)	Date, Time Functions	
		TODAY()	TODAY()
COUNT(range)	COUNT(A4:A7)		
	COUNT(G4:G7)	NOW()	NOW()
	COUNT(H4:H7)		
		DATE(year, month day)	DATE(98,10,13)
COUNTIF(range,"condition")	COUNTIF(A4:A7,">40")	TIME(hour, minute, second)	TIME(8,15,0)
		Financial Functions	
		PMT(rate, nper, pv)	PMT(10%/12,5*12,-3000)
		FV(rate, nper, pmt)	FV(10%/12,5*12,-150)
		PV(rate, nper, pmt)	PV(10%/12,5*12,-200)

Figure 20.5b Spreadsheet Functions as Formulas

A function is used in an expression to specify how to find the value of a specific cell. Formulas are used whenever possible to reduce the amount of data that has to be entered into the cells. For example, when several cells have the same constant value, the first cell will be given the constant, and the other cells will reference the first cell. In this way, when the value changes, the user has to change the value in only one cell instead of all the cells. Figure 20.6 presents some expressions that might be used in cells.

Mark any cells that need to be copied. Productivity increases when formulas can be copied. Copying cell formulas decreases data entry time and reliability of data entry. Two rules to increase productivity with spreadsheets are as follows: 1) reference as many cells as possible and 2) copy as many formulas as possible. Make sure you specify absolute cell references before copying the formula.

Step 6: Enter any tables required in the formulas.
Step 7: Develop any macros needed for use with the problem.
Step 8: Test your formulas as much as possible.

Putting It All Together

Figure 20.7 shows the development of a program for the amortization of a loan. Notice that the only cells containing data are the cells containing the values of the loan amount, the interest, and the time. All other cells are referenced to these three. Also notice the number of cells that are copied. Cells that have the same formula are typed only once and copied into the other cells. This can result in a big increase in productivity.

To add 2 cells together:
 +A5+A6

To sum a range of value (cells C3 through C19):
 SUM(C3..C19) ... indicates all cells
 between the first cell
 and the second cell

To average the same range of value:
 AVG(C3..C19)

To find the payment for a loan: where
 PMT(E1,E2/E4,E3*E4) E1 = Cell where the amount of loan is
 E2 = Cell where the interest is
 E3 = Cell where the number of years is
 E4 = Cell where the compound period (12/year, 24/year, etc.) is

To find % of a sum:
 +E5/SUM(E1..E10)*100
 E5 is what % of the sum of E1 through E10

To make a decision:
 IF(E1<E7,0,E7) when E1<E7 the cell value = 0
 otherwise the cell value = E7

To increase a number by a percentage:
 A10*(1+A4)
 A10 = number to be increased
 A4 = % (in decimal form) increase (to decrease a number
 this A4 would have to be negative)

To perform a table lookup: Formula Value

	Formula	Value		A	B	C
	VLOOKUP(3,A4..C7,1)	14.42	1	3	1	
	VLOOKUP(A1,A4..C7,B1)	14.42	2	2	2	
	VLOOKUP(2,A4..C7,2)	3.85	3			
	VLOOKUP(A2,A4..C7,B2)	3.85	4	1	12.92	1.75
			5	2	13.67	3.85
			6	3	14.42	5.95
			7	4	15.17	8.05

Figure 20.6 Sample Cell Contents of Formulas

ROW	A	B	C	D	E	F	G	H
	12	12	12	12	12	12	12	12
	COL. →							
	A	B	C	D	E	F	G	H
1		C L	Principal	=	C = $1,500			
2			Interest	=	P = 15%			
3			Years	=	= 1.			
4			PMT	=	C = PMT(E1, E2/12,12*E3)			
5	Period	Beg. Bal.	Interest	App. Prin.	E. Bal.			
6	1	C = E1	C = B6 * (E2)	C = (E4) – C6	C = B6 – D6			
7	2	= E6	Copy	Copy	Copy			
8	3	Copy	Copy	Copy	Copy			
9	4							
10	5							
11	6							
12	7							
13	8							
14	9							
15	10							
16	11							
17	12							
18								
19								
20								

Figure 20.7 Worksheet Planning Guide for a Loan Amortization (circled cells are absolute)

Scripts attached to buttons or other interactivity tools may be placed on any worksheet through the use of a language editor such as Visual Basic. Many times one sheet is designated as a control sheet that activates multiple tasks for use throughout the spreadsheet. This essentially automates the spreadsheet to increase productivity and user friendliness.

Spreadsheet packages allow the user to develop macros. *Macros* are small sets of instructions that facilitate the use of the spreadsheet. In reality, they are short programs. To use the capabilities of macros, the user must specify a set of keystrokes to accomplish a desired task. The user names the macro and tells the computer what the sequence of keystrokes is. From then on, the user implements the macro by its name. Macros increase productivity; however, some users avoid them because they do not understand or are afraid of them.

Keystroke macros replace a series of keystrokes that complete a task. For example, a user could develop a keystroke macro that would replace the keystrokes to set column width, to print the cell contents, or to erase the contents of a cell. Keystroke macros speed up the development of a worksheet by reducing the number of keystrokes. A keystroke macro template also can be developed, saved, and used later to prepare all new templates or worksheets. Any new macros that are developed are added to the macro template. The template holds all macros you use on a regular basis.

Another use for templates is as a structure for doing calculations for a user who knows little or nothing about spreadsheets and wants only the results. A macro can be written to enable the user to enter any data needed for calculations and then let the spreadsheet do the calculations. The template would be developed for problems that recur in an organization.

Macros are absolute. The user must be careful when inserting rows and columns after a macro has been developed. Any references to other cells will not be changed when the cells are moved. This can lead to serious errors. The best solution to this problem is planning. A well-planned solution involving macros should not require the insertion of new rows or columns.

Steps to Complete the Spreadsheet on the Computer

After the planning stage is completed on paper, it is time to go to the computer. These are the steps to follow at this time:

Step 1: Start with your macro template.

Step 2: Enter the cell contents you developed on your planning sheet. Copy as many formulas as possible. Be sure to designate all absolute references.

Step 3: Develop your template. A template is a worksheet that contains all labels, formulas, and values that do not change often; it contains everything except the data that change with each recalculation. A template is stored under a separate name from the worksheets containing the data. When the user needs to enter new values into the worksheet, he or she retrieves the template from storage and enters the new values. The worksheet is then recalculated and saved under a different name. It is never wise to save the completed worksheet under the same name as the template, or the template will not be available next time it is needed.

Templates make it easy to use a worksheet over and over again. A template can be very complicated or it can be simple. The aim of any user of a spreadsheet program should be to develop templates that will be reused later for similar problems.

Step 4: Test your answers. This is a very important step. If the answers in a spreadsheet are wrong, the results can be disastrous. Figure 20.8 shows the worksheet and the cell contents that result from the loan amortization problem in Figure 20.7.

	A	B	C	D	E
1			Principal	=	$1,500.00
2			Interest	=	15.00%
3			Years	=	1
4			Payment	=	$135.39
5	Period	Beg. Bal.	Interest	App. Prin.	End. Bal.
6	1	$1,500.00	$18.75	$116.64	$1,383.36
7	2	$1,383.36	$17.29	$118.10	$1,265.27
8	3	$1,265.27	$15.82	$119.57	$1,145.70
9	4	$1,145.70	$14.32	$121.07	$1,024.63
10	5	$1,024.63	$12.81	$122.58	$902.05
11	6	$902.05	$11.28	$124.11	$777.94
12	7	$777.94	$9.72	$125.66	$652.27
13	8	$652.27	$8.15	$127.23	$525.04
14	9	$525.04	$6.56	$128.82	$396.22
15	10	$396.22	$4.95	$130.07	$265.78
16	11	$265.78	$3.32	$132.07	$133.72
17	12	$133.72	$1.67	$133.72	($0.00)

C1: 'Principal
D1: ^=
E1: (C2) 1500
C2: 'Interest
D2: ^=
E2: (P2) 0.15
C3: (C2) 'Years
D3: (C2) ^=
E3: (FO) 1
C4: (C2) 'Payment
D4: (C2) ^=
E4: (C2) PMT (E1,E2/12,E3*12) ——→ Calculation of payment
A5: "Period
B5: "Beg. Bal.
C5: (C2) ^ Interest
D5: (C2) ^ App. Prin.
E5: (C2) ^ End. Bal.
A6: 1
B6: (C2) +E1
C6: (C2) +B6*E2/12 ——————→ Interest/month -- E2 must be absolute
D6: (C2) +E4–C6 ——————→ Applied Interest -- E4 must be absolute
E6: (C2) +B6–D6 ——————→ Ending Balance
A7: 2
B7: (C2) +E6
C7: (C2) +B7*E2/12
D7: (C2) +E4–C7
E7: (C2) +B7–D7
A8: 3
B8: (C2) +E7
C8: (C2) +B8*E2/12
D8: (C2) +E4–C8
E8: (C2) +B8–D8

A9: 4
B9: (C2) +E8
C9: (C2) +B9*E2/12
D9: (C2) +E4–C9
E9: (C2) +B9–D9
A10: 5
B10: (C2) +E9
C10: (C2) +B10*E2/12
D10: (C2) +E4–C10
E10: (C2) +B10–D10
A11: 6
B11: (C2) +E10
C11: (C2) +B11*E2/12
D11: (C2) +E4–C11
E11: (C2) +B11–D11
A12: 7
B12: (C2) +E11
C12: (C2) +B12*E2/12
D12: (C2) +E4–C12
E12: (C2) +B12–D12
A13: 8
B13: (C2) +E12
C13: (C2) +B13*E2/12
D13: (C2) +E4–C13
E13: (C2) +B13–D13
A14: 9
B14: (C2) +E13
C14: (C2) +B14*E2/12
D14: (C2) +E4–C14
E14: (C2) +B14–D14
A15:10
B15: (C2) +E14
C15: (C2) +B15*E2/12
D15: (C2) +E4–C15
E15: (C2) +B15–D15
A16: 11
B16: (C2) +E15
C16: (C2) +B16*E2/12
D16: (C2) +E4–C16
E16: (C2) +B16–D16
A17: 12
B17: (C2) +E16
C17: (C2) +B17*E2/12
D17: (C2) +E4–C17
E17: (C2) +B17–D17

Figure 20.8 Worksheet and Cell Contents for Loan Amortization Problem

Step 5: Recalculate and print for all different values of the data that change with each recalculation.

Step 6: Develop macros pertaining to this spreadsheet, graphs, and other types of analysis aids to help with the postanalysis.

Step 7: Analyze the results of the spreadsheet problem—this is the postanalysis. Problem solving for designing spreadsheets is important before, during, and after the user develops the worksheet. After completing the check of the worksheet, the user does a postanalysis, entering sample key values and recalculating the worksheet several times. Each result is evaluated and analyzed. This is the "what if" part of spreadsheets. Problems to determine the best bank for a loan or what investment to make are types of problems that require several worksheet calculations and analysis of each of the results.

The postanalysis should answer questions such as

- What happens to the profit when inflation increases and demand for our product also increases? (when completing a business analysis)
- How can we increase our savings? (using a budget analysis)
- Should we buy another business?
- Should we expand our store to include more square footage?
- How should we finance our car?
- How much profit can I project during the upcoming sale?
- How should we expand our product line?

Postanalysis is what spreadsheets are all about. Once the numbers are correct, they need to be analyzed to determine the outcome of the spreadsheet numbers. Postanalysis is a report summarizing the outcome along with recommendations resulting from the analysis. This analysis is difficult, and it requires practice to know what to look for and how to interpret the numbers. Graphs and summary sheets help in this analysis. Data modeling also helps. Data modeling is available only in the high-end spreadsheet package. Data modeling allows the user to set up and cross reference the data in multiple ways. Again, this is a tool to help in the final analysis.

Graphs and Presentation Graphics

Most spreadsheets have a means to graph the results. These graphs have two functions. First, they are useful for analyzing results. It is much easier to analyze information in graph form than in numeric form. The second use of these graphs is for presentations. Many spreadsheet packages include the capability to make presentation-quality graphs and to incorporate these graphs in a slideshow. Presentation quality graphs are attractive because of the use of special fonts, clip art, character enhancement, multidimensional graphs, and color. Stand-alone presentation graphics packages allow the user to import spreadsheet data to create slides and interactive slideshows that include text slides as well as graph slides. Presentation graphics is an expanding field. Using a multimedia notebook computer and a take-along projection system, a person can use an electronic slideshow to give a professional quality presentation.

There are several types of graphs. The most common graphs are the line graph, the bar graph, and the pie graph. The line graph shows movement, usually over a period of time. An example would be profit over the last five years. The bar graph shows comparisons, such as comparing the amount of corn produced in three states. The pie graph shows percentage of the whole. Pie graphs are usually used with budgets. Other types include a donut graph, a scatter graph, and three-dimensional graphs of the above. It is

important to choose the graph type that best represents the data. It is easy to become influenced by all of the fancy graph types; however, many times a simple graph portrays the data the best. Remember the graph should enhance communication of the data.

Putting It All Together

Figure 20.9a–b is an example of the development of a spreadsheet. The spreadsheet is a budget. A budget is a very good problem to solve using a spreadsheet. The spreadsheet is set up in three sections:

1. Spreadsheet headings
2. Income
3. Expenditures

	A	B	C	D	E	F	G
1	**My Budget**						
2							
3	Income (after deductions)						
4		January	February	March	Total	% of income	
5	Ron's Salary	1400	+B5	©→	SUM(B5..D5)	+E5/E8	
6	Ann's Salary	1400	+B6	©→	©	©	
7	Interest	100	+B7	©→			
8	Total Income	@SUM (B5..B7)	©→				
9							
10	Expenditures						
11		January	February	March	Total	% of income	% of Exp.
12	Rent	600	+B12	©→	SUM(B12..D12)	+E12/E8	+E12/E25
13	Gas/Electricity	100	+B13	©→	©	©	©
14	Telephone	75	+B14	©→			
15	Gasoline	100	+B15	©→			
16	Car Payment	250	+B16	©→			
17	Food	250	+B17	©→			
18	Entertainment	200	+B18	©→			
19	Clothes	100	+B19	©→			
20	Vacation Fund	300	+B20	©→			
21	Insurance	300	+B21	©→			
22	Car Expenses	150	+B22	©→			
23	Miscellaneous	300	+B23	©→			
24	Savings	+B8−SUM(B12..B23)	©→				
25	Total Expenses	SUM(B12..B24)	©→				

Figure 20.9a Example of a Planning Guide and Report for a Budget

My Budget

Income (after deductions)

	January	February	March	Total	% of income
Ron's Salary	1400	1400	1400	4200	48.28%
Ann's Salary	1400	1400	1400	4200	48.28%
Interest	100	100	100	300	3.45%
Total Income	2900	2900	2900	8700	100.00%

Expenditures

	January	February	March	Total	% of income	% of Exp.
Rent	600	600	600	1800	20.69%	20.69%
Gas/Electricity	100	100	100	300	3.45%	3.45%
Telephone	75	75	75	225	2.59%	2.59%
Gasoline	100	100	100	300	3.45%	3.45%
Car Payment	250	250	250	750	8.62%	8.62%
Food	250	250	250	750	8.62%	8.62%
Entertainment	200	200	200	600	6.90%	6.90%
Clothes	100	100	100	300	3.45%	3.45%
Vacation Fund	300	300	300	900	10.34%	10.34%
Insurance	300	300	300	900	10.34%	10.34%
Car Expenses	150	150	150	450	5.17%	5.17%
Miscellaneous	300	300	300	900	10.34%	10.34%
Savings	175	175	175	525	6.03%	6.03%
Total Expenses	2900	2900	2900	8700	100.00%	100.00%

Figure 20.9b Example of a Planning Guide and Report for a Budget

The finished planning guide in Figure 20.9a includes

1. The labels, including the spreadsheet heading and the row and column headings
2. The values for January
3. The formulas for the totals, the percentages, and the references from January
4. The cells where the formulas will be copied

Figure 20.9b shows results of the planning guide after the spreadsheet has been entered into the computer.

Summary

Spreadsheet packages allow the user to develop a worksheet of labels, values, and formulas for the computer to calculate. The spreadsheet is used when a problem requires a matrix of numbers and formulas. Spreadsheets increase productivity by allowing the computer to calculate and recalculate the worksheet. Once a worksheet is developed, the user can be assured that the results will be correct time and time again because the computer is very accurate.

A spreadsheet consists of the worksheet seen on the screen, the status area (also on the screen), the cell contents, and the commands. The worksheet is made up of rows and columns. The place where a row and column meet is called a cell. The cell name is designated by the column letter followed by the row number. The cell can contain a label, a value, or a formula. Labels are strings that tell the user what is in a cell, a row, or a

column. A value is a numeric constant. A formula is an expression. The resultant of the expression becomes the value of the cell as seen on the screen. Formulas contain constants, cell names, arithmetic operators, and functions.

Templates and macros increase productivity. A template contains the spreadsheet labels, the values that do not change, and the formulas. A template is saved and can be retrieved to enter new data time and time again. A macro saves keystrokes by setting up an instruction set so that it takes only one keystroke instead of many to perform a task. A problem that is solved with a spreadsheet must be well planned to be efficient. The steps to design a spreadsheet are

Step 1: Use the PAC to glean the given data, the required results, and the needed formulas.
Step 2: Create the overall design of the spreadsheet.
Step 3: Place the labels on the worksheet.
Step 4: Enter the constant data.
Step 5: Enter the formulas.
Step 6: Enter any tables required in the formulas.
Step 7: Develop any macros you will be using for the problem.
Step 8: Test your formulas as much as possible.

The steps to finish your spreadsheet on the computer are

Step 1: Start with your macro template.
Step 2: Enter the cell contents you developed on your planning sheet.
Step 3: Develop your template.
Step 4: Test your answers.
Step 5: Recalculate and print for all different values of the data that change with each recalculation.
Step 6: Develop macros pertaining to this spreadsheet, graphs, and other types of analysis aids to help with the postanalysis.
Step 7: Analyze the results of the spreadsheet problem—the postanalysis.

New Terms

cell	range name
cell protection	range of cells
command	spreadsheet package
default value	worksheet

Questions

1. What is the difference between a spreadsheet and a worksheet?
2. Define the term *cell*. Name three things that can go into a cell.
3. Name the parts of the status area.
4. Name six features of a spreadsheet.
5. Define *template*.
6. Define *macro*.
7. How is a cell named?

Problems

1. A student has to take out a loan to cover school expenses. He has gone to five banks, and each has a slightly different policy for student loans. The differences are in the areas of interest rate, payback period, and due date for the first payment. Design a template that will help the student choose the best bank for the loan.

2. a. Design a template for a budget for three months so that the amount of each variable category is a percent of the total amount brought in each month. Any money left over would go into savings.

 b. Modify Problem 2a to ensure having a minimum balance of $200 each month. When the balance is less than $200, borrow enough to bring the balance up to $200. When the balance is greater than $200, repay the loan.

3. A new father wants to save money for his child's college education. In 18 years he would like to have $20,000. He can get 8% interest for 18 years compounded monthly. Write the formula you would enter into a cell to find the amount of money the father has to deposit monthly in the account.

4. Evaluate the formulas listed below using the following data:

A1: 100	B1: 7	C1: 15	D1: 250
A2: 200	B2: 10	C2: 15	
A3: 500	B3: 6	C3: 12	
A4: TOTAL			

 a. IF(A1 < A2,B1,B2)
 b. SUM(A1 . . . C1)
 c. VLOOKUP(D1,A1 . . . C3,1)
 d. COUNT(A1 . . . A4)
 e. IF(C1,C2,B1,B2)

5. Dave's Bookstore is going to have a five-day sale on book bags. At present, the price of these bags is $12.00. The bookstore paid $7.00 for the bags, which means they make a profit of $5.00 on each bag they sell. The manager wants to see what the profits will be for the five-day sale, under the following assumptions:

 a. Each day of the sale, the price should be reduced by 10% of the previous day's price.
 b. The sale price should never fall below $7.00.
 c. They expect the sales to be as follows:

Monday	12 bags
Tuesday	15 bags
Wednesday	19 bags
Thursday	25 bags
Friday	40 bags

Follow the steps to design a worksheet for this problem.

6. Design a spreadsheet to calculate your grade point average for all of the classes you have taken so far in college.

7. Design a spreadsheet to calculate the number of square feet that needs to be painted per room in a house and the amount of paint needed per room. The number of square feet per gallon of paint will be a variable number. There are no more than three windows per room in the house. Also include the total number of square feet and the total number of gallons of paint needed.

8. Design a spreadsheet to calculate the charges to customers of the Good Taste Water Company. The total charges should be calculated also. The charges are as follows:
 a. A flat rate of $15.00 for usage up to and including 1,000 cubic feet.
 b. $0.0175 per cubit foot for usage over 1,000 cubic feet and up to and including 2,000 cubic feet.
 c. $0.02 per cubic foot for usage over 2,000 cubic feet and up to and including 3,000 cubic feet.
 d. A flat rate of $70 over 3,000 cubic feet.

Chapter 21

Document Processing

Overview

The Six Elements of a Computer System
Information Processing
Suites
Sharing Data
OLE

Objectives

When you have finished this chapter, you should be able to:

1. Develop data flow solutions for problems requiring the use of multiple software packages and the theory of information processing.
2. Understand how to increase productivity through the use of suites.
3. Understand the concept of sharing files.
4. Understand and develop solutions for problems using OLE.

The emphasis in the preceding chapters has been individual types of software, including presentation graphics, word processing, desktop publishing, spreadsheets, and relational database management systems. In reality, these products often are used together to develop a solution to a problem. You may want to design a letter for a client that includes a graphic and an imported spreadsheet. You may want data imported into a spreadsheet from a relational database management system. Keep in mind why each type of software was created:

- the word processor to efficiently accomplish text editing
- the desktop publisher to combine graphics and text on a page efficiently
- presentation graphics to create electronic slide shows easily
- spreadsheets to enter, calculate, and analyze numeric data easily
- relational database management systems to organize and report large amounts of text and numeric data easily

The Six Elements of a Computer System

An understanding of a complete computer information system is important to see the complete picture of computing. The definition of a computer system includes *all components required to complete a task.* A **computer information system** consists of six components. The first component is the hardware, which is the physical components of the computer. A computer system must have all the hardware parts needed to complete the tasks required of the business or home. This would include the system unit, input devices such as a keyboard and a mouse, output devices such as a monitor and a printer, secondary storage devices such as a hard disk drive and a floppy disk drive, and communications devices such as a modem. The second component is software. A system includes the software products required to complete the activities of the system. The third component is data. A system needs data to use with the hardware and the software. The fourth component is people. A system needs people to use the other components in the system. The fifth component is procedures. Procedures include how to use the software and the hardware (manuals and tutorials). The sixth component is facilities. The system must be placed somewhere. Furniture and square footage are required for every system. If any one of these six components is missing the computer information system is not complete.

The Johnson Lumber Yard will be used as an example throughout this chapter. Along with the regular applications such as using a database management system for inventory, using a word processor to write letters and prepare reports, using a spreadsheet for costing a project, and using an accounting package for all of the general accounting, Mr. Johnson would like to be able to write a letter to a customer. This letter would include a list of everything the customer ordered along with a price and the total cost. Mr. Johnson would like this to be done by the computer without re-entering data.

The computer system for the Johnson Lumber Yard would include the six components. The hardware would include three system units networked together, each with a keyboard, a mouse, a monitor, a hard disk drive, and a floppy disk drive. A printer would be included as part of the network. Software would include a word processor, a spreadsheet package, a graphics package, a database management system, and an accounting package. The data would be supplied from inventory lists, sales receipts, customer order forms, and other source documents. The people using the system would include the office staff and Mr. Johnson. Procedures would include the manuals that came with the hardware and software and the written procedures for the users. The facilities include the main office and Mr. Johnson's office, and the sales desk.

Document processing uses all components of the computer information system. This is an important concept. Document processing uses data from source documents created by hand and entered into the computer system using one of many software packages. Hardware and software are used to enter data, process data, store data, retrieve data, and create the final document. People complete the entry of the data and create the process of how to create the final document. People need manuals to learn how to use the software and the hardware. People need a place to put the hardware components, and a place to work, hence the need for facilities.

Information Processing

The use of more than one type of software to create a solution to a problem is called **information processing** or **document processing.** It incorporates the sharing of data

between software packages. For example, a letter developed in a word processor may require sharing data from a graphics package and from a spreadsheet software package. If this sharing of data were not possible, you would be required to enter data into more than one package, assuming that each package even has the capability to accept the data. Re-entering data is not only time consuming, but it also increases the possibility of errors.

When you are creating a solution to a problem using document processing, you should consider how each software package contributes to the final solution. Identify the package that will do the best job for each part of the solution. This is true information processing. In addition, keep in mind that the best solution may be to use only one type of software. There is no requirement that multiple software packages must be used to complete every solution.

For example, if you have a problem that requires numerical analysis on data, you should first consider using a spreadsheet. However, if the problem requires a letter that contains the spreadsheet, then you should consider using the spreadsheet to analyze the data. You can import it into a word processor because a word processor is the best type of software to create the letter. You could use the table feature in a word processor to analyze and display the data. However, a word processor does not accomplish numerical analysis as well as spreadsheet software.

Another example of information processing is costing a project. A contractor may use a relational database management system to store the data on the components of building a house. When he is creating a bid to build a specific house, he would import data from the database into the spreadsheet package to analyze the costs. He would then import the spreadsheet data into a word processor to complete the bid. He would be using the software ideal for accomplishing each portion of the bid process.

Suites

Software that has been presented in this text comes together in a package called a **suite.** A suite is three or more software productivity packages, such as word processing, spreadsheets, presentation graphics, and relational database management systems, that are packaged together along with other related software such as clipart, paint or draw software, internet software, and so on. The software company may have several combinations of these packages and identify their suites with slightly different names.

When you purchase a suite, examine the software that is included to ensure you have the desired combination. Many times a suite is bundled with the hardware that you purchase. If the bundled suite does not include all of the software packages that you need, plan to upgrade the suite to the one you desire.

At times you may be working on a computer that does not have all of the desired software. For example, perhaps the relational database management system is missing. You will need to decide which of the software packages you have will be the best replacement. In this case, the spreadsheet software package may be a good substitute for the RDBMS software. If you do not have a desktop publisher, a word processor is normally a good substitute. Very rarely will you have all of the desired software packages on your computer. A good computer skill to learn is how to choose the best software to accomplish the current job from the software packages that are available. When you have learned this skill, along with good planning techniques, you should improve your productivity and decrease your frustrations when using the computer.

In addition to multiple software packages, a suite has features that also increase your productivity, such as consistency of menu items, look-alike screens, and similar tool bars across packages. Another feature that saves time is the ability to **drag and drop** across applications. You may have both your spreadsheet package and your word processing package open and displayed side by side on your desktop. You may select items in your spreadsheet to drag and drop them into your word processing document, or perhaps drag and drop text from your word processor to a cell in your spreadsheet. There is no need to cut and paste.

Another feature of a suite is the file compatibility between software packages. This is sometimes called cross-application integration. There is no need to worry about saving in a compatible format because the suite is designed to accept all formats from other software packages included in the suite. Cross-application integration also includes the ability to use one package from within another package. For example, you may create a word processed report using the report form design in a RDBMS from the RDBMS software package. Or you may create a spreadsheet chart from the word processing software package.

The use of a suite is convenient in many ways. First, it is usually less expensive than purchasing each package separately. Second, because of the features, a suite increases your productivity. Third, a suite is easier to learn because of the commonalties. Last, for most problems in business, there are very few companion software packages that you will need beyond the suite.

Sharing Data

Information processing requires file sharing, including drag and drop across applications, cut and paste, and importing and exporting of files. One document may share all or part of a file. To share part of a file, you are required to use either the drag and drop feature or the cut and paste feature. Realize that cut and paste also means copy and paste. It is best to copy and paste. These features allow you to select all or part of another file for inclusion in the document in which you are working.

Importing files means having the computer copy all of the information in one file into the active document. When the imported information is inserted in the text, it starts at the point where the cursor is located even if the cursor is on another page. You should be sure to check the location of the cursor. If the imported information is imported as an floating object, then the object is inserted somewhere on the page you are viewing. A floating object may then be moved around on the page. If the object is not floating, then you move the object as you would text by using the Space Bar and the Enter key.

Exporting files means that you are saving the file in format other than the default format. Exporting files includes using the Save As feature to save files for use in other versions of the same program, in other software packages, and in other formats, such as RTF (Rich Text Format), HTML (Hypertext Markup Language), and PDF (Portable Document Format) formats. RTF format saves your file in a file format that retains very little of the character, paragraph, and document formats. HTML is the format used for Web page development.

PDF is the format used for electronic printing. It is one way to make your file into a format that almost anyone can read and print. PDF files may be exported from many software packages, but very few allow you to import PDF files. Adobe Acrobat allows

you to view and print PDF files. Adobe Acrobat may be downloaded from the Adobe Web site. PDF files are becoming a common way to distribute information to many people when you do not know the software available on their computers.

OLE

When you import data files, you will use **OLE,** object linking and embedding. Embedding a file means copying the contents of the file you are importing into the file in which you are currently working, called the active file. An active file is a file that is open, viewed on the monitor, and has the title bar of the software package colored (not grayed out). If you modify the original imported file at a later date, the data in the file to which it was first imported is not modified. After importing the file, there is no connection between the two files.

However, there are times when you want to maintain a connection to the imported file. When this is true, you should link the imported file. This will provide an ongoing connection between the imported file and the file to which it was imported. When you link the imported file, the location of the file is embedded instead of the complete file. This embedded location points to where the file can be located. The linked imported file is temporarily copied to the document each time it is viewed or printed. A low resolution copy may be embedded to retain the space and to show what the file looks like in case the link is lost. If the computer can find the linked file, the higher resolution graphic replaces the low resolution copy. A link is lost or broken when the embedded location of the imported file becomes incorrect. If the imported file is moved, the location of the imported file changes. The embedded location no longer points to where the import file is located. If you have saved the imported linked file on a hard drive of your computer at home and then come to school to view your document, you will have broken the link since the computer at school is not connected to your home computer. Figure 21.1 shows examples of OLE.

To ensure minimal problems with broken or lost links, create a file folder for your project. Save or copy all files to be imported to this folder before you import them. Save your document to this folder also. Import your files from this folder. When you change locations of your files, such as from your hard drive to a floppy disk or a zip cartridge, copy the entire folder from one place to the other. In this way all linked files and project files remain in the same folder as when they were created. Do not move any of these files from this folder, or you will find you have broken some of your links. If you do have broken links, your software program has a feature to update broken links.

There are three circumstances in which you may want to link files. First, when you want to modify the imported file at a later date and want the modifications to be made in the file to which it was imported. There are times when you want your file to change when the imported file is modified, such as a slide show presentation in which you have a picture that you want to change each time you are giving a presentation. If the imported picture file is linked to the slide show, whenever the picture file changes, the picture in the slide show will also change.

Second, you should link your files when there are many places in your document to which a file needs to be imported. Graphics tend to be large files. If the same file is embedded many times, your document file becomes very large. You can decrease its size by linking to the file. Because only a low resolution copy is placed in the document, the total document size decreases. For example, if you have a file that is three megabytes

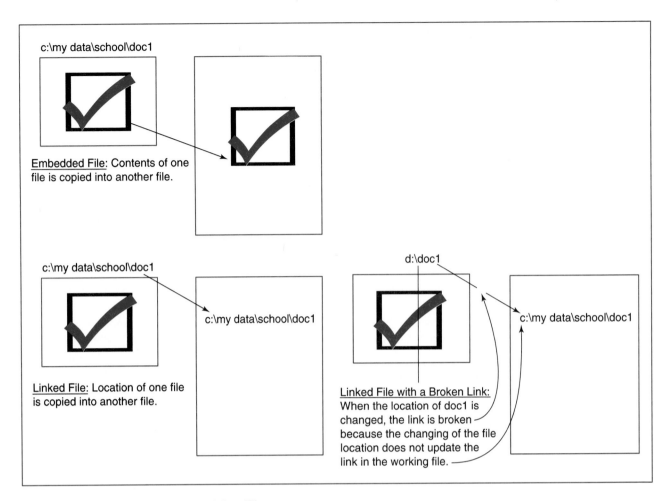

Figure 21.1 Linking and Embedding Files

in volume and you import it 10 times, it will add 30 megabytes to your document. If the low resolution copy is 0.25 megabytes, it will only add 2.5 megabytes to your file when the file is linked.

Third, you may want to link files when your file becomes very large. Because graphics tend to be large files, when you import many graphics files your working document becomes very large. The larger the file, the longer it takes to move from one part of the file to another and the bigger the chance of problems creeping in during the processing. When you link the files, you decrease the size of the imported graphics because only the location of the file and a low resolution image is embedded. Thus, the size of your slide show file will be decreased.

Linking files is necessary many times. However, links can cause problems, especially when you do not understand how linking works. The link simply points to where the imported file is located. If the file is not in that location, it cannot be imported. People tend to embed most files; however, you should link files under the three cases above.

The use of multiple packages, along with other components of a computer information system, to produce one document is called **document processing.** Figure 21.2 shows the flow of data from one package to another to produce the required document for the Johnson Lumber Company. The inventory file contains the product prices needed to develop the cost of the project. The customer file contains all customer

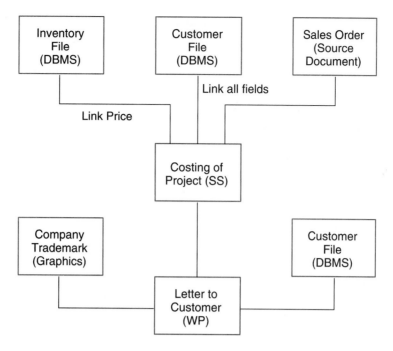

Figure 21.2 Data Flow to Produce Customer Letter

information. These two files are developed through the use of a database management system. The sales document is the source document; it contains the project information. The data from these two files are imported into a spreadsheet along with keying the data from the source document. A report on the total cost of the project is completed in the spreadsheet package. A letter is produced in a word processor by bringing in the company trademark as a letterhead (produced by a graphics package), the customer information from the customer file, and the cost of the components of the project along with the totals from the spreadsheet. This is document processing. There is no re-entering of data.

Information processing is very important to the development of good projects. The end user should understand the functions of the different application packages and decide on the best tool for the job to be completed, whether it is a word processor, spreadsheet, DTP, presentation graphics, other graphics software, RDBMS, languages, or any other support software. The efficient use of these packages' integrated capabilities should produce a better project in less time.

Summary

Information processing is the use of multiple software packages for a solution to a problem. It involves the use of the best software package to complete each part of a solution, a word processor for text editing, a desktop publisher for combining graphics and text, presentation graphics to create a slide show, a spreadsheet for numerical analysis, a

RDBMS to organize and report data, and so on. A suite is a group of several software packages that have commonalties, drag and drop capabilities, and other integrated features.

Information processing requires file sharing in the form of drag and drop, cut and paste, and importing and exporting. When a file is imported, it can be embedded or linked. Embedding an imported file copies the contents of the file into the working document. Linking an imported file copies the location of the file into the working document. The three reasons to link files are when a modified imported file should be modified in the working document, when there are multiple imports of the same file, and when the file becomes too large.

New Terms

Document processing	Information processing
Drag and drop	OLE
Exporting	Suite
Importing	

Questions

1. What is information processing?
2. What is a suite?
3. What are three features of a suite?
4. How can a suite increase your computer productivity?
5. What is the two processes that you can import a selected part a file?
6. What is OLE?
7. What is a problem with using linked files?
8. How can you minimize the problem of lost links?

Problems

1. The Exceptional Consultant Company has just finished a marketing survey for one of their clients. They would like to deliver a letter to the client that will include a summary report of the results, a statistical analysis of the results, and two charts of the results. Create a data flow diagram for this problem.

2. Joan Green has just been promoted to head of sales at the Best Computer Company. She knows that one of her duties is to make a presentation to the board of directors about the sales of the past year and projections for the future. She would like to keep information about her sales reps, their clients, and the amount of sales for each sales rep and for each client. She plans to develop a slide show to accompany her presentation. Create a data flow diagram for this problem.

3. You are planning to purchase a new computer and would like to compare four different brands. Find four different advertisements on computers to use for this problem. You must borrow the money for the new computer and need to write a letter to your banker to obtain the loan. Create a data flow diagram for this problem.

Unit Six

Supplementary Exercises

Develop a complete solution to the following problems:

1. Design a flyer to advertise the class you are taking.

2. Design a page for a training manual. The requirements are as follows:
 - page numbers in upper outside corners
 - header with the name of the chapter
 - footer with the name of the training manual
 - a margin large enough to make notes
 - a line separating the outside margin from the text

3. **a.** Design a form letter to use with the mail merge function of a word processor to send to friends inviting them to a party. Include graphics as necessary.
 b. Design the data file that would accompany the letter in Exercise 3a.

4. You are going to buy a car and need to borrow \$10,000 for five years. Design a spreadsheet template that will calculate the following for four different lending sources:
 - monthly payment
 - total amount paid
 - total interest

5. Develop a spreadsheet template that will calculate your grade average. (Use the IF function to find the grade points.)

6. The Buycheap Department Store is going to have a five-day sale. Each day the price will drop 10%. Develop the template to solve this problem. The columns should include the item number, name, wholesale price, retail price, and sale price for each day. The sale price should not drop below the wholesale price.

7. Evaluate the following formulas:
 a. $+A1 + B1$ where $A1:7$
 $$B1:10$$
 b. $IF(A1 < B1, B1 - 7, B1 + 7)$ where $A1:25$
 $$B1:15$$

(Exercise 7 continued on next page)

c. VLOOKUP(C10, A50 . . C54,1)

where

	A	B	C
50	0	.1	300
51	5	.7	200
52	15	.2	100
53	27	.6	0
54	35	.9	200

where 1. C10: 25
 2. C10: 27

8. Imagine yourself as a microcomputer specialist for a small business in a field of your choosing.
 a. How would you use word processor, spreadsheet, and database application packages to best serve your company? List and explain specific uses for each package.
 b. List and explain several situations in which the packages could be used in combination.

9. You are planning to buy a new computer and would like to compare four different brands. Find four advertisements for computers to complete this problem.
 a. Using a word processor, create a document that would compare the four systems.
 b. Using a spreadsheet, create a spreadsheet that would compare the four systems.
 c. Using a DBMS, create a database and a report to compare the four systems.
 d. Which of the applications packages was the best one to use for this problem? Justify your answer.
 e. You would like to write a letter to your bank to obtain a loan for your computer system. How would you use a word processor, a spreadsheet, and a DBMS in combination to accomplish this tasks? What pieces of data would you designate as linked data?

Appendix A

Formulas Commonly Used in Business Applications

The formulas found in Appendix A are commonly used in business applications. It is not necessary for the programmer or the user to know how to develop the formulas; however, it is very important to know how to use each of them. Almost every application will use one or more of these formulas. Included are the formulas to calculate principal and interest, commission, discount price, mark-up price, percent of a sum, economic order quantity, depreciation, break-even analysis, present value, future value, net present value, internal rate of return, average, payment, and others.

Equation	Explanation
$A = P\left(1 + \dfrac{I}{T}\right)^{NT}$	**Interest (A)** A = Amount of Principal + Interest P = Principal I = Interest rate N = Number of years T = Number of payments per year
$C = R\,S$	**Commission (C)** R = Rate S = Sales
$D = P - PR$	**Discount Price (D)** P = Price R = Percent discount
$C = P + PR$	**Mark-up Price (C)** P = Wholesale price R = Mark-up price
$\% = \left(\dfrac{P}{T}\right)100$	**Percent of a sum (%)** P = Item T = Total of all items
$E = \sqrt{\dfrac{2DC}{PH}}$	**Economic Order Quantity (E)** D = Average weekly demand C = Cost of placing an order P = Unit purchase price H = % cost of purchase price for inventory overhead
$D = \dfrac{P - S}{N}$	**Straight-Line Depreciation (D)** D = Depreciation P = Purchase cost S = Salvage value N = Useful life in years
$B = \dfrac{F}{S - V}$	**Break-Even Analysis (B)** B = Break-even point F = Fixed costs V = Variable costs S = Expected sales
Spreadsheet Function	**Present Value (PV)** This is the present value of known amount of money in the future.

Figure A.1 Commonly Used Business Equations *(continued on page 485)*

Equation	Explanation
Spreadsheet Function	**Future Value (FV)** This is the future value of a known amount of money today.
Spreadsheet Function	**Net Present Value (NPV)** This is the net return on an investment. It is the present value of cash inflows minus the present value of cash outflows.
Spreadsheet Function	**Internal Rate of Return (IRR)** Percentage of return on an investment.
$A = \dfrac{S}{N}$	**Average (A)** S = Sum of the items N = number of items
Spreadsheet Function	**Payment (P)** Given an amount of a loan, the interest rate, and the number of years to payback, will calculate the payment. L = Amount of loan R = Monthly Interest N = Number of payments
$P = \dfrac{N}{S}$	**Profit Margin (P)** N = Net profit after tax S = Sales
$ROA = \dfrac{N}{S}$	**Return on Assets (ROA)** N = Net profit after tax A = Total assets
$R = \dfrac{A}{L}$	**Current Ratio (R)** A = Current assets L = Current liabilities
$C = \dfrac{A}{S}$	**Average Collection Period (C)** A = Accounts receivable S = Sales per day

Figure A.1 *(continued from page 484)*

Appendix B

Sort, Merge, and Search Methods

Appendix B contains the algorithms and the flowcharts of popular methods of sorting data, merging files, and searching for a single record.

- Selection/exchange sort (Figure B.1)
- Bubble sort (Figure B.2)
- Shell sort (Figure B.3)
- Merge of two lists (Figure B.4)

Definitions:

N = Number of elements in list
I, J, K = Index values of arrays, counters
Min = Minimum value
A(I) = Element I of List A

Algorithm	Flowchart

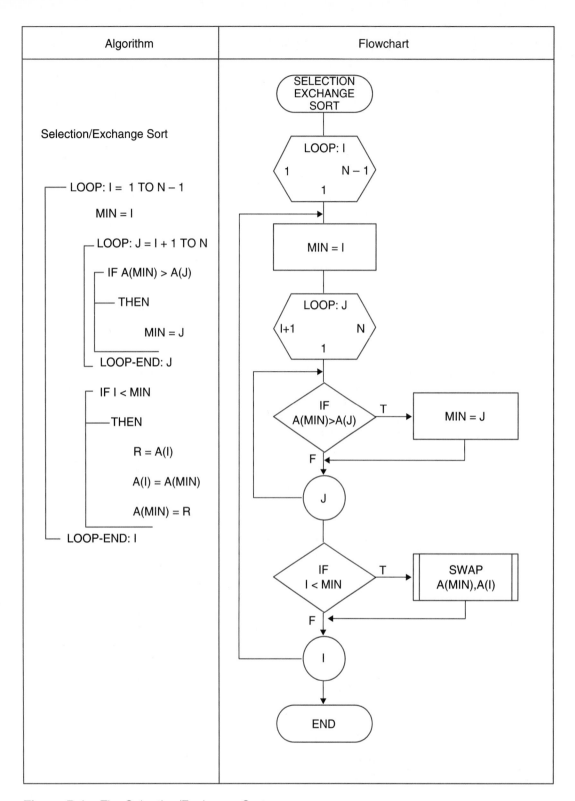

Selection/Exchange Sort

```
— LOOP: I =  1 TO N – 1
       MIN = I
     — LOOP: J = I + 1 TO N
       — IF A(MIN) > A(J)
         — THEN
               MIN = J
       — LOOP-END: J

     — IF I < MIN
       — THEN
             R = A(I)
             A(I) = A(MIN)
             A(MIN) = R
— LOOP-END: I
```

Figure B.1 The Selection/Exchange Sort

Algorithm	Flowchart
Bubble Sort 1. FLAG = 1 2. E = N − 1 3. ⌐WHILE FLAG = 1 FLAG = 0 ⌐LOOP: J = 1 TO E ⌐IF A(J) > A(J + 1) ─ THEN SWAP A(J),A(J + 1) FLAG = 1 └LOOP-END: J E = E − 1 └WHILE-END EXIT	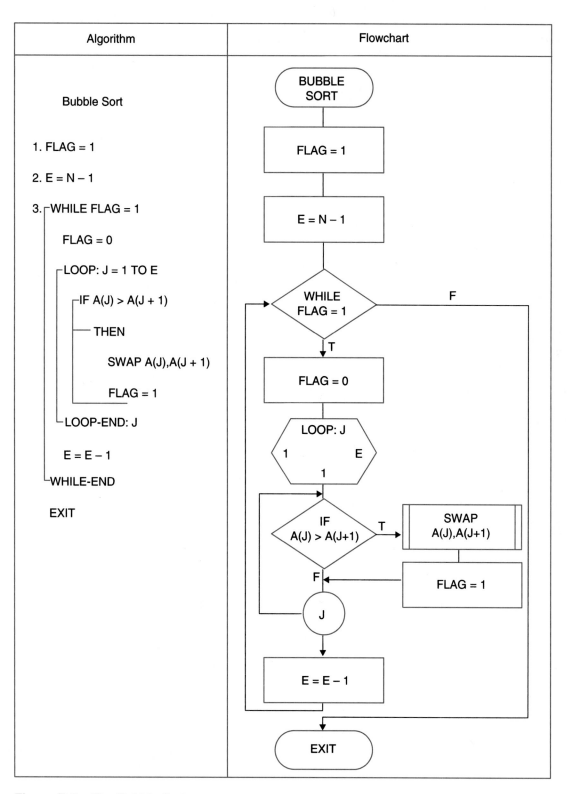

Figure B.2 The Bubble Sort

Algorithm	Flowchart
Shell Sort 1. G = INT(N/2) 2. ⌐ WHILE G > 0 　　M = N − G 　⌐ LOOP: J = 1 TO M 　⌐ IF A(J) > A(J+G) 　└ THEN 　　　L = J 　　　FLAG = TRUE 　└ WHILE FLAG 　　　SWAP A(L), A(L+G) 　　　L = L − G 　⌐ IF L >= 1 　　THEN 　⌐ IF A(L) <= A(L+G) 　└ THEN 　　　FLAG = FALSE 　└ ELSE 　　　CONTINUE 　└ ELSE 　　　FLAG = FALSE 　└ WHILE-END 　└ LOOPEND: J 　　G = INT(G / 2) └ WHILE-END 3. EXIT	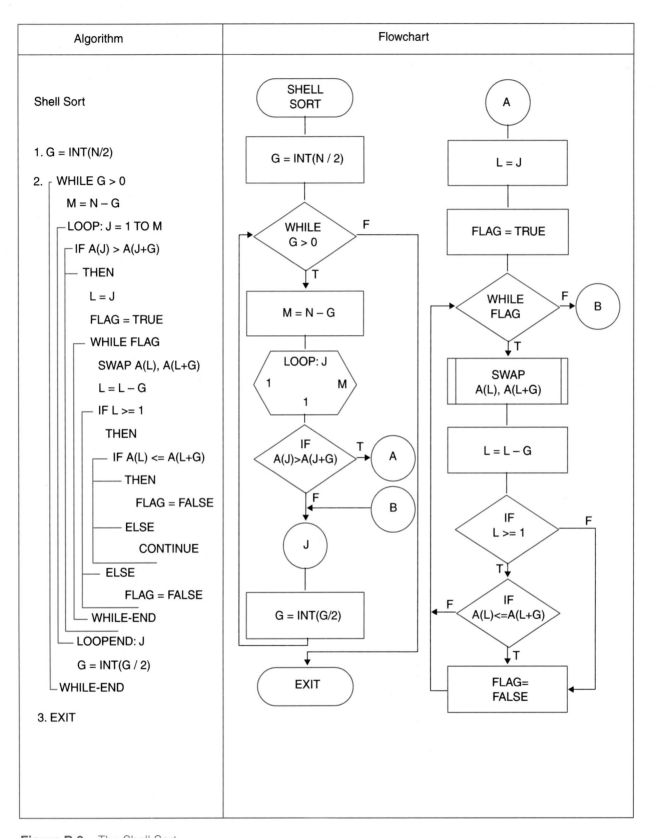

Figure B.3　The Shell Sort

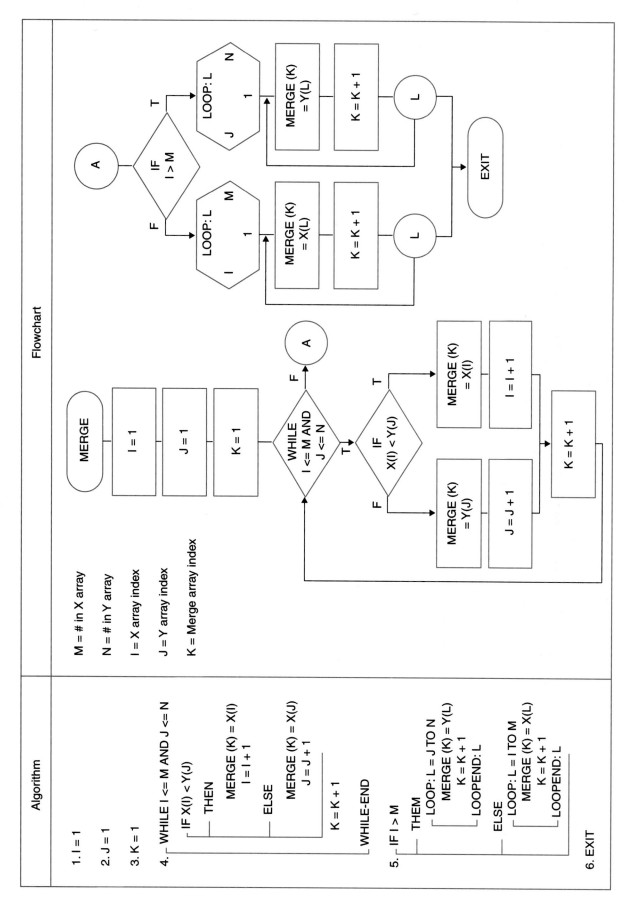

Figure B.4 Merge of Two Lists

Appendix C

ASCII and EBCDIC Codes for Data Representation

Character	EBCDIC Code Binary Code	Decimal Value
0	1111 0000	240
1	1111 0001	241
2	1111 0010	242
3	1111 0011	243
4	1111 0100	244
5	1111 0101	245
6	1111 0110	246
7	1111 0111	247
8	1111 1000	248
9	1111 1001	249
Space	0100 0000	64
.	0100 1011	75
<	0100 1100	76
(0100 1101	77
+	0100 1110	78
&	0101 0000	80
!	0101 1010	90
$	0101 1011	91
*	0101 1100	92
)	0101 1101	93
;	0101 1110	94
,	0110 1011	107
%	0110 1100	108
-	0110 1101	109
>	0110 1110	110
?	0110 1111	111
:	0111 1010	122
#	0111 1011	123
@	0111 1100	124
'	0111 1101	125
=	0111 1110	126
"	0111 1111	127

Character	EBCDIC Code Binary Code	Decimal Value
A	1100 0001	193
B	1100 0010	194
C	1100 0011	195
D	1100 0100	196
E	1100 0101	197
F	1100 0110	198
G	1100 0111	199
H	1100 1000	200
I	1100 1001	201
J	1101 0001	209
K	1101 0010	210
L	1101 0011	211
M	1101 0100	212
N	1101 0101	213
O	1101 0110	214
P	1101 0111	215
Q	1101 1000	216
R	1101 1001	217
S	1110 0010	226
T	1110 0011	227
U	1110 0100	228
V	1110 0101	229
W	1110 0110	230
X	1110 0111	231
Y	1110 1000	232
Z	1110 1001	233
a	1000 0001	129
b	1000 0010	130
c	1000 0011	131
d	1000 0100	132
e	1000 0101	133
f	1000 0110	134
g	1000 0111	135
h	1000 1000	136
i	1000 1001	137
j	1001 0001	145
k	1001 0010	146
l	1001 0011	147
m	1001 0100	148
n	1001 0101	149
o	1001 0110	150
p	1001 0111	151
q	1001 1000	152
r	1001 1001	153
s	1010 0010	162
t	1010 0011	163
u	1010 0100	164
v	1010 0101	165
w	1010 0110	166
x	1010 0111	167
y	1010 1000	168
z	1010 1001	169

Figure C.1 EBCDIC Codes

ASCII in Mnemonic	Binary Code	Decimal Code	Octal Code	Hex Code	Remark
NUL	00 00 00 00	0	000	00	Null, tape feed.
SOH	00 00 00 01	1	001	01	Start of heading; also SOM, start of message.
STX	00 00 00 10	2	002	02	Start of text; also EOA, end of address.
ETX	00 00 00 11	3	003	03	End of text; also EOM, end of message.
EOT	00 00 01 00	4	004	04	End of transmission (End); shut off TWX machines.
ENQ	00 00 01 01	5	005	05	Inquiry (Inqry); also WRU.
ACK	00 00 01 10	6	006	06	Acknowledge; also RU.
BEL	00 00 01 11	7	007	07	Ring the bell.
BS	00 00 10 00	8	010	08	Backspace; also FEO format effector.
HT	00 00 10 01	9	011	09	Horizontal tab.
LF	00 00 10 10	10	012	0A	Line feed or line space (New line); advance paper to next line.
VT	00 00 10 11	11	013	0B	Vertical tab (VTAB).
FF	00 00 11 00	12	014	0C	Form feed to top of next page (Page).
CR	00 00 11 01	13	015	0D	Carriage return to beginning line.
SO	00 00 11 10	14	016	0E	Shift out; changes ribbon color to red
SI	00 00 11 11	15	017	0F	Shift in; changes ribbon color to black
DLE	00 01 00 00	16	020	10	Data line escape.
DC1	00 01 00 01	17	021	11	Device control 1; turns transmitter (reader) on.
DC2	00 01 00 10	18	022	12	Device control 2; turns punch or auxiliary on.
DC3	00 01 00 11	19	023	13	Device control 3; turns transmitter (reader) off.
DC4	00 01 01 00	20	024	14	Device control 4; turns punch or auxiliary off.
NAK	00 01 01 01	21	025	15	Negative acknowledge; also ERR, error.
SYN	00 01 01 10	22	026	16	Synchronous file (SYNC).
ETB	00 01 01 11	23	027	17	End of transmission block; also LEM, logical end of medium.
CAN	00 01 10 00	24	030	18	Cancel (Cancl).
EM	00 01 10 01	25	031	19	End of medium.
SUB	00 01 10 10	26	032	1A	Substitute.
ESC	00 01 10 11	27	033	1B	Escape.
FS	00 01 11 00	28	034	1C	File separator.
GS	00 01 11 01	29	035	1D	Group separator.
RS	00 01 11 10	30	036	1E	Record separator.
US	00 01 11 11	31	037	1F	Unit separator.
Space	00 10 00 00	32	040	20	
!	00 10 00 01	33	041	21	
"	00 10 00 10	34	042	22	
#	00 10 00 11	35	043	23	
$	00 10 01 00	36	044	24	
%	00 10 01 01	37	045	25	
&	00 10 01 10	38	046	26	
'	00 10 01 11	39	047	27	Apostrophe.
(00 10 10 00	40	050	28	
)	00 10 10 01	41	051	29	
*	00 10 10 10	42	052	2A	
+	00 10 10 11	43	053	2B	
,	00 10 11 00	44	054	2C	
-	00 10 11 01	45	055	2D	Hyphen.
.	00 10 11 10	46	056	2E	
/	00 10 11 11	47	057	2F	

Figure C.2 ASCII Codes *(continued on pages 495–496)*

ASCII in Mnemonic	Binary Code	Decimal Code	Octal Code	Hex Code	Remark
0	00 11 00 00	48	060	30	
1	00 11 00 01	49	061	31	
2	00 11 00 10	50	062	32	
3	00 11 00 11	51	063	33	
4	00 11 01 00	52	064	34	
5	00 11 01 01	53	065	35	
6	00 11 01 10	54	066	36	
7	00 11 01 11	55	067	37	
8	00 11 10 00	56	070	38	
9	00 11 10 01	57	071	39	
:	00 11 10 10	58	072	3A	
;	00 11 10 11	59	073	3B	
<	00 11 11 00	60	074	3C	
=	00 11 11 01	61	075	3D	
>	00 11 11 10	62	076	3E	
?	00 11 11 11	63	077	3F	
@	01 00 00 00	64	100	40	
A	01 00 00 01	65	101	41	
B	01 00 00 10	66	102	42	
C	01 00 00 11	67	103	43	
D	01 00 01 00	68	104	44	
E	01 00 01 01	69	105	45	
F	01 00 01 10	70	106	46	
G	01 00 01 11	71	107	47	
H	01 00 10 00	72	110	48	
I	01 00 10 01	73	111	49	
J	01 00 10 10	74	112	4A	
K	01 00 10 11	75	113	4B	
L	01 00 11 00	76	114	4C	
M	01 00 11 01	77	115	4D	
N	01 00 11 10	78	116	4E	
O	01 00 11 11	79	117	4F	
P	01 01 00 00	80	120	50	
Q	01 01 00 01	81	121	51	
R	01 01 00 10	82	122	52	
S	01 01 00 11	83	123	53	
T	01 01 01 00	84	124	54	
U	01 01 01 01	85	125	55	
V	01 01 01 10	86	126	56	
W	01 01 01 11	87	127	57	
X	01 01 10 00	88	130	58	
Y	01 01 10 01	89	131	59	
Z	01 01 10 10	90	132	5A	
[01 01 10 11	91	133	5B	
]	01 01 11 00	92	134	5C	
]	01 01 11 01	93	135	5D	
^	01 01 11 10	94	136	5E	
–	01 01 11 11	95	137	5F	Underline.

Figure C.2 *(continued from page 494)*

ASCII in Mnemonic	Binary Code	Decimal Code	Octal Code	Hex Code	Remark
'	01 10 00 00	96	140	60	Accent grave.
a	01 10 00 01	97	141	61	
b	01 10 00 10	98	142	62	
c	01 10 00 11	99	143	63	
d	01 10 01 00	100	144	64	
e	01 10 01 01	101	145	65	
f	01 10 01 10	102	146	66	
g	01 10 01 11	103	147	67	
h	01 10 10 00	104	150	68	
i	01 10 10 01	105	151	69	
j	01 10 10 10	106	152	6A	
k	01 10 10 11	107	153	6B	
l	01 10 11 00	108	154	6C	
m	01 10 11 01	109	155	6D	
n	01 10 11 10	110	156	6E	
o	01 10 11 11	111	157	6F	
p	01 11 00 00	112	160	70	
q	01 11 00 01	113	161	71	
r	01 11 00 10	114	162	72	
s	01 11 00 11	115	163	73	
t	01 11 01 00	116	164	74	
u	01 11 01 01	117	165	75	
v	01 11 01 10	118	166	76	
w	01 11 01 11	119	167	77	
x	01 11 10 00	120	170	78	
y	01 11 10 01	121	171	79	
z	01 11 10 10	122	172	7A	
{	01 11 10 11	123	173	7B	
\|	01 11 11 00	124	174	7C	Alt mode.
}	01 11 11 01	125	175	7D	
~	01 11 11 10	126	176	7E	
DEL	01 11 11 11	127	177	7F	Delete, rub out.

Figure C.2 *(continued from page 495)*

Appendix D

Forms to Use in Problem Solving

Given Data	Required Results
Required Processing	**Solution Alternatives**

Figure D.1 Problem Analysis Chart

Figure D.2 The Interactivity Chart

Input	Processing	Module Reference Number	Output

Figure D.3 The IPO Chart

Algorithm	Flowchart

Figure D.4 Algorithm and Flowchart *(continued on page 501)*

Annotation	Test	Internal Documentation	External Documentation

Figure D.4 *(continued from page 500)*

Item	Variable Name	Data Type	Module	Scope	Pseudonyms/Module	Error Checks

Figure D.5 Data Dictionary

Record Data Dictionary

Field	Item Description	Field Name	Field Length	Data Type	# of Decimals	Comments

Record Layout

Figure D.6 Record Data Dictionary and Record Layout

Figure D.7 Screen Design

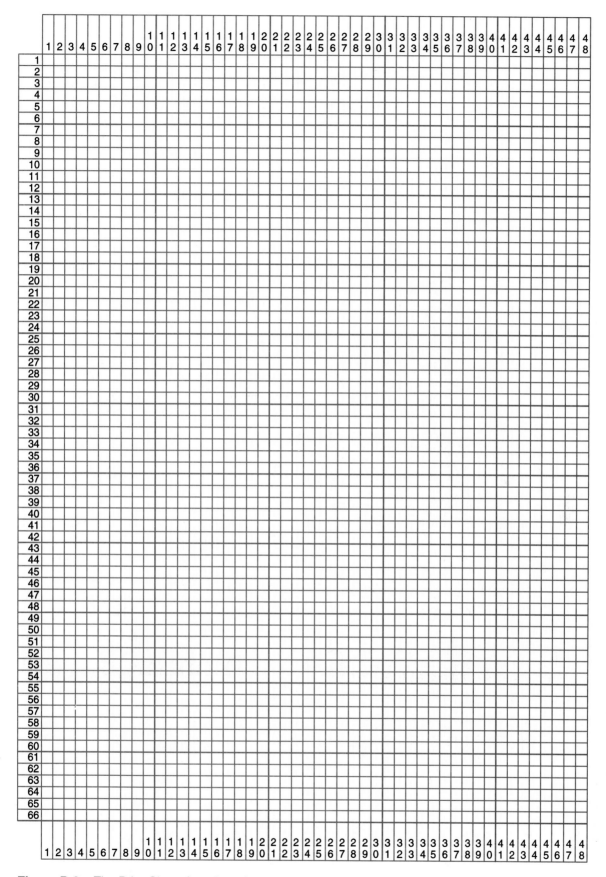

Figure D.8 The Print Chart *(continued on pages 506–507)*

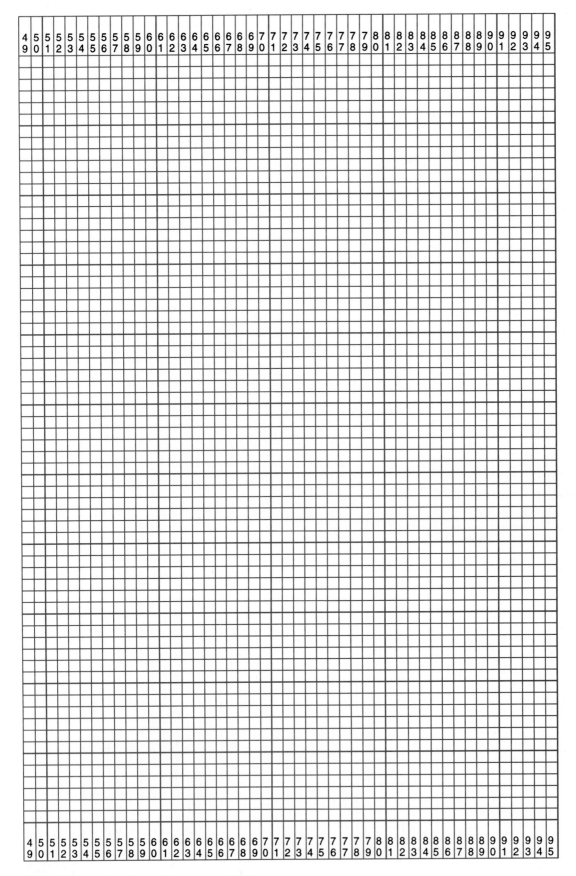

Figure D.8 *(continued from page 505)*

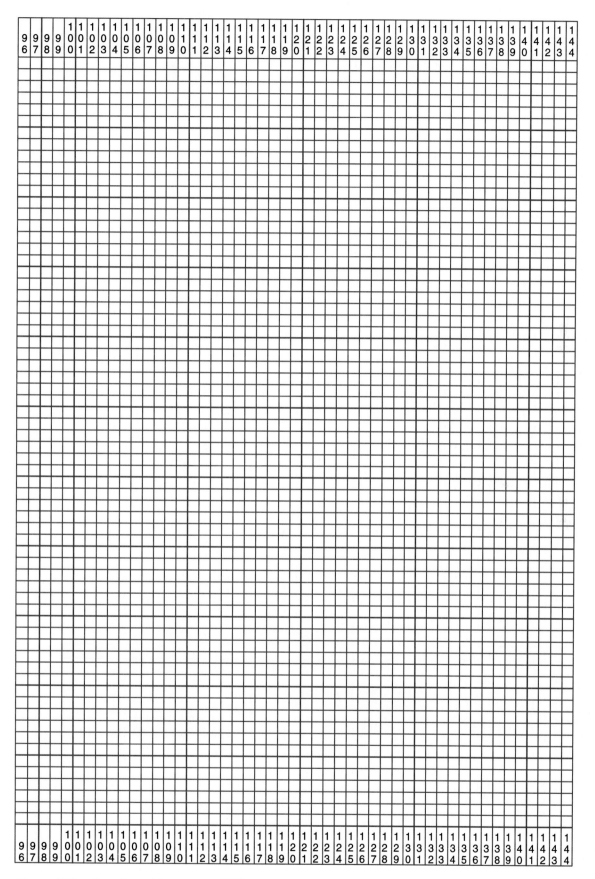

Figure D.8 *(continued from page 506)*

Header: _____

Left Margin: _____ Top Margin: _____ Right Margin: _____

Tabs:

Bottom Margin: _____

Footer: _____

	Other Information	
Paper Size: _____ Page Orientation: _____ Justification: _____	Graphics to Be Used:	Reminders: Mark Guidelines Mark Page Number Placement

Figure D.9 Page Planning Guide for Word Processing

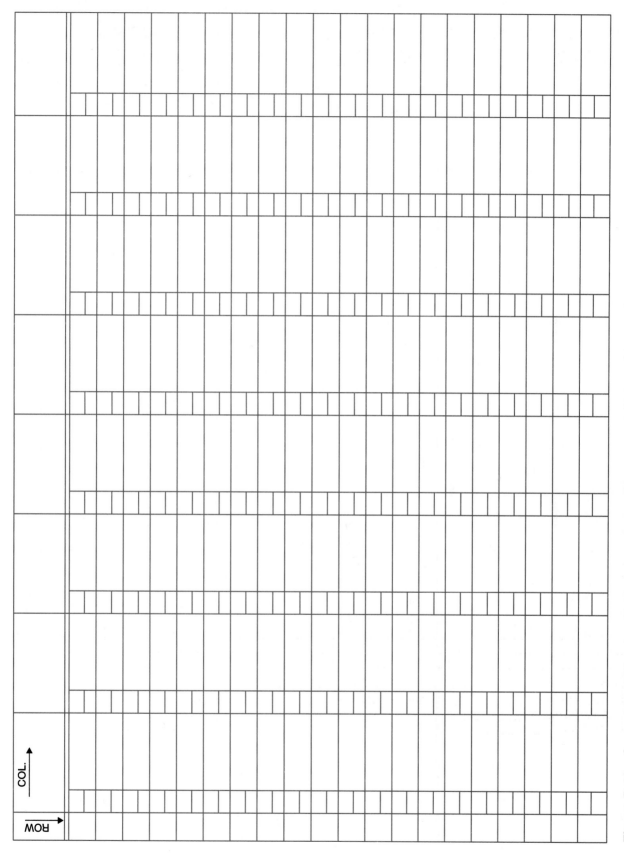

Figure D.10 Spreadsheet Planning Guide

Output Requirements	Input Requirements	Input Format

Figure D.11 Output and Input Requirements Chart

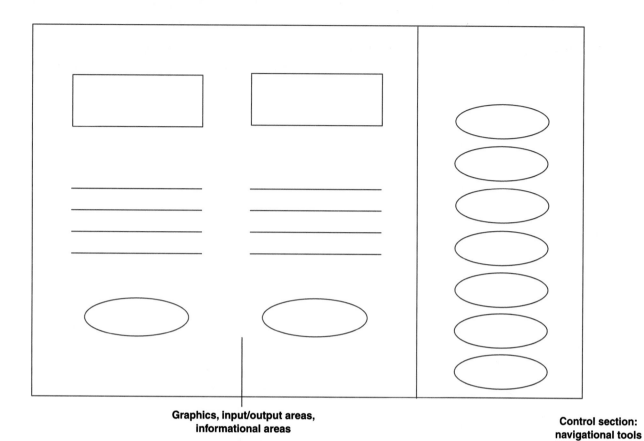

**Graphics, input/output areas,
informational areas**

**This is the area in which the user will input
data, read information, and receive output.**

**Control section:
navigational tools**

Figure D.12 The User Interface Screen Design Chart

Class	Properties	Functions

Figure D.13 Class, Properties, and Function Chart

Class: _____

Property	Attributes

Figure D.14 Attributes of a Class Properties Chart

Figure D.15 Storyboard

Object Name	Source/File Name	File Type/Extension	Screen Number(s)	H/I/O*	Script

*H = Hyper, I = Input, O = Output

Figure D.16 Object Dictionary

Appendix E

Other Problem-Solving Tools

Problem-solving techniques using seven basic tools have been presented throughout this book. There are other tools for problem solving that have not been presented. Among these tools are Nassi-Schneiderman charts and Warnier-Orr diagrams. Since these tools are replacement tools for the algorithm and flowcharts, they were not presented earlier in this text.

Both Nassi-Schneiderman charts (N-S) and Warnier-Orr diagrams are used to create structured solutions. Many people prefer one of these tools to flowcharts. Whichever tool is chosen, the use is the same. You should plan the solution, complete the problem-solving process, and end with a solution that is well done.

Nassi-Schneiderman Charts

Nassi-Schneiderman charts are sometimes called structured flowcharts. They are also referred to as N-S charts. These charts consist of a stack of rectangles one on top of the other and put together without flowlines. Processing flow always runs from the top to the bottom of the stack of rectangles. Figure E.1 shows the four logic structures as they would appear in an N-S chart. Note the similarity to algorithms as presented in this book. Figure E.3 shows an example problem and as compared to an algorithm, a flowchart, and a Warnier-Orr diagram. N-S charts fit better on one page. However, since there is no means to continue to the next page and there is no way to indicate a GOTO, the modules are required to be short and cohesive.

Warnier-Orr Diagrams

The theory behind Warnier-Orr diagrams is similar to an interactivity chart. The differences are:

1. The Warnier-Orr diagrams are sideways.
2. There are no rectangles.
3. They use a pseudocode for the instructions.

Sequential Logic Structure

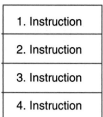

1. Instruction
2. Instruction
3. Instruction
4. Instruction

Decision Logic Structure

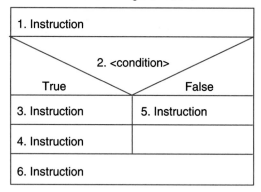

Loop Logic Structure

1. Instruction	
2. While<condition>	
	3. Instruction
	4. Instruction
	5. Instruction
	While-End
6. Instruction	

Case Logic Structure

1. Instruction	
2. Case of <variable>	
= const1:	3. Instruction
= const2:	4. Instruction
= const3:	5. Instruction
otherwise:	6. Instruction
7. Instruction	

Figure E.1 The Four Structures Using Nassi-Schneiderman Charts

A Warnier-Orr diagram consists of a series of left braces, pseudocode-like instructions, and a few special symbols. The symbols include:

{	The brace is used to enclose a zero logically related set of instructions.
(0,1)	A set of instructions is executed zero or one times. This is the notation for the decision and the case structure.
(0, *n*) or *n*	A set of instructions is executed zero to *n* times. This is the notation for the WHILE/WHILE-END loop structure.
(1, *n*)	A set of instructions is executed one to *n* times. This is the notation for REPEAT/UNTIL structure.
Blank or 1	A set of instructions is executed one time.
+	Exclusive OR. This is used with the (0,1) notation to show a selection.

Figure E.2 shows the four structures as written with Warnier-Orr diagrams. Notice the processes are broken up into modules just as in the structure chart. Most flow-charting templates have the Warnier-Orr bracket on one side. Figure E.3 shows the same problem developed using a Warnier-Orr diagram and a Nassi-Schneiderman chart.

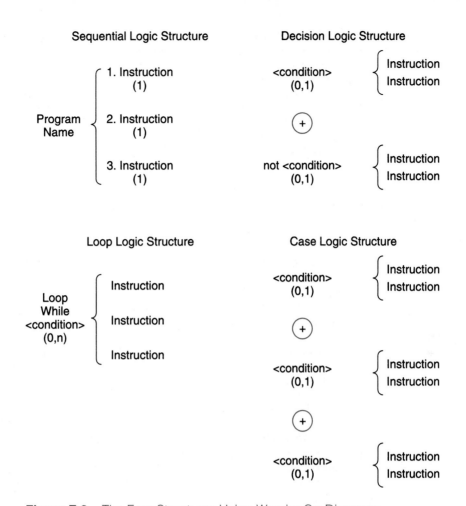

Figure E.2 The Four Structures Using Warnier-Orr Diagrams

Nassi-Schneiderman Charts

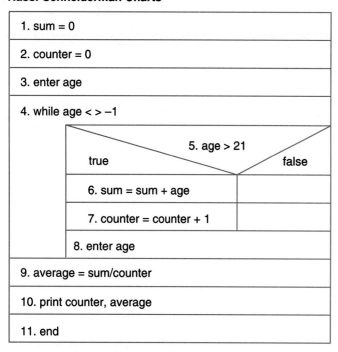

```
1. sum = 0
2. counter = 0
3. enter age
4. while age < > −1
        ┌──────────────────────────────┐
        │          5. age > 21          │
        │   true                  false │
        ├───────────────────────┬───────┤
        │   6. sum = sum + age   │       │
        ├───────────────────────┤       │
        │   7. counter = counter + 1 │   │
        ├───────────────────────────────┤
        │   8. enter age                │
9. average = sum/counter
10. print counter, average
11. end
```

Warnier-Orr Diagrams

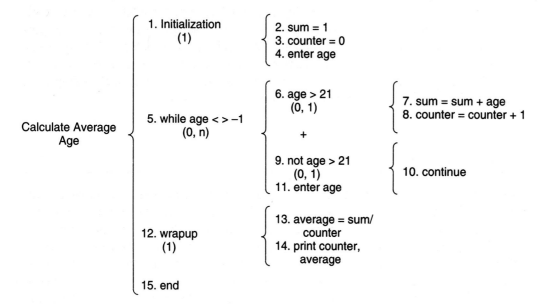

Calculate Average Age

- 1. Initialization (1)
 - 2. sum = 1
 - 3. counter = 0
 - 4. enter age
- 5. while age < > −1 (0, n)
 - 6. age > 21 (0, 1)
 - 7. sum = sum + age
 - 8. counter = counter + 1
 - +
 - 9. not age > 21 (0, 1)
 - 10. continue
 - 11. enter age
- 12. wrapup (1)
 - 13. average = sum/ counter
 - 14. print counter, average
- 15. end

Figure E.3 Comparison of a Nassi-Schneiderman Chart and a Warnier-Orr Diagram

Appendix F

Overview of UML
(Unified Modeling Language)

Unified Modeling Language (UML) is used to create a logical model of your solution. It uses various diagrams to describe the entire system. These diagrams help produce a model of the problem and lead to a better understanding of the requirements of a problem. Through a better understanding of a problem, a better solution can be developed using an object-oriented language. UML is increasingly being used in the programming industry to design and communicate a solution to a problem. This appendix presents an overview of the various diagrams and terminology that accompanies them.

The five diagrams defined in this appendix are use case diagrams, class diagrams, sequence diagrams, state chart diagrams, and activity diagrams.

Use Case Diagrams

Use case diagrams describe how a system functions from the user's standpoint. They present who may use what part of the system. There are basically two parts of a use case diagram: the actor and the use case. The actor represents the users of the system, including human and non-human entities. The use case describes the services required by the actor. It is represented by an oval with the service or function written inside. Communication is indicated by lines connecting the actor and the use case. The series of use cases required for this part of the solution are encased in a rectangle with the name of the system in the top part of the rectangle. See Figure F.1. Figure F.2 shows a use case diagram for the book system presented in Chapters 12 and 13.

The actors in Figure F.2 are the librarian and the client. The case uses are to add books, delete books, add clients, delete clients, list books, list clients, check out books, and check in books. There may be others, but these are some of the functions of the system. The client may access only the check in and check out case uses. The librarian may access all case uses. The assistant librarian can access everything except adding or deleting books and clients.

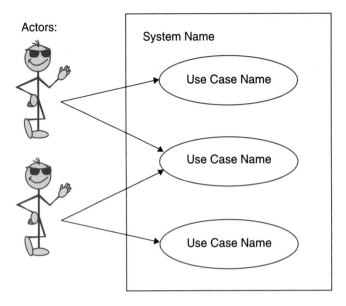

Figure F.1 Use Case Diagram

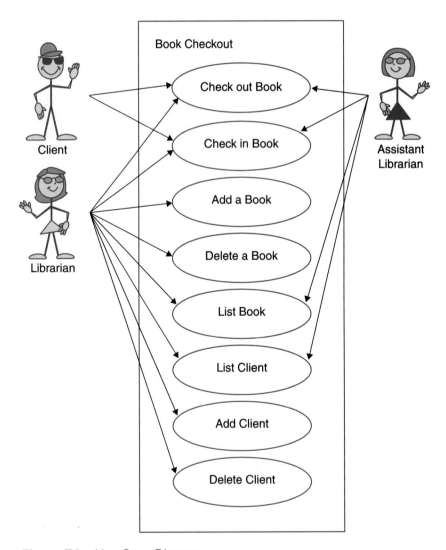

Figure F.2 Use Case Diagram

Class Diagrams

Class diagrams graphically describe how a class functions. Basically they describe the data structure of the solution. They describe the class, their attributes, and their operations. The simplest form is:

Class Name
Attributes
Operations

See Figure F.3 for a class diagram of the books class. The book class includes all books. The attributes for the book class include library ID number, the name of the book, its author, its ISBN, its publisher, year of its publication, whether it is checked out, the client to whom the book is checked out, and where in the library the book is located. The functions that involve the book class include checking out a book, checking in a book, listing all books, adding books, deleting books, and updating data on individual books.

Sequence Diagrams

Sequencing diagrams graphically present the interactivity between objects. The interactions are represented as messages in the diagram. The diagram shows the actions by the user and the subsequent actions that result between objects. See Figure F.4. Notice that the user interacts with the system by sending a message of what is to be accomplished to an object. The object may then send another message to another object to accomplish another task. The series may include any number of messages. The number of objects that are performing tasks and interacting with other objects also may be unlimited. See Figure F.5 for a sequence diagram of starting a car. The user sends a message to the ignition by turning key. The ignition then sends a message to the engine to start. The user sends a message to the engine to accelerate or not by pressing or releasing the gas pedal.

Book Class
Library ID number Name Author ISBN Publisher Year of publication Checked out Client checked out to Area in library
Check out book Check in book List each book data Add book Delete book Update book data

Figure F.3 Class Diagram of Book Class

Figure F.4 Sequence Diagram

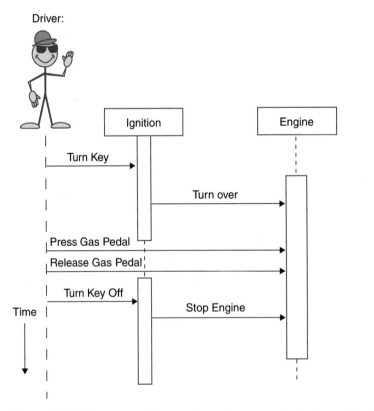

Figure F.5 Sequence Diagram for Starting and Turning off a Car Engine

The user sends a message to the ignition to turn off the engine by turning the key. The ignition then sends a message to the engine to stop running. This diagram is simplified. You may add brakes, gears, etc. to this diagram.

Statechart Diagrams

A statechart diagram describes what happens to a given value of an object as it proceeds through the system. See Figure F.6. An event is an activity that sends a message that causes something to happen. The action is the set of tasks or instructions that is the response to the event.

Activity Diagrams

An activity diagram describes the flow of activities. This type of diagram is similar to a data flow chart. It specifies what happens and when. The major difference is in the symbols used. Because you are representing general activities and not specific instructions, all activities use a flattened ellipse. See Figure F.7. You may add decision structures as necessary. The flow may be from top down or may be from left to right as in Figure F.8.

These UML diagrams illustrate how you can design and develop a solution using an object-oriented language. When you write the various functions, modules, or scripts, you will use the four logic structures presented in Chapters 4–8.

Figure F.6 Statechart Diagram

Figure F.7 Activity Diagram

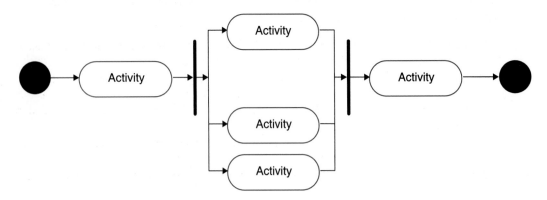

Figure F.8 Horizontal Activity Diagram

Glossary

Accumulating Keeping a running total by adding a current value in the form of a variable to a previous sum to find the current total. The instruction for accumulating is SUM = SUM + VARIABLE.

Activity file A file containing all of the updates to a master file.

Algorithm A sequence of instructions to solve a problem. A computer programmer writes a solution in the form of an algorithm before coding it into a computer language.

Algorithmic solution A solution to a problem that can be solved through a sequence of instructions.

Annotated flowchart A flowchart with explanations written beside the symbols.

Array Sometimes called a subscripted variable; a block of memory locations in the internal memory that is assigned to one variable name. Arrays may be one-dimensional, two-dimensional, or multidimensional.

ASCII file A file that contains the ASCII values of text. This is an unformatted file.

Automatic-counter loop One of three types of loop instructions. The automatic-counter loop enables the computer to repeat a set of instructions by using a variable as a counter that starts at a beginning number that is incremented by a given number each time the program executes the set of instructions. The loop stops when the counter is larger than a given ending number.

Backup file A file containing a master file as it was before the last update.

Base-one system A computer system in which array element numbers begin with one rather than zero.

Base-zero system A computer system in which element numbers begin with zero rather than one.

Batch processing Running successive sets of data through a program all at one time without human intervention.

Binary search A technique used to search for a single element in an array or a single record in a file.

Binary tree A list that has been created to have two linked fields—one to the right and one to the left to form a tree structure.

Buffer A small internal memory that is used to help process input and output operations.

Bug An error in a computer program.

Calculation module A process data module that does arithmetic calculations and accumulates, counts, or manipulates numeric data in some way.

Case logic structure One of the four logic structures for organizing the instructions to the computer that make up a program. The case structure allows the computer to select one set of instructions from among many, through data given by the user or calculated in the solution.

Cell protection A command used with spreadsheet packages to prevent the movement of the cursor into a cell.

Character data The data set of the character data type; it includes all symbols available on a computer.

Class A class is part of a solution of an object-oriented solution to a problem. A class models the properties or characteristics of a set of objects.

Codes Character data input by a user to direct the flow of processing, such as in the coded options in a record.

Cohesion A module that is written to be as independent of the rest of the program as possible.

Command An instruction to the computer in a productivity package or an operating system. Normally, a command is not part of a program.

Computer information system All components required to complete the tasks required to complete a job. These components include hardware, software, data, people, procedures, and facilities.

Concatenation Adding one piece of string data to another by placing the second piece immediately after the first.

Constant A value in a program that cannot change during processing.

Control-break A break in processing to print subtotals, or number of items, controlled by one field in a file.

Control module The module controlling the processing of all of the subtasks in a solution.

Coupling Techniques used to pass data from one module to another. Coupling techniques include the use of global variables, parameters, and function or module names.

Data Unorganized facts.

Data type The kind of data of a variable or a constant. The three basic data types are numeric, character, and logical.

Database A database is created through the use of Relational Database Management Systems software and contains a group of related tables of data, queries, forms, reports, and macros.

Database form A database form is the design of how one record of data will be viewed on the screen. A form may include other items, such as images and buttons for automation.

Database query A database query creates a new table containing selected fields and/or records from the database tables.

Database report A database report is a printed document created from data in a database or query tables.

Database table A database table contains data needed to create reports and forms for viewing information. A database table contains one or more records.

Debugging The process of correcting errors in a computer program.

Decision logic structure One of the four logic structures for organizing the instructions to the computer that make up a program. The decision structure selects one of two sets of instructions according to the resultant of a condition.

Decision table A programming aid to writing decision instructions for the computer; it consists of a rectangular grid divided into four parts, which specifies all possible actions the computer could take for each possible set of conditions.

Default value A value of a variable that is built into a program and that the computer automatically uses unless the value is changed by the user.

Document Text that is formatted and stored in a file.

Document processing Document processing refers to the use of multiple software packages in the development of a project.

Drag and drop Drag and drop is the ability to select a portion of a document and move it to another part of the document or to another document.

Dynamic array An array in which the maximum number of elements can change during processing.

END-OF-FILE (EOF) The marker in a computer solution indicating that there are no more records to be processed.

Equation A variable that is assigned the value of an expression, another variable, or a constant, as in $A = 5 + B$, $A = B$ (where B is a variable), or $A = 5$, respectively.

Event-driven object-oriented programming A solution to a problem in which the processing is controlled through events in the computer.

Exporting data The saving of a document in a format to be received by another software package.

Expression An operation or series of operations performed on variables or constants, as in $5 + B$.

External documentation Instructions to the user in the form of manuals or other written documents.

Field A field is one piece of data in a database table, such as a name or an address. A field is given a name and has the same attributes for all records in a table.

File A collection of related records.

File updating Adding, deleting, or changing data in records in a file to keep the records current.

Flowchart A graphic illustration of an algorithm.

Format In word processing, the size, shape, and appearance of a document, involving such factors as margins, indentations, line spacing, and type styles. In a spreadsheet, the way the contents of the cell will look on the screen.

Function A small set of instructions built into a computer language or applications that enables the user to complete a frequently repeated task with fewer keystrokes than would be required without the function.

Global variable A variable that can be accessed by all modules below and in line (in the interactivity chart) with the module that declared the variable.

Graphical User Interface (GUI) An interactive screen in which the user controls what happens in a program.

Hashing A method of using a predefined algorithm to designate the record number for fast retrieval of a given record.

Heuristic solution A solution to a problem that cannot be solved through a single sequence of instructions.

Hierarchy The order in which operations are performed for mathematical, relational, and logical operations.

Horizontal software Multipurpose software, such as word-processing programs, database management systems, and spreadsheets.

IF/THEN/ELSE instruction The instruction to the computer that implements the decision logic structure.

Import (Insert/Place) Importing a file refers to the use of all or part of one file within another file. Insert and place are additional names for import.

Incrementing The process of counting on the computer by adding a number to a previous number. The instruction for incrementing by one is COUNTER = COUNTER + 1.

Index file A file containing the key field and corresponding record number for all records in the master file. The file is in order of the key.

Indicator A value built into a solution by the programmer to redirect the flow of processing.

Information Organized facts.

Information processing Information processing is using the correct methods and software to process data into information.

Initialization module The module containing all of the processing that has to be completed only once and at the beginning of the solution.

Input mask An input mask specifies required characters when entering data into database tables. These characters are automatically entered into a field by the computer.

Installing software Setting up a software package so that it can be used on the computer for which it was bought.

Instance of a class An object.

Interactive processing Processing that involves user intervention to enter new data as needed, usually at a keyboard.

Interactivity The interaction between classes or between screens.

Interactivity chart A chart, also called a structure chart, showing all of the subtasks, or modules, in a program.

Internal documentation Remarks within a computer solution to explain the processing.

IPO chart Input-processing-output chart; describes the input, the output, and the processing needed to solve a problem on the computer.

Key The field of the records that is used to order a file or search for a record.

Linked list A list that has one field that contains the record number of the next record to be processed.

Local variable A variable accessed by the module that declared the variable. A single variable can be both local and global (see Chapter 4).

Logical data The data set consisting of TRUE and FALSE, used in making yes-and-no decisions.

Logical file number (LFN) In record processing, a number used to reference a file.

Logical operator An operator within an expression that uses logical data as operands and produces logical data as the resultant. The logical operators include NOT, AND, and OR.

Logic structure A structure for organizing the instructions to the computer that make up a program. There are four logic structures: the sequential structure, the decision structure, the loop structure, and the case structure.

Loop logic structure One of the four logic structures for organizing the instructions to the computer that make up a program. The loop structure enables the computer to process the same set of instructions repeatedly.

Macro A small set of instructions to the computer executed by a single command.

Master file The file in which all of the data needed for processing are kept current.

Mathematical operator An operator that uses numeric data as its operands and produces numeric data as the resultant. The mathematical operators include – (subtraction), + (addition), * (multiplication), / (division), \ (integer division), MOD (modulo division), and ^ (power).

Module A subtask of the larger task the computer performs to complete a whole solution. The modules include the Control, Initialization, Process, and Wrapup modules.

Multidimensional array A multidimensional block of elements with a single variable name. Each element is designated by element numbers representing row, column, page, and so on.

Multiple control-break A control-break on more than one field.

Navigational tools Tools including hot spots, hot objects, time, buttons of all types, and others, used to allow a program user to more through and interact with a computer program.

Negative logic A type of decision logic in which the action the computer is to perform follows from the FALSE side of the instruction.

Nested IF/THEN/ELSE A type of decision structure that nests one IF/THEN/ELSE instruction within another.

Normalize Normalizing data separates the data into related database tables.

Null file A file containing no records.

Numeric data The data type that includes all numbers, integers and real numbers, and is the only data type that can be used in calculations.

Object An item, or an instance of a class; an item on the screen when using an event-driven object-oriented programming language

Object-oriented programming (OOP) A language that supports object-oriented principles; a strategy of program design in which the data parts are the principal items, instead of the process.

OLE OLE is object linking and embedding.

One-dimensional array One column of elements with a single variable name. Each element is designated by an element number.

Operator A sign or symbol in an expression or equation telling the computer how to process the data. The three types of operators are mathematical, relational, and logical.

Page-break A break in the process of printing to finish one page and begin another.

Parallel arrays Two or more arrays in which the data in the same element numbers are related to each other.

Parameter A variable passed from one module to another module through the calling sequence of the module or function. There are two types of parameters, call-by-value parameters and call-by-reference. Call-by-value parameters cannot be changed in the module. Call-by-reference parameters can be changed and the changed value will be passed back to the calling module.

Pointer technique A programming technique that uses the value in one array to point to an element in another array.

Positive logic A type of decision logic in which the action the computer is to perform follows from the TRUE side of the instruction.

Primary key A unique key that is the major key for ordering a file or searching for a record.

Primer READ The first record that is read into the internal memory before starting a processing loop, in order to have values to process the first time through the loop.

Print module A process data module that prints the results of processing.

Problem analysis chart A chart used to begin to break down a problem into parts, including given data, results required, processing required, and solution alternatives.

Process Data modules The modules that process the data. They include Calculation, Print, Read, and File maintenance modules.

Productivity package Software designed to help the user solve problems, make better decisions, and increase productivity, efficiency, and accuracy. The most common types of productivity packages are word processing, database management, and spreadsheets.

Program A computer solution to a problem; a set of instructions in a given computer language that solves the problem.

Pseudocode A set of step-by-step instructions similar to that found in the programming language to be used.

Pushing and popping the stack Adding (pushing) and taking from (popping) a stack.

Query table A query table is the result of a database query. It contains selected records and/or fields from database tables.

Queue A list in which the next value used was the first one to be added. Data are added and used in a first-in, first-out basis.

Range name The name of a group or range of cells in a spreadsheet.

Range of cells A rectangular group of one or more cells in a spreadsheet.

Read module A process data module that enters data into the computer.

Record The set of data for one entity in a file or table.

Record management system A productivity package that performs many of the same functions as a database management system but that uses only one file rather than two or more.

Recursion Recursion occurs when a module or a function calls itself.

Relational Database Management System (RDBMS) A Relational Database Management System is a software program that allows the user to enter, organize, and report information.

Relational operator An operator within an expression or equation that uses numeric or string data as operands and produces logical data as the resultant. The relational operators include: = (equal to), < (less than), > (greater than), <= (less than or equal to), >= (greater than or equal to), and <> (not equal to).

REPEAT/UNTIL One of three types of loop instructions; this loop logic structure tests at the end of each loop to see if it should repeat; it repeats the instructions in the loop until the condition is TRUE.

Results The required answers to or output from a problem.

Schema The schema of a database is a graphic representation of the tables and their relationships.

Screen The image on the monitor in one instance.

Script A set of instructions attached to an object or a solution.

Secondary key A key, which may or may not be unique, that is used to order a file or to search for one or more records in a file. A secondary key is considered a minor key.

Sequential logic structure One of the four logic structures for organizing the instructions to the computer that make up a solution. The sequential structure processes the instructions one after another in a sequence.

Sequential search A method for locating a record in a file in which the computer starts at the first record and continues through the file, record by record, until the needed record is found.

Solution The set of instructions to the computer that will output the results; a computer program before it is coded into a computer language.

Sorting Arranging data in order; the order may be numeric or alphabetical.

Spreadsheet package A productivity program that allows the user to set up, calculate, change, and recalculate a worksheet.

Stack A list in which the value to be added and the value to be used are on the same end of the list. Data are added and used in a last-in, first-out basis.

Static array An array in which the maximum number of elements cannot change during processing.

Storyboard A design tool used to create screens.

Straight-through logic A type of decision logic that processes all decisions sequentially, one after another.

Structure chart A chart, also called an interactivity chart, showing all of the subtasks, or modules, in a program.

Suite A group of stand-alone software packages designed to work together. They have the same look and feel, as well as the ability to share and link data easily.

Syntax The rules of setting up instructions and commands in an operating system, a programming language, or an application program.

Systems flowchart Shows the overall flow of data throughout the entire computer system, including all peripheral equipment.

Template A shell, or structure, of a document or worksheet that can be reused repeatedly with new data.

Transaction file A file containing all of the updates that need to be made to a master file.

Tutorial A program provided with a software package to help the user learn how to use the package.

Two-dimensional array A block of elements with one variable name. Each element is designated by element numbers designating the row and column.

Variable A named memory location in the computer that is given a value that can be changed at any time during the processing.

Vertical software Software with a single purpose, such as dental accounting software, real estate management software, and the like.

WHILE/WHILE-END One of the three types of loop instructions; the loop logic structure that tests at the beginning of the loop and repeats as long as a condition is TRUE.

Word processing Editing text electronically through the use of a horizontal software package.

Worksheet The contents of the cells as seen on the screen when using a spreadsheet.

Wrap-up module The module containing all of the processing that has to be completed only once and at the end of the processing.

Index

L

Line Counters, 351
Linked Lists, 245
Logic Conversion, 115
Logical Data, 20
Logical File Number (LFN), 337
Logical Operator, 26
Logic Structures, 70, 161, 103, 137, 87
 Case Logic Structure, 70, 161
 Decision Logic Structure, 70, 103
 Looping Logic Structure, 70, 137
 Sequential Logic Structure, 70, 87
Looping Logic Structure, 70, 137

M

Macros, 439
Master File, 383
Mathematical Operator, 25
Merging Files, 491
Module, 48, 74
Multidimensional Array, 201
Multiple Control Breaks, 359

N

Nassi Schneiderman Charts, 517
Navigational Tools, 312
Negative Logic, 106, 113
Nested IF/THEN/ELSE, 107
Nested Loops, 151
Normalization of Tables
Null File, 370
Numerical Data, 17

O

Object, 310
Object Dictionary, 327
Object Oriented Programming (OOP), 307, 315
Object Linking and Embedding (OLE), 477
One-dimensional Array, 178
Operand, 25
Operators, 24
 Logical Operators, 26

Mathematical Operators, 25
Relational Operators, 26

P

Page Break, 351
Parallel Arrays, 178
Parameter, 22, 76
 Call by value, 78
 Call by reference, 78
Pointer Technique, 206
 Cross Tabulation, 211
 Frequency Distribution, 206, 207
Positive Logic, 106, 109
Primary Key, 234, 280, 337
Primer Read, 142, 345
Presentation Graphics, 466
Problem Analysis Chart, 46, 47, 92
Problem Solving, 3
Problem Solving Steps, 3, 4
Processing Information, 17
Product, 139
Productivity Package, 431
Program, 6
Pseudocode, 52

Q

Query, 287

R

Random Access Files, 336
Random Access File Processing and Updating, 347, 407
Range, 454
Range of Cells, 454
Real Numbers, 17
Recalculation, 456
Record, 45, 278
Record Data Dictionary, 341
Recursion, 156
Relational Database Management System, 277
Relational Operator, 26
REPEAT/UNTIL, 142
Resultant, 25
Results, 6